Roswell Revealed

The New Scientific Breakthrough into the Controversial UFO Crash of 1947

SUNRISE INFORMATION SERVICES

ROSWELL

REVEALED

*The New Scientific Breakthrough into the
Controversial UFO Crash of 1947*

SUNRISE

SUNRISE Information Services
Canberra

SUNRISE Information Services (SUNRISE)
PO Box 1468
Woden, Canberra ACT 2606 Australia
www.sunrisepage.com

ISBN 978 0 994282 63 7

Book cover design by Lyndsey Lewellen of Expert Subjects
www.expertsubjects.com

Typeset by SUNRISE

Printed by LightningSource/IngramSpark

Acknowledgements

Our sincerest thanks must go to Dr. Andrzej "Andrew" Calka, for his valuable time in providing copies of selected metal articles and giving his insights into the nickel-titanium alloy, which form the basis of this book's argument.

We are grateful to the late Colonel James Bond Johnson, Ph.D. (1926-2006), military reporter and representative of the *Fort Worth Star-Telegram*, for the details of what happened in Brigadier-General Ramey's office.

Thanks must also go to the University of Texas in Arlington, for their permission to reproduce the photographs taken by Mr. Johnson, and Major Charles A. Cashon of Fort Worth Army Air Field.

We appreciate the interest expressed by Mr. Anthony "Tony" Bragalia in our research when we first brought it to the attention of his friend, Mr. Thomas J. Carey (a former USAF anti-crypto specialist turned Roswell investigator), and his co-author, Mr. Donald R. Schmitt. When we contacted Mr. Carey and Mr. Schmitt by email on January 21, 2008 and precisely six months later began conversing with an unknown gentleman named Mr. Bragalia (and eventually we learned of his connection to Mr. Carey), our aim was to let Mr. Carey and Mr. Schmitt know about our research and to ask for any information they could provide to help complete our research prior to publication of our first edition book. At the time, no books on the Roswell case discussed the results we had found, so we thought these two U.S. investigators specializing in the Roswell case could bestow on us their expertise in this field. Since then, Mr. Bragalia has kindly provided us with additional information about the Battelle Memorial Institute, and gave us a copy of the 1972 report from Dr. Frederick Wang, and the *New York Times* obituary details of Prof. John P. Nielsen, the scientist mentioned in a number of important scientific articles we have found and used in this book.

Also, in this updated edition, we have included some information Mr. Bragalia has gathered about the world's top expert in chemistry, Dr. Linus Pauling. To ensure the information is accurate, we have examined the original communications between Pauling and the then-director of

Battelle, Mr. Clyde E. Williams, which occurred between January and February 1951, and have decided to include only what is relevant to this book. Any further controversial claims made by Mr. Bragalia about Pauling are covered in a footnote in Chapter 7, for those readers who wish to have them clarified in terms of what we do know about the work and interests of the scientist.

We would also like to thank Dr. Bob Rich of Wombat Hollow, Australia, for straightening out and editing an important aspect of Chapter 10 and for providing his keen skeptical eye, especially with regard to Chapters 4 and 5; senior engineer Imtiaz Siddiqu, from Kibin, for his brilliantly meticulous editing work of Chapters 3, 6 and 10; and Daniel T. Moore, also from Kibin, for carefully checking Chapters 1 and 2 (and a bit of Chapter 9 too for good measure). In this edition, updates to various chapters have also been edited by various editors at Kibin, including Crystal W., Michael S., Jane Jacobi, and Erin Sapre. For this we are forever grateful. And, of course, how can we ignore the fine editing work of Shawn Wells of Canton, USA, for the additional information we added in the previous edition about iron-based shape-memory alloys in Chapters 7 and 8, together with a final check by the talented Bernard Farber of Chicago, USA?

We are indebted to all of the original witnesses, especially Major Jesse A. Marcel, Bill Brazel, Lydia A. Sleppy and many others, who had the courage to reveal their experiences of events at the time. Without them, this book could not have been written.

Finally, to anyone else who we cannot or have not been able to acknowledge (either because we could not track them down, or they wish to remain anonymous), we would like to thank you for providing information to help with this research work.

"Toto, I've a feeling we're not in Kansas any more."

Dorothy, *The Wizard of Oz*

Contents

1	An Overview	1
2	A Brief History	29
3	The Witnesses Speak	109
4	The U.S. Government Speaks	133
5	An Australian Researcher Speaks	149
6	Nitinol: The World's Most Powerful Memory Alloy	155
7	The Official Scientific History of Nitinol	205
8	The Official U.S. Military History of Nitinol	285
9	Are There Alternative Materials?	353
10	Was the Object Alien- or Man-Made?	375
11	Conclusion	411
	Appendices	423
	Bibliography	489

CHAPTER 1

An overview

SOMETHING CRASHED to earth on the night of July 2, 1947, northwest of Roswell, New Mexico, and it left dozens of civilian and military witnesses baffled over what it was and where it came from.

William "Mac" Brazel, a local rancher, was the first to see the wreckage. Earlier that night, he was at home with his family when they heard an odd explosion in the midst of a severe thunderstorm sweeping over New Mexico. At first, there were frequent lightning strikes over one area of his property. No rain, just the lightning. William thought that perhaps underground mineral deposits may have been attracting the lightning, but soon after the explosion, the lightning died down and nothing more was heard that night.

The next morning, William rode on horseback to check on his sheep. He got to an escarpment where he could take in a greater view of the countryside. From there he saw his sheep cowered in the distance, as if trying to avoid something. He could see why. Lying in front of him, below the escarpment where he stood, was a large quantity of metallic foil and plastic-like beams and sheets scattered over a large area. Naturally curious, William got off his horse, walked down the escarpment, and took a closer look.

It didn't take long for William to realize the extraordinary toughness and extreme lightweight nature of all the materials. More importantly there was one peculiar observation he could neither ignore nor forget. It relates to the dark-grey foil, found in great abundance on the ground. Apparently, he saw a material that looked very much like it was a metal, described as being as thin as the foil in his cigarette pack, but that it could behave like plastic. In other words, this strange metal could be made to return to its original shape every time he tried to bend it and let it go.

William kept some of the larger foil pieces in a shed about 3 miles from the debris site and later visited friends in the town of Corona to discuss his discovery. No one knew what it was, and there were no indications in newspapers or on the radio to suggest the U.S. military authorities had been looking for something in the desert. The only thing mentioned was a financial reward[1] of $3,000 for the recovery of a "flying saucer" or parts of one. Given the unusual nature of the materials found on his property, William naturally wondered whether he had the remains of one of those elusive and enigmatic flying objects strewn about his pasture. If so, the money would certainly go a long way towards helping him and his family. Combined with his intense interest in the materials, it was clear he had to find out what had crashed on his property. In the end, his friends persuaded him to go to the authorities and tell them about his discovery.

Four days after William had discovered the wreckage, he drove to Roswell to notify the local sheriff, George Wilcox, who in turn contacted the military authorities stationed in the city. Two U.S. Air Force (USAF) representatives from the Roswell Army Air Field, Major Jesse Antoine Marcel (1907–1986) and Captain Sheridan Cavitt, went to investigate.

On returning to base with a couple of jeep loads of the wreckage, Marcel and Cavitt indicated to their colleagues and the base commander that the entire object, or at least part of it, had exploded above ground as it traveled at high speed and had scattered debris over a wide area. Marcel had a particularly in-depth knowledge of practically every aerial object flown in his country and abroad. Yet on this occasion, he was stumped. In fact, neither men, nor anyone else at the base who attended the meeting and tested the wreckage, could explain what the materials were or what had flown over the desert.

Whatever the object was, it was enough for the commanding officer at Roswell AAF, Colonel William Blanchard, to accept the view of his men who examined the wreckage and his own personal experience in handling the materials that a "flying disc" was recovered even though no direct evidence for the shape could be determined from the wreckage itself. The choice of words to explain the object is interesting because the military was clearly trying to link the unusual nature of the materials to those unexplained daylight metallic-looking symmetrical UFOs allegedly observed by a growing number of civilians and military personnel at the time as the best explanation for what was found. Apparently the unusual nature of the materials was enough for the military to reach the conclusion that whatever this flying object was that had dropped large amounts of artificial-looking pieces on the desert floor was definitely not man-made. So whatever flew in the air had to be something unknown.

The front-page of the Roswell Daily Record, July 8, 1947.

During the recovery operation, the public information officer at the Roswell Army Air Field (AAF) was ordered to issue a news release stating that a flying "disc" had been picked up to avoid curious media representatives checking the location of the crash site. This was no doubt an extraordinary claim if it was true, and even if it wasn't, it certainly received considerable front-page prominence in the *Roswell Daily Record* and several other newspapers before international media outlets eventually picked it up.

As the news began to captivate the public's imagination and light up the media's interest like a Christmas tree, the military was placed under considerable pressure to release more details. It didn't take long before top Air Force officials in the Pentagon realized the unsettling predicament they were in.

Did the media get more information? Yes, but not as they had expected.

Within a couple of hours of the initial news release, Brigadier-General Roger Maxwell Ramey (1903–1963), commanding officer of the Eighth AAF at Fort Worth AAF (later Carswell Air Force Base), Texas, received a message[2] at 2:00 PM Roswell time from his chief of staff, Colonel (later Brigadier General) Thomas Jefferson DuBose (1903-1992). The man who ordered DuBose to write and personally send the message to Ramey was Major General Clements McMullen (1892–1959), Vice-Commander (or Acting Director) of the Strategic Air Command (SAC) at Andrews AFB in Maryland. Later, Ramey received a separate top secret memo detailing the reason for what he had to do next by briefly mentioning the recovery of "victims" and a "disc" and how vitally important it was for him to clean up the mess brought on by the initial news release by issuing a new story claiming the "disc" explanation to the public was a mistake. The object was really the humble remains of an ordinary weather balloon.

As if anticipating the likely skepticism among the public and the media after the initial unfortunate word choice made in the news release to describe the shape of the object, Ramey presented promptly what he claimed was the evidence for the entire incident by way of a torn-up weather balloon, or at least the partial remains of such a device. The only problem was, it seems he did not feel particularly keen to have any members of the press coming in to ask a barrage of questions while inspecting the device all at once, or even one or two media representatives visiting the wreckage over an extended period of time. Ramey was more interested in a quick discussion with just one trusted reporter.

James Bond Johnson, a man with a long and distinguished military career who became a reporter, photographer and representative of the *Fort Worth Star-Telegram*, would make his name in the history books as the only person to view the carefully laid-out remains of the weather balloon in Ramey's office. While Johnson was there, a brief exchange of questions and answers ensued. There was a moment when another military officer was brought into the office and asked to confirm the legitimacy of the "weather balloon" explanation after examining the pieces. Then Mr. Johnson was permitted to take several photographs as an official record of what he had seen.

Again anticipating the possibility of reporters asking why Marcel was not in the pictures, Ramey ordered his public information officer, Major Charles A. Cashon, to take a couple of photos of Marcel holding a piece of the weather balloon. Before Mr. Johnson arrived, Marcel was told to leave the room and was kept in seclusion for the next 24 hours to avoid answering any questions from the reporter. Later, these additional photos were quietly slipped in to form the full set of official photos for Mr. Johnson's newspaper to use and distribute to other interested media agencies.

Despite the photographs and the presence of Major Marcel in a couple of them, a number of reporters still managed to find problems with the new "weather balloon" story. For example, how could a badly torn weather balloon, as revealed in the photographs, be described as a "disc"? Also, why couldn't dozens of military men identify the object as such when they examined it closely? After all, the officer who was brought into Ramey's office to identify it did so without hesitation.

In an attempt to quell the continuing skepticism, the USAF tried to tackle these and any other pressing questions on the minds of a number of eager reporters at the time. Unfortunately, in doing so, the military only succeeded in creating more questions than answers. It wasn't long before the media noticed further discrepancies in the responses.

One such discrepancy was the size of the object in terms of the amount of material found scattered over a large area compared with the size of the object shown in the photographs. Major Marcel, who was ordered by Ramey to support the USAF's position that a "weather balloon" had crashed near Roswell and had to be the spokesperson on the matter since the media knew he was initially in-charge of the preliminary investigation of the debris field and materials, managed to slip up when he told reporters the wreckage was spread out in a fan-shape covering up to a mile in length and 200 to 400 feet wide with most of the material found at the narrowest end. The photographs clearly could not support the observation given the limited quantities presented to Mr. Johnson when the balloon was shown to him in Ramey's office.

Again, Ramey had to intervene personally to clear up the confusion. He went on radio stating to the media that what the USAF had found was a particularly large weather balloon consisting of multiple metallic radar reflectors attached by cables to multiple balloons. Afterward, Ramey ordered another military spokesman named Major Kirton to become the new frontline for the USAF in handling media questions relating to the so-called Roswell incident, as if Ramey was not entirely confident in Marcel improving his public relations duty. Either that or Marcel was telling far too many truths, and these were embarrassing the military.

Still the USAF was not having a particularly good day, as Major Kirton slipped up by not knowing exactly how large the balloon was. In fact, neither Ramey nor Kirton could come up with a consistent figure describing the balloon's exact size. It is almost as if the USAF was admitting to the media of no knowledge and responsibility for what crashed on the rancher's property and was trying hard to cover up something of potentially far greater importance given the inconsistencies emerging on the nature of the downed object. Still the stuff ups from the USAF continued.

For example, in another discrepancy to make the rounds in the media was why the USAF claimed the balloon would not be analyzed. The experts at Wright-Patterson AFB, whose job it was to examine aerial objects of interest to the military, made it clear to ABC News that they were awaiting the arrival of the wreckage for analysis, as though some people in the military still could not be sure whether the materials were from a weather balloon. Either that or it was a foreign-type of weather balloon and the military wanted to know who built it. Or perhaps it was both? Whatever the object was or whoever built it, the public would never learn the results of that analysis, as if what was found remained top secret to this day.

Despite the glaring inconsistencies, there was no joy for the media in pursuing the case as the USAF eventually stopped providing details in the hope the media would go away. With nothing else to go on, no direct evidence to contradict the military's explanation, and no one willing (or able) to talk about the materials and flying object[3], the media did exactly that, and nothing more was heard. And it might have remained this way if it were not for the dedication of two U.S. researchers.

In 1976, the story resurfaced when William L. Moore and Stanton R. Friedman sifted through interview notes of two witnesses whom Friedman had met. Retired Air Force officer, Major Jesse Marcel, and Lydia A. Sleppy, a former employee of Albuquerque radio station KOAT, both claimed they had chilling firsthand accounts of the events[4], as well as details of the debris found. We will discuss this further in Chapter 2. At any rate, when the statements were combined with another remarkable story told by an experienced civil engineer named Barney Barnett to his close friends and boss, claiming to have seen a badly damaged "dirty stainless steel" (or dark-grey) "disc" along with bodies sometime in the summer of 1947, it made for an intriguing case.

Were all these witnesses referring to the same object? If so, what was this "disc"? Could this wingless object represent something alien (and hence something new to science)? Or was it a new type of man-made experimental flying machine?

The results of the four-year investigation led to the publication of a book in 1980 titled *The Roswell Incident* by Charles Berlitz and William L. Moore. Additional information from Stanton R. Friedman and his co-author Don Berliner appeared in *Crash at Corona* in 1992.

When *The Roswell Incident* was released, it rekindled a fresh wave of controversy. Why? Because something in the witnesses' descriptions of the materials made them certain that this mysterious flying object was too unusual to be considered a weather balloon or any other man-made object of any sort. Indeed, Major Marcel, the man who knew practically every conceivable aerial object there was to be known at the time, which is why he was sent to investigate the crash site by his base commander while Cavitt was there to help confirm his findings, claimed that the wreckage could not have come from any known missile or aircraft. He firmly believed the materials were not the standard issue for building any type of man-made aerial object known at the time or to this day.

Perhaps the object was a new type of experimental flying machine developed in secret by the USAF that was being tested for a range of new hi-tech and super-tough materials? If so, the dark-grey foil itself was not used to build new fighter jets after 1947 despite the relative case by which the USAF could make this foil, apparently in large quantities. Why? Certainly it is not because the foil was a terrible material to use for aerospace applications. Quite the contrary. Its toughness and extreme lightweight properties would have made it the world's best aerospace material, and made any top military official at the Department of Defence proud of the person(s) who invented this material. Yet for some reason, the USAF decided against using this foil. It is as if something else associated with the materials had frightened off the USAF from ever using them again, or at least not for a long time.

Now we may understand why. The observations by a civil engineer of human-like subjects lying on the ground would suggest that somehow the experiment also included the testing of odd-looking and short-statured pilots. In fact, the engineer had paid very close attention to the bodies, as if these pilots were too real in his own mind to be something artificially made (i.e., dummies). And he was not the only person at the scene to be looking at the bodies. It would appear a number of other witnesses were trying to assess whether any of the pilots were alive.

To further fuel the controversial nature of this case, according to Barnett, the bodies were described as being "too unusual" to be human. Sure, each body was described as having arms and legs, a torso and a head, and all in the right places to be thought of as human...but they were not human in his opinion. Talk by Barnett of these victims having unusually large heads, large eyes, and a small slit-like mouth attached to incredibly thin and small bodies seems to be the obvious giveaway on this unusual factor.

Perhaps the USAF had also developed a new type of highly realistic-looking, child-sized thin body and big-headed plastic dummy designed to possess these unusual alien-like features. Or else we have a bunch of well-shaven and highly intelligent monkeys with over-sized heads (most probably suffering a medical condition known as macrocephaly) who died in the crash. Or perhaps very thin and short Japanese prisoners of war?

Leaving aside the astonishing claims of small bodies (or plastic dummies) with large heads that may or may not be linked with the first crash site and focusing more on the wreckage itself at the rancher's property, it would appear from the original witnesses' reports, as we shall discuss in Chapter 3, that the flying object was made of extremely lightweight, but extraordinarily tough, metallic and plastic-like materials. Moreover, it possessed unusual physical properties that had not previously been observed in any aircraft, missile or weather balloon, or discussed in books or the media. For example, witnesses claimed the plastic-like structural beams and parchment sheets would not burn with a cigarette lighter. In the case of the newspaper-thin foil, described as metallic-looking by the civilians but a metal by the military personnel and the most abundant material found at the crash site, not even a blowtorch could melt it either, according to Roswell military personnel who tested all the materials. More importantly, at least in regards to this research, the foil was described as dark-grey in color and having the unmistakable ability of returning to its original shape.

As no man-made plastic in those days could withstand the high temperatures of a blowtorch, this is almost certainly a metal or alloy of some sort. But if it were a metal or alloy, even by today's standards, the public remains relatively unaware of this type of unusual metal. So imagine back then. There would almost certainly be no talk of shape-memory foils by Marcel and those at the Roswell AAF, let alone the unwitting members of the public.

For something the USAF feverishly believes is man-made, this unusual shape-memory foil must definitely be part of some secret military experiment involving the testing of new materials. Yet if we accept the official and latest finding of the USAF on the incident (as we shall explain in Chapter 4), the wreckage found near Roswell was nothing more than the remains of a flimsy and ordinary weather balloon made of aluminium foil, balsa wood and plastics. In other words, the USAF believes no exotic metals were used in the experiment, let alone whatever these tough plastics associated with the flying object were. Somehow, one gets the impression there is a problem with this claim.

This is quite odd. Here we have two opposing groups looking at the exact same materials in all their glory, and yet they have managed to come up with completely different descriptions. On the one hand, we have the USAF claiming the metallic foil is made of flimsy aluminium with no unusual properties; and, on the other hand, the witnesses claimed a dark-grey metal was observed as being virtually indestructible when subjected to the high temperatures of a blow torch and heavy blows from a sledgehammer. In terms of the shape-memory recovery effect of the foil in question, the USAF claims no such material was used, whereas the witnesses are certain they had seen foil pieces able to return to its original shape. As for the color, we are dealing with either a bright-silvery metal, according to the USAF, to explain the choice of a metal (in this case aluminium), or a dark-grey "dirty stainless steel" metal according to the witnesses. The only thing both can agree on is the fact that the foil is a metal or alloy. That is about it. Everything else is completely the opposite. How is this possible? Clearly, to see the exact same materials up close and personal, one has to be telling the truth in this matter, and the other is completely wrong.

So who is right?

If the witnesses are telling the truth and the object was not a weather balloon of any sort (it would be hard to imagine the use of so much high-tech and super-tough materials for a slow-moving object driven by the wind), and then assuming it is a man-made flying object, it leaves us with an aircraft or missile as the most likely explanation. Except it would have to be a new type of aircraft or missile since no one officially knew anything about a dark-grey and extremely lightweight shape-memory foil and other super-tough plastic-like materials. In addition, talk of a dark-grey "disc" would suggest it was symmetrical in design, making it more likely to be a secret military experiment (and probably an aircraft since we have dummies or pilots associated with it), although how such an object would fly through the air with stability has yet to be explained.

Certainly, a secret military experiment involving a new type of aircraft would help create a compromise between the two camps. The only thing is, the USAF must agree it was a *secret* military experiment and one that involved the testing of a new aircraft—which, in the early 1990s, it had not, assuming, of course, that it has nothing to hide after all this time.

In fact, over the decade following the publication of Berlitz and Moore's book, the military maintained an eerie silence over the incident. This was unusual considering the USAF never made a claim to the media of a secret experiment in July 1947. Why was the USAF doing everything in its power to avoid discussing or releasing documents about the case whenever members of the public requested further information during the 1980s and early 1990s?

This was something former New Mexico congressman Steven H. Schiff had experienced personally. So he asked the General Accountability Office (GAO) in Washington, D.C. to conduct a comprehensive audit on all information relating to the case. Being "an independent, nonpartisan agency that works for Congress"[5], the GAO agreed.

A letter was sent to then Secretary of Defense William J. Perry requesting details on the policies and procedures for documenting "weather balloon, aircraft, and other similar crash incidents" and a test of them by conducting a search for one case. While the GAO avoided mentioning the Roswell incident specifically, the USAF managed to obtain the name of the person requesting the audit directly from GAO investigators and soon learned that the audit was related to the Roswell case from previous written correspondence between Mr. Schiff and the department. Eventually Mr. Perry sent the information he received about Mr. Schiff and the GAO to all the USAF chiefs.

The word "Roswell" in the context of a mandatory audit was enough to stir the military's top brass into action. Rather surprising considering this was supposedly an ordinary non-secret weather balloon experiment of bygone days for the military. Already the stirring of the top echelon pot was revealing the hallmarks of a highly secret experiment. The question is, why the secrecy? What's so special about the balloon (or was it really an aircraft or missile)? At any rate, with the incident given high priority, the USAF looked far and wide for anything to explain the case. And, it didn't take long for the military to find a scientist and some documents dated in the right timeframe that might do the job.

The latest official explanation for the Roswell case, as the USAF would have us believe, is that it was definitely a weather balloon since the materials looked similar to the witnesses' descriptions of the wreckage (i.e., witnesses did mention a lightweight metallic foil, structural beams looking a bit like balsa wood with some flower-like designs painted on, and plastic sheets), except that this time the USAF revealed why it had been enthusiastically (for want of a better word) avoiding the subject with the public: the object was part of a *secret* military experiment known as Project Mogul. Which balloon specifically crashed on the rancher's property has not been categorically stated in the documents uncovered for Project Mogul or confirmed with reasonable certainty from reports written by the people at Wright-Patterson AFB, assuming, of course, the materials had been analyzed at this base.

In fact, being something so secret that many USAF personnel had no knowledge of Project Mogul, one would expect scientists at Wright-Patterson AFB to have analyzed the materials and figured out the object in due course. So there must be reports from the base. Unfortunately, the USAF is not willing to release these reports to confirm its new explanation.

In 1994–95, the USAF provided a somewhat copious report to support its new position and had it published in a book for the public to read. With no other information to go on after receiving the documents, the GAO had to accept the USAF's conclusion as the most plausible explanation.

Was this enough to end the controversy surrounding the Roswell case? Apparently not.

As the public began to question further the claims of *bodies* recovered from the object, the USAF again had to revise its report in 1997. In this latest report, the USAF stated that the alleged "victims" were probably "dummies" flown in certain high-altitude weather balloons.

To support this latest position, a press conference was held on June 24, 1997 in Washington D.C. Colonel John Haynes, the USAF spokesman for the case, wanted to "set the record straight" by presenting not just the latest 1997 report with its new cover design and "Case Closed" title and mentioning dummies as a probable explanation for the bodies, but also a video tape.

To understand the USAF's latest and allegedly final position on the Roswell case, let's explain in a little more detail the contents of the video tape as the media saw it.

The tape begins with a symmetrical object containing a white top and silvery metallic underside being unveiled from a large box marked "Viking Project Langley Research Center". In fact, this object is the precursor to the Mars Viking lander that was built after 1967. The purpose of showing this object was presumably to let the media know about the existence of symmetrical objects in weather balloon experiments, even if this object was not built in the late 1940s. In the case of this Viking Project object, it had presumably reached an undisclosed altitude by a balloon before it was jettisoned, allowing the object to fall to earth as shown in the video. Next, the media sources were shown a white-colored, disc-shaped object slowly spinning from three exhaust jets as it returned to Earth. Then another type of disc-shaped object is seen landing gently on the Earth's surface with a parachute. Several more pictures of symmetrical metallic objects[6] presumed to have been built by the USAF at around the time of the Roswell incident are shown, including one sitting upside-down on a flat desert floor. However, the USAF spokesman for the video did admit these objects were, in fact, recovered in 1972. Another scene shows five military personnel rolling up a high-altitude balloon or parachute during a recovery operation in New Mexico. The USAF spokesman said this occurred in the mid-1950s, all the while emphasizing the unusual nature of the payloads that could be construed as being disc-shaped. More pictures of non-symmetrical objects appear (perhaps an early attempt at building a disc-shaped object) dangling at the end of cables held by large cranes.

We finally get around to looking at some file footage of equipment used in Project Mogul in 1947. A picture of a single silvery box-kite reflector and single balloon floating into the air is followed by another scene of eleven large spherical-shaped balloons tethered to the ground, as if suggesting these could have been responsible for the Roswell incident.

At last, the USAF introduced scenes of anthropomorphic dummies used in certain experiments. We initially see two men dressing up a dummy in standard pilots' clothing. Another scene shows the dummy being levered down by a crane into a cockpit. Apart from one happy shot of the dummy with its arms around the shoulders of two men, another scene shows what was initially thought to be a dummy ejected from a USAF fighter jet that looked like it was built in the 1950s. In fact, it was a real person who was ejected.

We go back once again to another high-altitude weather balloon, this time with a transparent skin, being filled with lightweight gases. Another scene shows a non-symmetrical payload of one or two dummies being lifted into the air by this transparent balloon. According to the USAF spokesman, the dummies reached an altitude of 98,000 feet, where one was ejected. A camera attached to this dummy permitted the USAF to observe what was happening. It is presumed the dummy had survived the fall when the media saw another scene showing the dummy being held in the air by its parachute.

The USAF spokesman also mentioned how these dummies had landed all over New Mexico and were often seen by civilians, as if to help explain to the media how bodies could have been observed in early July 1947, even though these dummies were deployed in the 1950s. In fact, the USAF later confirmed after the video was shown that the first dummies ever used by the military were in 1953 and not earlier.

The video ended with a scene of a real test pilot getting strapped up and ready for an experiment where one can observe him jumping off a high-altitude balloon. We understand this had occurred on August 16, 1960 on the Excelsior III at a height of 102,800 feet, with the man presumably surviving the fall, although we do not see this by the time the video ended.

If all this testing of weather balloons and dummies did involve the Roswell event in some way and was meant to culminate with a man parachuting from a great height just to see if he would survive in 1960, that's a very long time for the USAF to figure this one out. Even if the USAF was a little unwilling to commit a human test subject to this experiment as the most likely reason for waiting, the existence of parachutes from World War II would have seen the USAF perform tests with the dummies prior to 1950. Furthermore, assuming Barnett had observed dummies in early July 1947, it is clear they landed without parachutes. Naturally, this would suggest the USAF needed to attach parachutes to the dummies. Since the technology of parachutes existed by 1947 and given the persistence of the USAF to solve the problem, it is unlikely it would have taken them until the 1950s to realize these dummies had to wear parachutes. As for a human test subject, why would it take until 1960 to see if it worked? Indeed, the ability of the

USAF to allegedly build unusually tough plastic and metallic foil, as in the Roswell flying object in early July 1947, should have shown the highly advanced knowledge and technology the USAF had in the use of high-tech materials to protect humans. It would have been easy for the USAF to build the equivalent of a space suit for the human test subject using these materials by no later than 1949.

There is only one slight problem, though. The video and Colonel Haynes somehow failed to mention that any kind of exotic materials had been used during these highly elaborate military experiments. Furthermore, the experiments show all this work to allow a man to survive a fall to Earth from a high-altitude balloon occurred in the 1950s culminating in the final test in 1960. That is a very long time for the USAF to figure things out and have the solution ready given the state of advanced materials available to the USAF by July 1947. Does this mean the Roswell case represents another military experiment not seen in the video presentation?

As for the timing of these experiments involving dummies, this raised the eyebrow of more than one reporter. As one media representative said during the video conference held on June 24, 1997:

> "How do you square the UFO enthusiasts saying that they are talking about 1947 and you're talking about dummies used in the 50s, almost a decade later?"

USAF spokesman Colonel John Haynes explained this as:

> "Well, I'm afraid that's a problem that we have with time compression. I don't know what they saw in '47, but I'm quite sure it probably was Project Mogul. But I think if you find that people talk about things over a period of time, they begin to lose exactly when the date was."

So while the USAF claims the witnesses were probably having time compression problems, consistent claims of an unusual shape-memory foil together with other tough materials by a number of witnesses would suggest the USAF might be suffering amnesia problems of its own given the "aluminium foil" explanation it has suggested for the Roswell case.

So who is really having the problem with remembering the finer details of the Roswell case?

Despite this, the media had to go quiet on the Roswell case once again, as they did in 1947. There was simply nothing more to go on, and certainly nothing in the materials to cast doubt on the official explanation.

So what now?

It was starting to look like the public had reached the end of the line for the Roswell case. In that case, there should be no further scrutiny of the case by anyone whatsoever. In fact, to make sure of it and so discourage others from looking at the case again, the USAF marked the updated 1997 report with the words "Case Closed".

The case was far from being closed.

Mr. Johnson, the man who took the photographs of the weather balloon in Ramey's office, was not entirely pleased with the way the USAF had conducted the investigation. As he stated in October 1996:

> "I was not impressed with the attitude of the Air Force Intelligence officer who interviewed me in 1994 in connection with the GAO investigation in that he seemed to be just going through the motions and not too concerned with finding the true facts—his mind seemed already made up!"[7]

Further supporting his view was the way the USAF had analyzed one of his photographs. He had mentioned to the military investigators that in one of his photos, Ramey had inadvertently held part of the top-secret contents of the memo facing toward his camera. Since the camera he used at the time had a high-quality lens, he wondered whether it was possible to read the text in the memo.

When the USAF learned of this, the photograph was sent to an unnamed national-level organization for independent analysis. The results came back negative from the experts, saying it was too hard to read the text.

Not convinced, Mr. Johnson took it upon himself to conduct his own investigation. This time he assembled a team of experts to analyze the original photographic plate to see if anything could be read. Remarkably, Mr. Johnson and his team claimed certain words were readable. More importantly, the words revealed a different story than the one officially provided by the military.

Given the potentially explosive nature of Mr. Johnson's findings, other researchers tried subsequently to reproduce the results with varying degrees of success.

Then in 2000–01, an optometrist turned researcher, Mr. David Rudiak, made further progress into the analysis of the memo by performing his own digital enhancements to the blown-up image, counting the number of letters in each word, being able to identify the letters that were readable and leaving aside those letters considered more contentious. The difference this time, compared to his fellow researchers, was that Mr. Rudiak added one other tool: an electronic dictionary. Eventually, after a long and laborious analysis of the memo and making sure the sentences made reasonable sense, not only were other researchers and he able to clearly see and agree on certain words —including "victims", "disc", and "weather balloons"—but he was also able to hone in on the most likely words for other portions of the text, including "Rawin demo crew" and "new find". While these words on their own do not prove conclusively that the Roswell object is anything exotic in the sense that it is of extraterrestrial origin, their position within the memo and choice of words does provide another level of controversy to the case.

For instance, it is clear that the word "disc" is still maintained in Ramey's secret memo. For some reason, the Pentagon thought it was important to keep using the word "disc". Perhaps it was merely in reference to the original news release? Except the Pentagon also wanted to include the word "victims" in the final sentence of the previous paragraph just before the word "disc" is mentioned in the first sentence of the next paragraph, as if there is meant to be a connection between the two. Even if there is no connection, using the word "victims" is highly unusual because the word was never mentioned in the original news release, or by any of the witnesses of the initial crash site. This is something that the USAF invented after the wreckage was recovered from the rancher's property.

This implies that the USAF had discovered bodies out in the middle of the desert.

Or were these actually dummies?

Well, the USAF cannot be referring to "anthropomorphic dummies". The word "victims" points strongly to a biological discovery that was undeniably once alive. Seriously, how can a dummy be described as a "victim" after crashing to earth? It has no feelings, so it should not be called a "victim". This is reinforced by the fact that the words "demo Rawin crew" (and "weather balloon") in the final paragraph are separated significantly from "victims", as if no direct link is meant to be present. Such a separation distance indicates that genuine yet unknown (or perhaps simply unnamed) individuals (human, or some other terrestrial "human-like" species) had died in the crash and the military wanted to cover up the situation by suggesting the use of "dummies" as a good enough explanation in case the media somehow learns about these "victims".

In other words, the early part of the memo was meant to explain to Ramey the actual reason for the cover up by mentioning the recovery of a disc-shaped object and "victims" without going into too much detail as to whether we were dealing with something extraterrestrial or some other foreign source. It is only in the final paragraph do we learn about the possible use of dummies (i.e., "demo Rawin crew") and in presenting a weather balloon explanation as the best way to handle the slightly more troublesome questions relating to the apparent recovery of "bodies" and the so-called "disc" from the public and media representatives.

But why? No one ever mentioned bodies (or even dummies) in the context of the initial crash site. Yet, somehow, the USAF had obtained information to make it believe "victims" were found and was anticipating the possibility of the media learning about them and was prepared to give an alternative explanation if it would keep the media quiet.

As for maintaining the word "disc" in the memo—again separated appreciably from the term "weather balloons" at the end, as if suggesting no direct connection—it is pushing this Roswell incident toward a different story from the one being told to the media. Since "victims" were not mentioned in the initial news release, the memo is pointing to another type of secret military object, possibly shaped like a disc, which the USAF does not want the public to know about (i.e., most likely a highly advanced aircraft carrying pilots and possibly some passengers), except the "weather balloons" explanation would be used as a decoy.

In fact, the continual persistence in mentioning the word "disc" in the memo at a time when the military would have had ample time to examine the materials at Wright-Patterson AFB and, presumably, the rest of the original object (it must have been found some distance away together with the "bodies") must support the idea of a "disc". The choice of the word must be because either the object was shaped like a "disc", or the USAF assumed a "disc" of the type seen in those UFO reports because of the unusual nature of the bodies, and possibly materials.

But even if it is not something exotic, what kind of man-made disc-shaped object was it?

A disc-shaped weather balloon? Unfortunately, the super-tough materials claimed by practically all the firsthand witnesses would suggest a much faster flying object. Perhaps a disc-shaped aircraft or missile? However, with "victims" involved, it would have to be an aircraft of some sort, right?

Otherwise, the only other alternative is to say the object does represent something unknown to the military in terms of a new technology that was built by someone else. However, no one from the scientific community or elsewhere in the world has come forward to take responsibility for the object that crashed in the New Mexico desert, explaining the unusual materials they had used and who piloted the "disc-shaped" aircraft (or missile). This raises the prospect that the creators of this object could be a little more unknown, perhaps unusual in the sense that Barnett could be right in his claims of something extraterrestrial.

Of course, this would bring considerable debate into the scientific realm if we introduce the extraterrestrial hypothesis. Yet genuine scientists cannot discount the possibility. Certainly, it would explain the subsequent behavior shown by the military after recovering the super-tough materials and "victims" and the choice of using the word "disc" to describe the object. The word "disc" had to be used because the materials were indeed unknown to the military, and the bodies made them realize it was probably one of those flying saucers talked about in the media.

Sounds reasonable. But is it true?

Whatever the truth, one thing is certain: no Project Mogul balloon can ever be described as "disc-shaped". The photograph of Dr. Charles B. Moore (the scientist who formerly worked on the secret project) taken by New York University staff on behalf of the USAF investigating the incident in 1993–94, could not reveal the disc-shape nature of a typical Mogul reflector as he stood in front of the camera holding up the reflector itself by the cables. Nor has the USAF satisfactorily explained the reason for choosing yet another unfortunate word choice through "victims" in the secret memo held by Ramey and in its 1997 report. As for the materials used to construct the object, virtually every firsthand civilian witness and military personnel at Roswell AAF who inspected the materials remained baffled by the materials' high strength, hardness, and high-temperature resistance[8]. The weather balloons discussed by the USAF in its official 1994 report were too flimsy to support these claims. In addition, there is no indication of any Mogul balloon having a new high-tech metallic foil capable of returning to its original shape.

The case clearly cannot be "Case Closed".

Of course, all this could still be explained as another secret military experiment.

Is there a way to resolve this controversy?

Well, one thing is absolutely certain. Despite the different views and inconsistent information coming out of the USAF, there is universal agreement on one simple fact: something crashed near Roswell in early July 1947. As the *USAF Executive Summary of the Report of Air Force Research Regarding the "Roswell Incident"* published on September 8, 1994, stated:

> "There is no dispute, however, that something happened near Roswell in July, 1947, since it was reported in a number of contemporary newspaper articles; the most famous of which were the July 8 and July 9 editions of the *Roswell Daily Record*. The July 8 edition reported 'RAAF Captures Flying Saucer On Ranch In Roswell Region', while the next day's edition reported, 'Ramey Empties Roswell Saucer' and 'Harassed Rancher Who Located Saucer Sorry He Told About It'."[9]

And we can also add to this agreement the view that a metal or alloy was found for the Roswell foil. Otherwise the USAF would not have chosen "aluminium" as its preferred explanation for the foil. And it must also take into account the blowtorch test performed on the Roswell foil by the military witnesses, as well as the fact that virtually all the military witnesses were confident in describing the foil as definitely a metal or alloy.

Together with a large body of testimonial evidence from civilian and military witnesses and the official news release, this isn't a question of "Did the crash occur?" but rather "What crashed near Roswell in early July 1947?"

And indeed, what type of materials were found? And what's the composition of these materials to explain the incredible toughness and extreme lightweight characteristics?

Here lies the crux of the problem. Are we dealing with a man-made flying object (probably from another undisclosed secret military experiment) operated by the USAF and testing new types of exotic aerospace materials, or is there meant to be more to this case than meets the eye?

Then, in the early 1990s, an Australian researcher noticed, within the original witnesses' statements, a particular observation never before analyzed at the time. It concerns the dark-grey foil found at the Roswell crash site. According to the witnesses, this dark-grey foil had the claimed ability to return to its original shape after applying a force to deform it, either by hand or with an object, such as a sledgehammer. On the face of things, this may seem rather insignificant, except that an alloy having this distinctive dark-grey color and memory behavior had been identified in scientific literature, and is considered the world's most powerful shape-memory alloy (SMA) and the first to start serious interest in SMAs by the scientific community.

The "distinctly dark-grey" SMA at the centre of this research is NiTi (or nitinol), containing nickel and titanium in roughly equal quantities.

Not expecting to find anything unusual, the researcher made a study into the alloy's properties (see Chapter 6) and history (see Chapters 7 and 8) only to discover some remarkable information about the relationship between the USAF and SMA research, including NiTi. Not only was the alloy studied at Wright–Patterson AFB (the place where the Roswell wreckage ended up for analysis) at nearly the right time (just after 1947, or to be more precise, in early 1948), but at least three, and potentially four, SMAs were studied: TiZr, NiTi, NiTiCo and NiTiCr.

What is more astounding is the realization that NiTi is not only the world's most powerful SMA, but it is also the world's lightest titanium-based SMA. And with titanium making up a significant proportion of its composition, it is also one of the toughest alloys around. In view of the witnesses' statements of how lightweight and tough the Roswell foil was, the dark-grey SMA would be the perfect candidate, assuming, of course, the USAF wanted to build an extremely lightweight and strong secret flying object of some sort, using whatever technology was available prior to July 1947 to make a NiTi-like material.

Of course, this is not to say the Roswell foil is NiTi. Perhaps it was. Or it could be NiTi with a little bit of cobalt. Or perhaps there is another dark-grey SMA? Certainly, TiZr is not dark-grey—it is more bright silvery in color. Unfortunately, from the available official military reports on titanium at the time, only those alloys containing NiTi (if the Ni and Ti elements are in sufficient quantities) are likely to possess this distinctive and natural dark-grey color. Furthermore, the scientific literature has not revealed another equally powerful, let alone more powerful, SMA composed principally of titanium that is dark-grey like NiTi. NiTi (and any derivatives of it) is the only dark-grey SMA we know of to this day. Clearly, NiTi is the alloy we should be looking at in regards to the Roswell case.

Whatever the USAF had found in NiTi during its secret studies, research has indicated that NiTi, NiTiCo and NiTiCr (where Co and Cr are in small quantities compared to Ni and Ti) are currently the closest titanium-based SMAs we have to match the observations of the Roswell foil. And these are the ones looked at by the USAF at around the right time. What other NiTiX (where X is another element) could there be?

Additionally, NiTi's phenomenal hardness with continual cold-working (e.g., bending) and in the "shape-memory" activated crystalline state upon heating or application of an electric current would have been ideal for aerospace applications (and would certainly explain the alloy's ability to resist the appearance of dents after heavy blows from a sledgehammer).

Furthermore, the alloy has the necessary high-temperature resistance (as well as high strength, and hardness, especially in the activated shape-memory response state) to support the witnesses' claims of a super-tough metallic foil at the time. And if the USAF had indeed built a weather balloon or something else using a SMA, revealing this information would have aided them in explaining away the crash debris as man-made. Just tell the public that the military had been using NiTi or something similar (well, it must be of a dark-grey color with shape-memory properties, and an alloy of some sort) in its secret experiments using weather balloons (or more likely something else, such as an aircraft) and that's it. End of story. Then we can all move on with our lives.

But they haven't. Why?

Just to add to the intriguing discovery, the principal high melting point metal in question—titanium—had a major and sudden technological status change in the U.S. after 1947, made evident by the way Battelle scientists, under contract with the USAF, had suddenly discovered the useful engineering properties of highly pure titanium in alloy form after 1947. This was happening nowhere else in the world. In fact, it was only after 1947, when the USAF at Wright-Patterson AFB requested that the Battelle Memorial Institute in Ohio study the metal and at least one SMA of interest to the USAF known as NiTi, with emphasis on high purity in the metal for producing these sorts of alloys, did U.S. scientists become suddenly and uniformly enthusiastic about the potential of the metal in alloy form according to a *Scientific American* article dated April 1949.

Yet before this time, no one in the scientific community was officially expressing such gusto towards the metal. In fact, no scientist in 1947 was aware of the importance of high purity work in developing new types of titanium-based alloys with unusual and highly useful engineering properties, such as a shape-memory effect. NiTi wasn't even known to be a SMA (in fact, the term hadn't been invented). Only the USAF was showing great interest in the alloy and titanium at some point (presumably in 1947) and eventually had to ask Battelle to study these materials after 1947. It was as if the military had discovered something special, and wanted to know how the alloy worked, and that the high purity of the alloy was paramount to revealing this. However, the military had to ask for outside scientific assistance to solve this problem. Furthermore, the request for help came after 1947.

This is rather odd.

You see, for a secret military experiment involving the use of a titanium-based SMA to occur by early July 1947, the USAF would have already acquired all the essential knowledge and developed the technology to manufacture pure alloy in newspaper-thin sheet form for assembling into a "disc". Yet the military had to ask for scientific assistance after 1947 to study a number of titanium-based alloys, including NiTi, as well as increase the quantities of high-pure titanium available.

How can this be possible? Didn't the USAF know how to do all this work by no later than July 1947?

And why NiTi? Was this because the USAF hadn't worked out the knowledge behind how SMAs worked? Or is it something else? But if it was something else, then we would have to realize that the purity and quantity needed to reveal the shape-memory effect in NiTi (a critical factor) was not there until after 1947 (in fact, it was not until the 1950s did the USAF have enough titanium to consider making large quantities of a highly pure titanium-based alloy to build a titanium aircraft) when scientists became officially enthusiastic about the metal. So why wasn't the interest there sooner?

And even when the enthusiasm was (finally) there, there was nothing in 1948 or 1949 to mass produce titanium-based alloys in highly pure form to manufacture the amount of NiTi-like foil used in the Roswell object.

Even if this is untrue, a question remains. Where is the location of the secret titanium manufacturing plant to support the USAF manufacturing of the secret flying "disc"? Because officially speaking, recently declassified documents suggest the work into all titanium-based alloys recommenced on May 18, 1948. And from these documents, we discover NiTi and a select number of other titanium-based SMAs. Never before do we learn of any scientist or the USAF ever making a discovery of a titanium-based SMA. The only other time we learn about the existence of titanium alloys in association with any USAF study was in early 1947 through Project Air Force RAND. However, nothing in the conclusion of the March 1947 report indicated the need to produce high-purity titanium-based alloys considered critical to making SMAs. Only after 1947 do we see high-purity titanium alloys given their deservedly serious importance by the USAF, of which no less than three are notable titanium-based SMAs (including NiTi), and to find ways to mass produce those titanium-based alloys of interest to the USAF.

This raises a serious question: Assuming we can rely on the official evidence from these declassified USAF/Battelle documents and Project Air Force RAND, to ask for scientific assistance on the grounds of purity and finding ways to manufacture titanium-based alloys after 1947 was tantamount to saying the military didn't have the technical knowhow to produce a titanium-based SMA for itself in 1947.

As further evidence, we find from the historical records how the first official titanium military aircraft was built in 1951–52. Why? On further reviewing the evidence, it seems the reason was because there wasn't enough titanium in commercial production to meet the demands of the U.S. military. Well, didn't the USAF have enough titanium by July 1947 to build one large flying object (apparently to carry a few dummies or pilots)? Even more unusual is the decision by the USAF to stop all work into titanium and its alloys in 1947. This is despite the fact that the USAF had already discovered, and was allegedly an expert in developing, a unique dark-grey titanium-based SMA on its own (as there are no reports in existence to support "outside assistance" from Battelle or anyone else). It is also despite the fact of USAF having built an allegedly secret military object containing a significant amount of this SMA covering at least the outer skin component. Then, for some reason, the USAF decided to come back to study titanium after 1947 where a dark-grey SMA and several others of its kind were mentioned for Battelle to look at (we do have the reports for this "outside" help). And finally the USAF decides to build another titanium aircraft at such a late stage in the game without using a SMA of any sort.

Something doesn't make sense.

If all this is untrue, where is the report dated prior to the Roswell event to support this secret work on titanium-based SMAs by the USAF? Well, surely, the military must have known something about SMAs. The witnesses spoke of a foil that could return to its original shape, and with emphasis on being an alloy of some unknown composition at the time (although one military general has suggested the alloy probably contained titanium, just to make things more interesting). The USAF is claiming the event is man-made. And we know the materials ended up at Wright-Patterson AFB for analysis where the shape-memory property would have been easily observed and later documented in the USAF/Battelle reports after 1947 of several titanium-based SMAs of interest to the USAF. So there must be a report dated before July 1947 somewhere in the bowels of the military. And it must be a report showing how a SMA using titanium had been discovered. Was it done on a trial-and-error basis? Or did it involve a lengthy process of analyzing the scientific literature and coming up with a new scientific theory for how a SMA works, and then applying the theory to a titanium-based alloy?

The report should also explain how the USAF managed to build a titanium-based flying object by early July 1947. It is clear that special equipment must have been developed by the USAF or someone else under contract with the military for this specific purpose of extracting titanium in highly pure form, combining an additional element or two to create an NiTi-like alloy, and then manufacturing this alloy in a large "newspaper-thin" sheet to at least cover the skin of a 9-meter diameter, presumed to be disc-shaped, flying object. The report must show how this was achieved in considerable detail.

And if such a report does exist, why bring Battelle back in to study titanium and its alloys after 1947? Indeed, why is there continued secrecy in the USAF over titanium-based SMAs to this day? Clearly, the USAF needs to write another report to explain the continuing secrecy over SMAs and the object itself. Did the military forget? Or is the USAF hiding something else?

Indeed, the secrecy must extend to the Roswell case as a whole since it is clear from Mr. Johnson's photographs that a crinkled bright silvery aluminium foil was shown. Why crinkled? Apart from aluminium foil not showing the required high-temperature resistance and other toughness tests, a SMA would not show a crinkled surface. It had to be smooth just as the witnesses reported[10] (and at room temperature).

Again, something doesn't add up.

Unless the USAF had forgotten or did not locate the right documents to explain the highly secret work it did on SMAs prior to July 1947, it could mean the USAF is concerned the public would discover something else in association with SMAs that might seriously question the "weather balloon and dummies" theory, or even the man-made explanation, should evidence be found to suggest otherwise.

Could this be because of the alleged "victims" found associated with the "disc"? Is there something sensitive about the "victims" and the USAF does not want the public to know who these victims are?

Well, the secrecy cannot be about the SMAs, per se. As we shall see, SMA research is not exactly new to science—certainly not today, and not even 50 years ago. Okay, the work into this branch of science wasn't exactly on the minds of every scientist working outside the U.S. military in 1947 or 1948. One or two scientists at Battelle who were contracted to work in secret with the USAF might have known. Anyone else at the institute was probably kept in the dark. The only other people who might have known were a handful of scientists working at Wright–Patterson AFB. The only hint the rest of the scientific community had of a potentially new branch of science was a couple of ordinary alloys exhibiting relatively weak "pseudoelastic behavior". Of course, we now understand this behavior to be the "shape-memory effect". Yet it was not enough to warrant a new scientific classification (and, therefore, an all-out study) into this class of alloys. In fact, a close study into the history of SMAs (see Chapter 6) will reveal how these two pseudoelastic alloys were essentially ignored as scientific oddities. With the scientific community quiet on the subject, there was nothing in scientific literature to predict the discovery of a more powerful and tougher SMA using titanium as the principal metal in its composition. Only after 1958 (or 1961 when the news officially reached the public on the world's most powerful dark-grey SMA called NiTi from U.S. Navy metallurgists) did the official study of SMAs finally take off—some 12 years after the Roswell crash.

Leaving aside how the presence of a NiTi-like material at the crash site can be explained, the secrecy must be about something else. And it must relate in some way to the "victims". Is there a way we can find out?

Before we reveal these intriguing new findings, let us start from the beginning. As with any ground-breaking research work, it is important that we look at the *essential history* of the Roswell crash, paying particular attention to the early statements made by the original witnesses, as mentioned in official affidavits, Berlitz and Moore's 1980 book, and any later interviews with the original witnesses, to help paint the clearest picture of what was witnessed. Of course, no research is truly balanced unless we also present the official view of the U.S. government, according to the USAF in 1947 and in the most recent study on the case conducted in 1994 and published in 1995 (and again updated in 1997).

Finally, we will compare the two opposing theories to explain the Roswell case, using the latest information from the scientific literature and historical insight to see how well the witnesses and the U.S. government fared in the analysis. This is the kind of intriguing analysis that could, at last, lift the shroud of secrecy still surrounding the Roswell incident and show, on the basis of probability, what the USAF found.

CHAPTER 2

A Brief History

A UFO is seen heading toward a large thunderstorm

ACCORDING TO Berlitz and Moore, on the night of Wednesday July 2, 1947, at about 9:50 PM, a brightly glowing disc-shaped object flew at high speed over Roswell heading in a north-westerly direction toward a large thunderstorm that was sweeping over New Mexico.

Observed by Dan Wilmot, a local hardware dealer, and his wife while sitting on the front porch of their home at 250 South Penn Avenue in Roswell, *The Roswell Daily Record* on Tuesday, July 8, 1947, claimed Mr. Wilmot's description of the UFO was "like two inverted saucers faced mouth to mouth".

Mr. Wilmot said:

> "All of a sudden a big glowing object zoomed out of the sky from the southeast. It was going northwest at a high rate of speed."

Curiosity clearly got to the better of the Wilmots. After noticing the UFO's unusual symmetrical shape, emitting a white glow over its surface, and moving silently at low altitude, the observation was compelling enough to beckon the couple to run out into their front yard to take a better look.

Mr. Wilmot, an old man at the time, had probably reached a point in his life where his hearing was not at its best. As a result, he said the object was silent, while Mrs. Wilmot claimed it was silent except for a brief moment when a swishing sound could be heard as it passed directly overhead. At its closest point, the couple thought there was a glow coming from *inside* the object. They also had the impression the object was large. In forty to fifty seconds it flew out of sight.

The Wilmots did not speak to anyone about the sighting for fear of being ridiculed until someone else was brave enough to report on the sighting. No one did. Time passed and eventually, on July 8, Mr. Wilmot's firm conviction of what he saw made him pluck up the courage to personally report it to the *Roswell Daily Record*. Luckily for Mr. Wilmot, his story came within a matter of minutes before a news release from the Roswell Army Air Field (RAAF) had arrived claiming a flying disc was recovered from a rancher's property somewhere near Roswell.

The local reporter who heard Mr. Wilmot's story did not know what to make of it and was not sure if it should be published. All he had was the testimony of Mr. Wilmot—later confirmed by his wife. The next thing he knew, the military stationed in the city had announced the recovery of a flying disc on a rancher's property. Were the two stories connected? He did not know. Certainly Mr. Wilmot's timing in releasing his story was impeccable. The only question was, could the story be considered sufficiently credible to put into the newspaper?

The reporter checked with locals and discovered that the Wilmots were "...one of the most respected and reliable citizens in town..."[1]. Realizing these were not the sort of people to make up a flying saucer tale for the sake of it, a decision was made by his editor to publish the story.

Apart from the timing of the sighting and the direction the UFO took, it remains unclear to this day whether this was the UFO that had dropped pieces northwest of Roswell and eventually crashed at another location.

The odd explosion

A short time later, a mysterious explosion was heard, at a place about 127 kilometers northwest of Roswell, near the Foster sheep ranch, managed by William Ware "Mack" Brazel[2] (1899-1963). As Brazel and his family described it, the explosion was not like ordinary thunder, but different.

William's son, Bill, a geoseismologist in Alaska's North Slope oil region and an employee of Texas Instruments, recalled some details of his father's involvement. Bill said that his family—his father, mother and the two children—saw a terrible lightning storm[3] come up from the east. There was not much rain, only lightning.

Bill said:

> "He [his father] said it seemed strange that the lightning kept wanting to strike the same spots time and again, almost as if there was something attracting it to those spots; he thought maybe underground mineral deposits or something. Anyway, in the middle of this storm there was an odd sort of explosion, not like the ordinary thunder, but different. He said he didn't think too much about it at the time because the storm was so bad that he guessed it was some freak lightning strike, but later he wondered about it."[4]

The lightning quickly died down following the explosion, the thunderstorm moved away, and the family was able to take a hard-earned rest from the previous day's chores.

Despite things getting back to normal, William was not entirely comfortable with what had transpired. He couldn't shake the feeling that perhaps some of his sheep could have been injured from all the lightning strikes. As he prepared to go to sleep he decided that he would check on his sheep the moment there was daylight.

The first crash site

William awoke at dawn, as usual. He sat on the edge of the bed, stretched out his arms, and gave out an almighty yawn that any American man could be proud of. He got up, washed and dressed in his

American cowboy clothes. He went into the kitchen to have a healthy helping of bacon and eggs and some coffee. Before he left, he told his wife Maggie[5] (1902–1975) of his plans to check on his sheep and go down to pick up Timothy "Dee" Proctor, the seven-year-old son of a neighboring rancher, whose sharp "eagle eyes" would help him find his animals.

As he headed out the door, he donned his cowboy hat. He walked outside and looked in the general direction of where he would investigate his property. Nothing unusual he thought. Clearly his sheep were not anywhere to be seen, suggesting to William that the animals must either be congregating near the watering hole or hiding in another part of his property. He decided to check the area described as the Hines pastures for any wounded animals, and later to look elsewhere should he find his sheep grazing in another part of the property. It would likely be a long day ahead.

He took a deep breath to enjoy the brief moments of cool air outside. He could see the sky was clear and it was going to be another hot day. He carried the saddle to his horse, strapped it up, jumped on and made his way toward the Proctors' home on what he thought would be another uneventful day.

This was the morning of July 3, 1947. Everything was suggesting this was going to be another typical day in the life of an American cowboy managing his property and animals, but this was not going to be an ordinary day for him.

As he and Timothy rode gingerly on horseback over the property, it didn't take long[6] before they reached an escarpment where they were able to look down and over a wide area of the property.

As they reached this new vantage point, they were quickly distracted from their search by something else. What they saw was not the sheep they had come to find, but something far more intriguing. Lying before them and scattered far and wide was wreckage from an unknown flying object.

Bill stated:

> "Anyhow, the next morning while riding out over the pasture to check on some sheep, he came across this collection of wreckage scattered over a patch of land about a quarter mile long or so, and several hundred feet wide. He

said to me once that it looked like whatever this stuff had come from had blown up [above the ground]."

The horses were getting agitated at this time, as if neither animal wanted to go down the escarpment to get amongst the debris.

One of the reasons for bringing Timothy along was his excellent eyesight. No doubt the young budding cowboy proved his worth yet again when Timothy said he could see William's sheep at a good distance away, apparently avoiding the debris site for some reason. No indications of any injured animals lying on the ground as far as Timothy could see. William was grateful for this. Realizing the animals were probably fine and a quick glance in the direction of the sheep seemed to confirm this, William's focus was drawn back to the debris. He jumped off his horse and told Timothy to stay put. He walked down and picked up the nearest piece. It was a metallic foil, dark-grey in color, and the most abundant of all the materials he saw. From the way it was machined into a smooth sheet and with a slight curve to it, he thought the foil must have come from a large object. Unfortunately, he could not tell *what* it was or precisely *how big*. In fact, there was nothing in what he saw to indicate the size or shape of the object, just a lot of debris scattered everywhere.

The color of the foil reminded him of lead, except that was about as far as the resemblance would go. Everything else about this foil suggested to him that this was definitely not lead. For a start, it was too lightweight. And its incredible thinness would disguise what was essentially a very tough material with an unusual property. What he had in his hands was a sheet with no creases. The foil looked flimsy because of its weight—it was like holding a feather—and remarkable thinness, but it was not. William tried to do everything he could to damage it without success. The most the foil would show was the ability to bend up to a point as he tried to scrunch it up in his hands. But as soon as he let go, the foil suddenly returned to its original shape. He was left astounded by what he saw. It was almost like a plastic of some sort, except it was definitely a metal. It was a metal he had never seen before, or see later. Surprise quickly turned to excitement as he realized the unusual nature of his discovery.

Noticing there was plenty of this stuff lying around, he dropped the foil he had in his hands quite casually and looked around some more. It didn't take long before he noticed a couple of different materials. One looked like a sheet of plastic and another was a tough plastic piece of broken beam reminiscent of balsa wood because of its color and extreme lightweight nature. Again these weren't the normal plastic or wood he was familiar with. These materials could not be torn or broken with his hands, nor would they burn with his cigarette lighter. Again he was baffled.

Timothy was sitting patiently on his horse watching from a distance. He could see that William was getting more excited with each passing minute he spent with the debris. After he saw William trying to burn what appeared to be a piece of the parchment with his cigarette lighter, William promptly returned to his horse with a handful of this unbreakable plastic sheet and put it in his saddlebag. He told Timothy he was going to visit his parents' house to seek his neighbor's opinion of what he had found.

Was there any indication of the direction in which this mysterious object that had lost some wreckage was heading? William's son, Bill, revealed:

> "He [Bill's father] also said that the way this wreckage was scattered, you could tell it was traveling an airline route to Socorro, which is off to the southwest of the ranch."

The second crash site

One morning in the summer of 1947, Grady Landon "Barney" Barnett (1892–1969), a former Lieutenant in the U.S. Army who became a civil engineer for the U.S. Soil Conservation Service, was in the Socorro region not far from the Plains of San Agustin. It is a region lying approximately 400 kilometers from Roswell and 200 kilometers in a westerly direction from the Foster Ranch.

As he drove, he noticed light reflecting off a large metallic object in the desert. At first he thought a plane had crashed. However, his initial assessment of the object was quickly dashed upon arriving at the scene. Now he could see the object was not a plane, but a large disc-shaped dark-grey or "dirty-stainless steel"[7] metallic object about 9 meters in diameter.

The object looked like it had been "burst" open by an explosion or impact (although almost no wreckage was scattered around, as if the materials were tough). Where the object was intact, he observed no signs of riveting on the surface to show the skin was composed of multiple sheets. It looked as though it was a single large sheet that broke apart. Also, there were bodies lying about the wreckage. The bodies were small and slender with large heads, no hair, each wearing a one-piece and smooth skin-tight grey metallic suit without zippers, buttons or belts.

Not long after Barnett arrived at the scene, a university archaeological research team—about four or five people, including a professor—also allegedly arrived at the place where they thought a plane had crashed.

Barnett thought the bodies all looked like males, but he couldn't confirm it[8]. No one could say exactly who these "victims" were. There was no indication at close range that Barnett and the others were looking at dummies from some military experiment conducted in the area. If they had, it is unlikely they would have continued their focus on the victims for as long as they did. Something was telling him and the others that these bodies were real and each one had to be checked in case any were alive.

Unfortunately, there was not a lot of time for Barnett and the others to determine if there were any survivors. In a matter of minutes, a military officer drove up in a truck and told everybody to get out of the way. Before the witnesses left, they were told that this was a classified project and that it was their patriotic duty not to talk to anyone about what they had seen.

Was this enough to keep everyone quiet?

The university group certainly did as they were told, and for a while Barnett did as well. However, Barnett was already an old man when he eventually spoke for the first and only time of his remarkable experience. He mentioned his unusual sighting to his wife, Ruth, his boss, and his close friends, Mr. L. W. "Vern" Maltais and his wife, Sean, in February 1950. He cautioned them not to repeat it. To make sure of it, it seems Barnett decided not to reveal the exact date of when it happened.

This is the only time we would learn about this second crash from civilians.[9]

Clearing the first debris site

By late morning, William took Timothy home to his parents—Floyd (1903-1985) and Loretta (1892-1979) Proctor—who lived about 10 miles away from his ranch and about 20 miles from the debris site. Before he left, he took another look at his sheep. They had quietly moved over to another spot, but well away from the wreckage. Indeed, William couldn't help noticing the way the sheep was refusing to cross the Hines pasture, where the wreckage was scattered, to get to the watering station nearly a mile away.

Floyd and Loretta greeted William on his arrival that afternoon. He didn't waste any time mentioning his discovery. He eagerly got off his horse and took a fragment out of his saddlebag to show the couple. Loretta later recalled hearing William say something about the metallic foil he found on his property as, "...if crumpled and let loose, regained its original shape again."[10]

The sample William took out of his bag was the extremely lightweight brown plastic sheet. He explained it couldn't be cut with a knife or burned with a cigarette lighter. He even tried to give a demonstration.

Loretta could not get direct confirmation for the odd behavior of the metallic foil by seeing a sample as William was, at that precise moment, more excited by the unusual toughness of the plastic sheet he found.

In an interview with William Moore in June 1979, Floyd recalled what had happened:

> "[William] Brazel had come over to my place late one afternoon all excited about finding some sort of wreckage on his ranch. He wanted me to come over with him and look at it, and described it as 'The strangest stuff he had ever seen'. I was tired and busy and just didn't want to bother going all that way over there right then. You know he tried, he really tried to get us to go down there and look at it."[11]

It was clear in Floyd's mind how excited William had been about the materials. He could not stop talking about it, which was unusual for a man who preferred to keep quiet. As Floyd recalled:

"He [William] was in a talkative mood, which was rare for him, and just wouldn't shut up about it. He described the stuff as being very odd. He said whatever the junk was, it had designs on it that reminded him of Chinese and Japanese designs. It wasn't paper because he couldn't cut it with his knife, and the metal was different from anything he had ever seen..."[12]

Loretta came into the room and on hearing the interview gave further details to William Moore. In an affidavit signed on May 5, 1991, Loretta said:

"In July 1947, my neighbor William W. 'Mac' Brazel came to my ranch and showed my husband and me a piece of material he said came from a large pile of debris on the property he managed. The piece he brought was brown in color, similar to plastic. He and my husband tried to cut and burn the object, but they weren't successful. It was extremely light in weight. I had never seen anything like it before."[13]

Was Loretta familiar with weather balloons and did she think the material she saw came from one? In response to this she said in her affidavit:

"The piece of material I saw did not resemble anything from a weather balloon. I had seen weather balloons before. I had never seen anything like this."

In other words, she did not claim she was an expert in all aerial objects, but she knew what a weather balloon was and confirmed this wasn't part of one.

William persisted in trying to get the Proctors to return with him to look at the wreckage. All manner of persuasion was attempted, but the Proctors declined. Loretta explained:

"We should have gone [to view the debris field], but gas and tires were expensive then. We had our own chores, and it would have been twenty miles."

37

Instead, the Proctors advised William to go to the authorities and report what he had found, especially as there "is usually a reward for this sort of thing."[14]

William returned home. Far from keeping quiet or feeling dejected by the limited interest from his neighbor, he went on and on about his amazing discovery to his wife and daughter. For a moment the family must have thought these materials were the most exciting things he had ever seen. Perhaps not the sort of thing a wife would want to hear even at the best of times, but here was a man who was clearly moved by what he had seen. And he just wouldn't stop talking. The description he gave, the sample materials he showed, and his excitement was enough for the family to go with him to the crash site and collect some more debris. As stated in the 1997 book by Saler et al., *UFO Crash at Roswell - Genesis of a Modern Myth*, Brazel, his wife and their daughter went out to the site and filled several feedbags of the debris on July 4, 1947. Still neither his wife nor their daughter could give any hint of what the debris was or where it came from.

On the morning of Saturday July 5, 1947, nearly two-and-a-half days after William first found the debris on his property, he decided to carry some of the larger pieces of the wreckage and store them in a cattle shed at the Hines House, a small bunkhouse located approximately three miles from the debris site.

In the afternoon, William visited another of his neighbors about 10 miles from his home—Lyman and Marian Strickland—where he showed them a piece of the wreckage. At first glance the piece of material seemed uninteresting but they politely agreed to take a look at it. The piece was played with for a bit and soon the Stricklands expressed amazement at what they saw. Again, like the Proctors, they remained just as ignorant as William about what it was, and where it came from. William did not show them the foil's peculiar shape-memory characteristics, but their daughter would confirm with investigators the behavior later after seeing the foil from Bill Brazel during another visit.

Bill, who knew of his father's interest in the unusual material, managed to pick up a few pieces of the material including the dark-grey metallic foil from the site about a week after it was discovered, and again after "a good rain" had exposed some more pieces. He kept them in secret, just long enough to show to the Stricklands, before he spoke too much in public only to have his collection confiscated by the U.S. military. In a televised interview in 1991 on *UFOs, a Need to Know*, Marian Strickland said:

> "The time that he [Bill] brought the sample of what he had picked up, he was at the corral. My daughter and two sons and husband were at the corral, and they saw it. My daughter says that it could be crumpled up and straighten right back out."

The foil made an impression on the daughter—Sally Strickland Tadolini (1936–2010). In a signed affidavit dated September 27, 1993, Sally stated:

> "A week or so after all the excitement, Mac's son Bill...showed us a piece of the thing his father had found, and he asked us not to say anything about it....What Bill showed us was a piece of what I still think of as fabric [because of its behavior and thin appearance]. It was something like aluminium foil, something like satin, something like well-tanned leather in its toughness, yet it was not precisely like any one of these materials....It was about the thickness of very fine kidskin glove leather and a full metallic greyish silver, one side slightly darker than the other[15.] I do not remember it having any design or embossing on it."

The foil's behavior was unmistakable. It was something she would never forget:

> "Bill passed it around....I did a lot of sewing, so the feel made a great impression on me. It felt like no fabric I have touched before or since. It was very silky or satiny, with the same texture on both sides. Yet when I crumpled it in my hands, the feel was like that you notice when you crumple a leather glove in your hand. When it was released, it sprang back into its original shape, quickly flattening out with no wrinkles. I did this several times, as did the others..."

On the evening of Saturday July 5, 1947, William drove into the nearest town, Corona, to do some shopping. His uncle, Hollis Wilson,

39

and a friend from Alamogordo (with whom he later met up at Wade's Bar) mentioned stories about flying saucers and the rewards (around US$5,000) being offered to recover one. William proceeded to tell them of the strange wreckage he had found on his property. Not being sure what it was, his uncle and friend strongly advised him to go to the authorities.

Notifying the authorities

Not the kind of man to let the materials be forgotten, early the next morning, at around 7:30 AM on Sunday July 6, 1947, William got into his truck, rounded up his two kids, and made the long arduous drive southeast to Roswell via Tularosa, where he stopped off and left the kids with their mother. It would be nearly three-and-a-half hours before he reached the city.[16]

Upon arriving at the city around 11:00 AM, William made sure he didn't make his usual shopping trip, followed by visiting a few friends for a long and enjoyable drink or two at a local bar and listen to a few good yarns to pass away the time. He understood right there and then the importance of what he had found and immediately visited the local sheriff of Chaves County named George Matt Wilcox (1894-1961). It seemed the excitement was too much for William.

William came in with two boxes containing some of the debris. He explained that he'd found strange materials on his ranch and had no idea what they were or where they came from. Wilcox looked somewhat interested, so he asked if he could look at a sample. Sure enough, William offered one. About the only material William did not show to others was the foil, so he decided to take out a piece of the strange metal and showed it to the sheriff. He was particularly eager to show Wilcox the metal's unusual shape-memory behavior and asked if Wilcox knew what it was.

Wilcox stood there in total disbelief. Here was a piece of foil looking too thin, flimsy and lightweight not to have disintegrated or collapsed under the pressure of a man's hand and stay that way, but there it was doing an amazing trick of returning to its original shape right in front of his eyes. Nor could he tear the foil apart with his hands. This was no ordinary foil.

At last, Wilcox could see what William was talking about and why he got so excited.

Wilcox thought long and hard about what he had in his hands. He looked intently at the other materials in the box in the hope of solving this problem. Still, all his experience and time spent looking at the materials and manipulating them would not give him any clues as to what he was dealing with. In the end, he had no choice as he too was curious of what these materials were and the type of object these materials came from, so he contacted Roswell AAF (as it was known in those days, later called Roswell Air Force Base, or Roswell AFB). If anyone knew what type of flying object it was, the guys down at the Army Airfield would.

Like William, Wilcox also didn't waste any time.

In fact, there was nothing in the call from Wilcox to the U.S. military to suggest these materials came from an ordinary weather balloon. Nor was it a call to suggest:

> "Well, whatever it was, it's not that important. But after I
> have my lunch, I will give the guys at the local Air Field a
> ring to see what they think."

It was a call that had to be made virtually immediately and with a sense of overwhelming excitement to the point where both men needed to know what it was that had lost its pieces and what type of metal could return to its original shape.

Wilcox dialed the number that would connect him to the military base stationed in his city.

The call was answered by Major Jesse A. Marcel, an intelligence officer of the 509th Bomb Group at Roswell AAF while he was having lunch. He listened intently. At first Marcel thought the description of the wreckage was reminiscent of the remains of a downed weather balloon, or something similar, once he heard the sheriff mention words to the effect of a thin foil, plastic sheets and beams being recovered in the desert. It is altogether likely the rancher had mistaken the pieces for a humble weather balloon and needed reassurance from him that this was all it was. But why didn't the sheriff know what a weather balloon looked like? There was something in the sheriff's voice that told him he had to come over right away. So he quickly finished his lunch and drove

41

to the sheriff's office, arriving at 12:50 PM, to inspect the samples of debris.

Marcel gave more than a cursory examination to the metal fragments and other materials lying in the boxes. He had a very good look at all the materials, making sure the foil pieces were the same and behaved in the same way, as well as trying to figure out the type of plastic materials making up the sheets and beams. All the pieces were indeed extremely lightweight, just as Wilcox and the rancher had mentioned. Certainly it *looked* like the sort of material you could use to build a weather balloon, he thought. However, they were too incredibly tough. The materials that went in to making weather balloons of any type as he knew them at the time were a lot more flimsy. And the behavior of the metal fragments was something he had never seen before. This was no ordinary weather balloon. Far from alleviating both the sheriff and rancher's concerns by stating it was a weather balloon, or some other natural or man-made object, Marcel quickly realized the importance of the find.

Marcel returned with haste to his base with one of the boxes and reported to his commanding officer, Colonel William Hugh "Butch" Blanchard (1916-1966).

Blanchard was sitting in his chair behind his desk when Marcel almost burst into his office. "This had better be important," Blanchard said. Marcel assured him it is. Without wasting any time, Marcel proceeded to explain to him the wreckage he had in the box he was carrying.

Blanchard listened carefully. He looked into Marcel's face as he spoke and noticed how unusual it was for Marcel to be excited by some materials. Indeed, from where he sat, the materials looked fairly unassuming. Yet Marcel was making them out to be incredibly important in his opinion. Then Marcel started to explain some of the unusual physical properties of the materials. "Interesting," Blanchard thought.

Being a rational man trusting of his own eyes, Blanchard had to see these materials and their properties in action for himself. So he inspected some of the pieces in the box. He played with the metal as Marcel reaffirmed what he said about its unusual shape-memory property. Blanchard noticed the way the metal fragments were behaving as he tried to bend and let them go back into shape, in agreement with

Colonel William H. "Butch" Blanchard (1916–1966)

Marcel's claims. He then looked at the plastic-like sheets and the artificial wood-like beams. Still maintaining interest, Blanchard tried to rip one of the larger sheets and break one of the beams with his hands —he couldn't.

"What is this stuff?" Blanchard probably wondered. "And where did it come from?"

Blanchard's own experience in aerial matters could not give him any answers. Knowing Marcel's considerably greater experience in aerial objects, he asked Marcel for his opinion. Marcel was just as perplexed as Blanchard, despite his extensive knowledge and years of experience in virtually every known aerial object produced by his country and those overseas.

Marcel mentioned that more of the fragments remained littered on the rancher's property, and there was another box just like what he had brought in sitting at the sheriff's office. Blanchard thought for a moment. Not knowing precisely how much material there was, he assumed that it could be collected in less than a day by a couple of men, so he ordered Marcel to find someone from the Counter Intelligence Corps (CIC) to accompany him and the rancher to the crash site, and to carry the debris back to Roswell AAF where a decision could be made on what to do with these strange materials. Later, military police arrived at the Sheriff's office to pick up the second box provided by the rancher for personal delivery to Blanchard's office. Hopefully this was sufficient manpower and time to collect everything.

As Marcel left the office, Blanchard called Brigadier-General Roger Maxwell Ramey (1903-1963), commanding officer of the Eighth AAF at Fort Worth AAF (later Carswell Air Force Base), Texas, to which the 509th Bomb Group at Roswell AAF belonged, and informed him of the unusual discovery. Ramey, in turn, contacted the Pentagon in Washington with the interesting news.

Nothing more was heard from top U.S. military officials until the following day.

Collecting debris from the first crash site

Marcel notified the sheriff that he would return and asked William if he would take him and a colleague he would bring with him to the debris

44

field for collection. William agreed. Marcel left the base around mid-afternoon. He was accompanied by the highest-ranking CIC agent on the base, Captain Sheridan Cavitt from West Texas. Cavitt was dressed in plain clothes at the time compared to Marcel in his standard military uniform; and drove a Jeep Carry-all, while Marcel followed William's pickup truck in his own blue-colored 1942 Buick Roadmaster convertible.

The remoteness of the site and the time spent traveling there made it necessary to stay overnight at the Hines House. The next day, Marcel and Cavitt had considerable time to closely inspect the debris scattered only 3 miles from the house. Until then, the men settled down for a cold meal of Campbell's Pork n' Beans. Before bunking down for a warm and windy night, Marcel and Cavitt took the opportunity to inspect the cattle shed—located a stone's throw away from the house (or roughly 100 feet)—where William said he'd stored some of the larger pieces. Unfortunately, none of the materials inspected in the shed gave any tantalizing clues as to what had flown over the rancher's property, or how big it had been.

The next morning, William left the truck behind at the house to jump on horseback. He led the way as Marcel and Cavitt followed him to the location in their better-equipped cross-country vehicles.

In an interview with investigators in 1979, Marcel recalls the morning of Monday July 7, 1947:

> "...[I] saw a lot of wreckage but no complete machine. Whatever it was had to have exploded in the air above ground level..."[17]

He said to investigators on December 8, 1979, just three years before his death:

> "When we arrived at the crash site, it was amazing to see the vast amount of area it covered. It was nothing that hit the ground or exploded on the ground. It's something that must have exploded above ground, traveling perhaps at a high rate of speed...."[18]

Marcel had every reason to be in-charge of investigating the debris. Blanchard, the head of the Roswell AAF, showed an innate confidence in Marcel's ability to identify the object. Cavitt's role was merely to confirm Marcel's initial findings, and help with the clearing operation. However, on this occasion as Marcel recalled, both he and Cavitt were stumped as to what had exploded in the air. And none of the materials looked at would reveal anything to suggest these were ordinary materials.

Based on his analysis of the materials, Marcel ruled out familiar man-made flying objects such as a weather balloon, aircraft and missile. The most he could figure out from the way the pieces landed on the ground and were distributed that the object was heading to the southwest.

As Marcel said:

> "...It was quite obvious to me, familiar with air activities, that it was not a weather balloon, nor was it a plane or missile....It was all evenly spread out, as if something had exploded in the air and fallen down on the ground. What impressed me was that one could clearly see which direction it came from and where it was going. It flew from northeast to southwest. One could see where it began, and also where it ended..."[19]

William watched on horseback from the top of an escarpment as Marcel and Cavitt walked through the field to see how far the materials were scattered, examined a number of the materials along the way, and then filled the rear of Marcel's Buick and Cavitt's Jeep Carryall with as much of the material as possible. He later left the area to return home, leaving the two men to investigate and clear the wreckage in their own time.

The furthest extent of the debris field, as far as Marcel could observe, seemed to cover nearly one mile in length and several hundred feet wide. Ignoring the few extraneous pieces lying further away, the bulk of the materials were within the three-quarters of a mile by 200 to 400 feet wide range, distributed in a fan-shape with the heaviest concentrations at the narrowest end.

Also the materials found were extremely lightweight, very strong and hard, yet flexible. They consisted of small beams that looked like balsa wood, with pink or purple hieroglyphic-type writing on them, a brown parchment-like substance that was extremely strong and lightweight, and a great quantity of a newspaper-thin lead-colored (or dirty stainless steel-colored) foil.

By dusk, Marcel and Cavitt realized there was too much of the material to be collected by them alone. A decision was made for Marcel to return to Roswell AAF and notify Blanchard of the situation, while Cavitt stayed at the site for a while longer. Later, Cavitt would return to base with his load of the materials.

Before Marcel made his way back to base, at around 2:00 AM, he went home at 1300 W. 7th Avenue where his family lived. His obvious excitement awoke his wife Viana and their eleven-year-old son.

In an interview conducted years later, Marcel Jr. (1936-2013), the son of Jesse Marcel and now a retired Colonel, recalls seeing some of the material in the house and in the jeep:

> "My father was so excited about the debris that he drove by our house to show my mother and me the material before delivering it to the air base. There were several boxes of it in our car but we emptied the contents of just one of these boxes on our kitchen floor. He wanted to see if he could piece some of the fragments back together."[20]

In an affidavit dated March 6, 1991, Marcel Jr. described the materials as follows:

> "There were three categories of debris: a thick, foil-like metallic gray substance; a brittle, brownish-black plastic-like material, like Bakelite; and there were fragments of what appeared to be I-beams."[21]

He recalls his father successfully putting together a few pieces on the kitchen floor, but not enough to tell him what had crashed, or exactly how large it was. Certainly it was large, but not enough to give a definite size. Whatever this mysterious flying object was, it clearly had affected his father in a deep and profound way. For someone who had an intimate knowledge of virtually every aerial object in the sky at the time, this was a surprise. Marcel Jr. commented in his affidavit:

> "My father said the debris was recovered from a crash site northwest of Roswell. He felt it was very unusual and may have mentioned the words 'flying saucer' in connection

with the material. He was certain it was not from a weather balloon."

Marcel Jr. did not recall seeing the unusual shape-memory property in the dark-grey metallic foil as described by William Brazel and his father. Apparently, he thought he needed to have a sledgehammer to see the effect. As Marcel Jr. said:

"I didn't try to bend it nor did I witness the 'metal with a memory' that some have described. I do recall my dad saying a colleague had tried to bend one of the larger pieces with a sledgehammer without denting it."[22]

The best Marcel Jr. could remember of the materials were the wood-like beams containing the unusual hieroglyphics painted or embossed onto them. Marcel Jr. said:

"The one I remember best was about twelve to eighteen inches and was made from a very lightweight material. When I held it up to the light I could see what appeared to be symbols printed or embossed along the length of the beam. They looked at first like hieroglyphics, but on closer scrutiny appeared to be geometric designs."

When his quotes were later published, some people tried to question the integrity of Marcel Sr. as if trying to prove his observations were not real. On seeing the reaction, Marcel Jr. wrote a book discussing in further detail what he saw along with the history of his father in the U.S. AAF. As he explained the reason for writing his book *The Roswell Legacy*:

"I've grown tired of hearing other people speak for me, and for my father, and of hearing some people deny what I know to be true. It's difficult to believe an incredible tale, especially when you don't know the storyteller.

My hope is that by reading this book, you'll get a better feel for who Jesse Marcel, Sr. was, who I am, and what has guided all of us who have been involved in the legacy borne of the Roswell crash.

In the larger scheme, we may all be mere footnotes in history, but we were so much more than that. We were (and are) living, breathing human beings who, through no choice of our own, have been caught up in a whirlwind of controversy and wonder. It is my fondest hope that you, the reader, might end up feeling as if you know the people a little bit better, might have a better understanding of the controversy, and might share with us the same sense of wonder that has so filled our lives."[23]

Preliminary testing begins on the strange wreckage

Marcel arrived at the base sometime after three in the morning. He stored some of the material in Blanchard's office and the rest was left in his Buick.

Relaxing over a cup of coffee, Marcel saw Cavitt drive in with his load of the materials. Together, the two men sat down to discuss the events of the day, in particular what they had discovered and what to do with the stuff as they waited for Blanchard to arrive at his office at precisely 6:00 AM.

By dawn, on Tuesday July 8, Marcel, Cavitt and several other officers at the base got together to inspect the materials more closely. One man began testing the grey metallic foil using a sixteen-pound sledgehammer and a blowtorch.

In a personal interview with investigators in November 1989, CIC agent Lewis S. "Bill" Rickett (1910-1993), claimed to have been present to witness the preliminary tests:

> "Using a sixteen-pound sledgehammer, they tried to dent one of the larger, metallic chunks. They couldn't even scratch the surface. They tried to burn it using a [blow]torch, but the metal seemed to dissipate the heat. There was no sign that the torch had touched it. Their tests seemed to prove one thing: The material was not the flimsy stuff it appeared to be...."[24]

Based on his personal experience of metal foils, Rickett made it clear that he had never seen anything like it:

"I never saw a piece of metal that thin that you can't bend. The more I looked at it, I couldn't imagine what it was."[25]

An emergency meeting was held at 7:00 AM to discuss the results of the tests. Both Marcel and Cavitt attended the meeting with Blanchard's regular staff. Pieces of the wreckage were passed around for everyone to examine. Despite all the military personnel present and their in-depth collective knowledge and experiences of many different types of aerial objects known at the time both locally and abroad, all agreed the material collected was unknown. The only further details some of the men could contribute to the discussion was the possibility that one of those mysterious flying discs talked about in various newspapers and radio broadcasts had exploded in the air and disintegrated. There was nothing to suggest the object remained in the air and crashed elsewhere as far as Marcel and Cavitt could tell. Their opinion was that the entire object came to grief in the spot where Marcel and Cavitt had been clearing the area.

Realizing the potentially important nature of these materials and how likely its origins might have been outside of the U.S., Blanchard knew he had to act promptly.

The meeting abruptly ended when Blanchard ordered his men to clear the property of the debris. Later he contacted his superior, Ramey, for the final official word on what to do with the material. Ramey later informed Blanchard to send the material to his headquarters based on the information he received from Dubose. However, before Blanchard could make this crucial second call to higher headquarters, a report was prematurely released to the press.

In an attempt to keep the curious public away from the debris field, Blanchard ordered Second Lieutenant Walter Haut (1922-2005), a public information officer (PIO) for the 509th Bomb Group at Roswell Air Force Base, to immediately release to members of the press a report[26] stating that a flying disc had been picked up. The order came through at 9:30 AM.

By around 11:00 AM, the telex report was personally delivered by Haut to the local newspapers—*The Roswell Daily Record* and the *Roswell Morning Dispatch*—and two radio stations in town, KGFL and KSWS (where it was seen as a newsworthy item and so was quickly broadcast to the public by 12 noon).

```
DXF54

     MORE FLYING DISC (DXR53)
                  - O-
     THE INTELLIGENCE OFFICE REPORTS THAT IT GAINED POSSESSION OF
THE "DIS:" THROUGH THE COOPERATION OF A ROSWELL RANCHER AND SHERIFF
GEORGE WILCOX OF ROSWELL.
     THE DISC LANDED ON A RANCH NEAR ROSWELL SOMETIME LAST WEEK.
NOT HAVING PHONE FACILITIES, THE RANCHER, WHOSE NAME HAS NOT YET
BEEN OBTAINED, STORED THE DISC UNTIL SUCH TIME AS HE WAS ABLE TO
CONTACT THE ROSWELL SHERIFF'S OFFICE.
     THE SHERIFF'S OFFICE IN TURN NOTIFIED A MAJOR OF THE 50 9TH
INTELLIGENCE OFFICE.
     ACTION WAS TAKEN IMMEDIATELY AND THE DISC WAS PICKED UP AT THE
RANCHER'S HOME AND TAKEN TO THE ROSWELL AIR BASE.  FOLLOWING
EXAMINATION, THE DISC WAS FLOWN BY INTELLIGENCE OFFICERS IN A
SUPER- FORTRESS TO AN UNDISCLOSED "HIGHER HEADQUARTERS."
     THE AIR BASE HAS REFUSED TO GIVE DETAILS OF CONSTRUCTION OF THE
DISC OR OF ITS APPEARANCE.
     RESIDENTS NEAR THE RANCH ON WHICH THE DISC WAS FOUND REPORTED
SEEING A STRANGE BLUE LIGHT SEVERAL DAYS AGO ABOUT THREE O'CLOCK IN
THE MORNING.
          J241F T/S
```

Official news release of July 8, 1947 (with typo in the word "disc" as in the original) from Second Lieutenant Walter Haut, the public information officer for the 509th Bomb Group at Roswell AAF.

Unfortunately all this work to notify the media was done without the explicit permission from Blanchard's military superior, Ramey. As a result, Haut was punished for his actions, and Blanchard was put on leave and replaced with another commander. When the public learned about this, the USAF described the change in authority as standard procedure.

Just prior to the news release going out to the media, CIC agents Captain Sheridan W. Cavitt and Master Sergeant Lewis S. Rickett arrived at the Foster Ranch and cordoned off the area[27]. A number of military trucks came in with soldiers. Their orders were to clear the property of the debris, leaving behind nothing. This was not a simple and straightforward recovery operation as one would expect for an ordinary weather balloon if it was one. Either that or it was a particularly special type of weather balloon built with new and expensive materials. Or was it a new type of foreign hi-tech aircraft or missile that crashed? At any rate, over a dozen military personnel had to collect and clear the material in the minutest of detail. It involved turning the soil and raking it. The debris was placed inside large wooden crates and loaded onto several trucks.

It was late in the afternoon before the bulk of the debris arrived at Roswell AAF, but not without one or two military personnel involved in the clearing operation allegedly keeping a piece of the unusual grey metallic foil as a souvenir. Of course, whether they were successful at keeping it for long enough is another question. Whatever the situation, statements from several military witnesses who claimed to have seen others keep a souvenir would eventually emerge for investigators to record.

Did the USAF discover something else?

During the clean-up operation, it quickly became apparent to the USAF of another discovery at the rancher's property. The discovery prompted the USAF to quickly locate the rancher and interrogate him. Brazel, who had just left Walt Whitmore's house, was tracked down and brought to the RAAF base guesthouse.

Far from having a friendly chat over a cup of coffee, Brazel was kept at the base for five to six days against his will. While he was there, Brazel was asked a barrage of terse questions about what he knew, whether he saw anything else on the property, was given an embarrassing head-to-foot army physical examination (was the USAF trying to find places where Brazel could have stashed away some of the debris?), and told not to speak about the incident other than being instructed to tell the media that what he found was a weather balloon. To do this, he was required to visit *The Roswell Daily Record* accompanied by a USAF security officer, and later on his own (when the USAF saw that he complied with the order) to radio station KGFL.

According to Carey & Schmitt's book *Witness to Roswell*:

> "Six witnesses have testified to seeing Brazel escorted through Roswell under military guard, as he was taken to *The Roswell Daily Record* and *Morning Dispatch* newspapers and KGFL and KSWS radio stations to personally retract his claims of discovering the remains of a flying saucer."[28]

As a result of the interrogation and being accompanied by a USAF security officer, Brazel's original story began to change as can be seen in the July 9, 1947 article in the *Roswell Daily Record* titled "Harassed Rancher who Located 'Saucer' Story sorry he told about it". As reported in the article (and later used by the USAF in 1994 for its *Roswell Report*):

> "Brazel related that on June 14 he and an 8-year old son, Vernon, were about 7 or 8 miles from the ranch house of the J. B. Foster ranch, which he operates, when they came upon a large area of bright wreckage made up of rubber strips, tinfoil, a rather tough paper and sticks.
>
> At the time Brazel was in a hurry to get his round made and he did not pay much attention to it. But he did remark about what he had seen and on July 4 he, his wife, Vernon and a daughter, Betty, age 14, went back to the spot and gathered up quite a bit of the debris.
>
> ...Monday he came to town to sell some wool and while here he went to see sheriff George Wilcox and 'whispered kinda confidential like' that he might have found a flying disk....
>
> Brazel said that...it might have been about as large as a table top. The balloon which held it up, if that was how it worked, must have been about 12 feet long, he felt, measuring the distance by the size of the room in which he sat. The rubber was smoky gray in color and scattered over an area about 200 yards in diameter.
>
> When the debris was gathered up the tinfoil, paper, tape, and sticks made a bundle about three feet long and 7 or 8 inches thick, while the rubber made a bundle about 18 or 20 inches long and about 8 inches thick. In all, he estimated, the entire lot would have weighed maybe five pounds.
>
> There was no sign of any metal in the area which might have been used for an engine and no sign of any propellers

of any kind, although at least one paper fin had been glued onto some of the tinfoil."[29]

Whether this was yet another example of standard procedure by the USAF when handling civilians with different points-of-view, or the result of finding something else in the middle of the desert which prompted more serious action[30] is still a matter of debate. The sudden quietness of Brazel over the discovery this time around was assumed by the reporter from the *Roswell Daily Record* to be the intense publicity he received:

> "W. W. Brazel, 48, Lincoln county rancher living 30 miles south of Corona, today told his story of finding what the army at first described as a flying disk, but the publicity which attended his find caused him to add that if he ever found anything else short of a bomb, he sure wasn't going to say anything about it."[31]

But one thing was certain. For a long time afterwards, Brazel would not explain what happened to him or his discovery after his poor treatment in the hands of the military, except on two occasions: (1) in total defiance of his military captors, he said to a reporter it was no weather balloon; and (2) years later he mentioned some details of the event to his son Bill. To people who knew him, it was thought Brazel had been threatened with imprisonment (or something far worse) if he spoke out.

As part of his show of defiance to the military, the following quote reveals Brazel's true thoughts about the incident:

> "I am sure what I found was not any weather observation balloon....But if I find anything else besides a bomb, they are going to have a hard time getting me to say anything about it."[32]

Subsequently, the USAF made routine inspection of the debris field after it rained and asked if William had found any more pieces.[33]

Controlling the media

To add to the interesting military behavior over what is presumed to be a man-made object, Walt E. Whitmore, Sr. (1921–1995), the man who had tried to hide William in his house for the purpose of getting an exclusive interview for his majority-owned KGFL Radio, had managed to successfully record the story and was about to get it on the air when someone wanted to put a stop to the broadcast. He began broadcasting over KGFL a preliminary release explaining there was an interview made with the rancher, but before he could reveal details of the interview to the public, he received a call from a man calling himself "Slowie" claiming to be the Secretary of the Federal Communications Commission in Washington, D.C. In a firm and uncompromising tone, this person told Whitmore to cease transmission of the story or face a revocation of his FCC broadcasting license. Whitmore realized his interview with the rancher was having considerable impact on the authorities.

Suddenly another call from Washington came through from Senator Dennis Chavez of New Mexico, and former chairman of the Senate Appropriations Committee. Chavez suggested that Whitmore should obey the FCC directive from Slowie. Fearing he would lose his license, Whitmore complied.

Whitmore visited the site of the debris several days after the USAF had cleared it. But since then, he has faithfully kept quiet until his retirement when he told his story to his son Walt Whitmore, Jr.

In interviews with investigators, Walt Whitmore, Jr., remembered his father saying he saw some of the materials brought in by the rancher: a tough metallic, grey foil and small wood-like beams with hieroglyphic-type writing on them. The largest was a piece of metallic foil about four or five inches square, weighing practically nothing.

He also recalls his father saying he visited the crash site several days after the area was thoroughly cleared by the USAF. He claims a stretch of about 175 to 200 meters of land had been uprooted in a fan pattern with most of the damage at the narrowest end.

Johnny McBoyle is another reporter claiming to have been restricted from performing his duties by the authorities. At the time the official news broke of the recovery of a so-called disc, McBoyle had high hopes of reaching the debris field to learn more. Unfortunately, he was stopped by the military at a road block northwest of the city. Far from letting him go and asking him to return home as there was nothing to see or wasn't worth his time, it was necessary for the military to quickly seize him and return to Roswell under military guard. He arrived at the base at around 4.00 PM where he was kept until the clearing operation was completed.

While at the base, McBoyle was not completely isolated from everyone else. He could observe the results of the recovery operation in progress and he tried to listen in on some of the conversations. Then, at some point he managed to make out words to the effect of "little green men" as if something biological had been picked up at the ranch. Because while he was left unattended, McBoyle decided to make an urgent and necessary call to his parent radio station KOAT in Albuquerque (the other station he partly owned was KSWS in Roswell) asking his teletype operator, Lydia Sleppy, to type the latest story he heard for transmission on the AP wire. He spoke with her with a sense of excitement.

According to Sleppy, McBoyle claimed to have said (perhaps exaggerating slightly):

> "Lydia, get ready for a scoop! We want to get this on the ABC wire right away. Listen to this! A flying saucer has crashed....No, I'm not joking. It crashed near Roswell. I've been there and seen it. It's like a big crumpled dishpan. Some rancher has hauled it under a cattle shelter with his tractor. The Army is there and they are going to pick it up. The whole area is now closed off. And get this— they're saying something about little men being on board....Start getting this on the teletype right away while I'm on the phone."[34]

However, as she tried to type the message, she was interrupted after a couple of sentences by the following message from an unidentified sender: "THIS IS THE FBI. YOU WILL IMMEDIATELY CEASE ALL COMMUNICATION."

The military police were quickly informed of what was happening and saw McBoyle talking to someone on the phone. He was pulled aside and his life was threatened. He later spoke with Sleppy saying bluntly, "Forget about it. You never heard it. Look, you're not supposed to know. Don't talk about it to anyone."

He never talked about the incident again.

Despite the effort to keep civilians and the media quiet, there was just one problem: the USAF had already sent a news release, and was reaching media in other states, and other countries. Part of the statement was published on the front-page of the *Walla Walla Union Bulletin* in Washington on Tuesday July 8, 1947. The next day the statement was published, in full, in the *San Francisco Chronicle*.

The news release created such considerable interest and controversy that it was quickly acted upon by reporters in the USA and some abroad. Soon the USAF was under pressure to release more details.

ROSWELL STATEMENT

Here is the unqualified statement issued by the Roswell Army Base public relations officer:

"The many rumors regarding the flying discs became a reality yesterday when the intelligence office of the 509th Bomb Group of the Eighth Air Force, Roswell Army Air Field, was fortunate enough to gain possession of a disc through the co-operation of one of the local ranchers and the Sheriff's office of Chaves county.

"The flying object landed on a ranch near Roswell sometime last week. Not having phone facilities, the rancher stored the disc until such time as he was able to contact te Sheriff's office, who in turn notified Major Jesse A. Marcel, of the 509th Bomb Group Intelligence office.

"Action was immediately taken and the disc was picked up at the rancher's home. It was inspected at the Roswell Army Air Field and subsequently loaned by Major Marcel to higher headquarters."

The official USAF news release of a flying disc recovered from a local rancher's property is published in full in the San Francisco Chronicle.

Top U.S. military brass finally stirred into action

It must have seemed like an eternity for some military officials, but, at last, Ramey's chief of staff, Colonel (later Brigadier General) Thomas Jefferson Dubose, received the all-important telephone call from the Pentagon at 2:00 PM Mountain Time Zone (MDT) (i.e. Roswell) on Tuesday July 8, 1947 from Major General Clements McMullen (1892-1959), Vice-Commander (or Acting Director) of the Strategic Air Command (SAC). A man with a long and distinguished military career, McMullen ordered Dubose to send the box containing the original Roswell fragments in a sealed container via Fort Worth to Andrews AAF near Washington, where he would receive it personally. Before ending the call, McMullen also allegedly mentioned in no uncertain terms to concoct a weather balloon story to the press to stop the growing interest in the debris.[35]

Introducing the weather balloon explanation

There was no time to waste for the USAF.

At approximately 4:30 PM on Tuesday July 8, the U.S. military authorities changed their story. There was no disc, and definitely no "little green men". This was the remains of either an ordinary "Rawin sonde" or "Rawin target" weather balloon.

The front-page article of *The Roswell Daily Record* of Wednesday July 9 titled "Gen. Ramey Empties Roswell Saucer" ran with this new story:

> "An examination by the army revealed last night that mysterious objects found on a lonely New Mexico ranch was a harmless high-altitude weather balloon—not a grounded flying disc. Excitement was high until Brig. Gen. Roger M. Ramey, commander of the Eighth air forces with headquarters here cleared up the mystery."

Another front-page spread—this time in the *San Antonio Express*—suggests the explanation from the USAF was generally accepted by the media.

Well, to be more accurate, not all reporters at the time were entirely convinced of the explanation. Indeed, some questioned how the USAF could have mistaken an unknown disc for a weather balloon? Didn't they know what had crashed in the desert before releasing the initial alarming report to the media? And how could a weather balloon create so much debris covering three-quarters of a mile and several hundred feet wide of a rancher's property? Unfortunately, with no other evidence to go by at the time, the media had to accept the latest official explanation.

The two Rawin weather devices, whichever of the two was allegedly responsible for the wreckage, were both constructed of lightweight materials made of thin (and rather flimsy) balsa wood frame covered with aluminium or tin foil and a plastic sheet, and were suspended by one or more helium-filled polyethylene or rubber balloons. In other words, if we were to accept the USAF explanation of what happened in early July 1947, the following pictures are what Marcel and the other witnesses should have seen on the ground (or at least parts of it).

A "Rawin sonde" weather device in flight over New Mexico (left) from the New Mexico Institute of Mining and Technology in Socorro, and a "Rawin target" weather device used in 1947 by meteorological stations in the area (right) from the U.S. Meteorological Service.

Furthermore, the USAF assumed the metallic foil picked up from the rancher's property had no shape-memory effects. It had to be aluminium foil. End of story.

Photographer Mr. James Bond Johnson

To lend weight to this new explanation, military reporter and representative of the *Fort Worth Star-Telegram* (since 1943), former U.S. Army Colonel James Bond Edward Johnson (1926-2006), was

Major General Clements McMullen (1892–1959)

granted an exclusive interview and photo opportunity with Brigadier-General Ramey in his office at 6:00 PM. As Mr. Johnson recalled:

> "I was assigned in mid-afternoon on July 8, 1947, by my city editor to go out to FWAAF (later Carswell AFB) and get a shot of a 'flying disc' the Air Corps had 'captured' and was flying in from Roswell AAF. I now think I arrived at General Ramey's office within less than an hour after the debris arrived and it was for some reason spread out on the carpet in the general's office."[36]

Thinking he was going to see a "flying disc", Mr. Johnson had to be briefed prior to the meeting on what it was he was going to photograph. In a taped interview, Mr. Johnson stated:

Second photograph from James Bond Johnson showing Brigadier-General Roger Ramey with the substitute wreckage (UTA No. AR406-6 #2026 10000684.tif).

> "I posed General Ramey with this debris. At that time I was briefed on the idea that it was not a flying disc as first reported but in fact was a weather balloon that had crashed."[37]

On entering Ramey's office, he saw Ramey, Dubose and the remains of a highly crinkled weather balloon already carefully laid out and displayed before him on the floor. He was permitted to take four black and white photographs[38] to confirm what he saw using his 4"x5" Speed-Graphic press camera.

Mr. Johnson later claimed:

> "As far as is now known I was the only photographer or reporter to see or photograph the debris in the general's office since no one else has been identified as such and no other photos have turned up."[39]

However, unbeknown to Mr. Johnson were an additional three photographs taken of the weather balloon debris. And they weren't from Mr. Johnson's own camera.

Before Mr. Johnson arrived, he was not made aware of the fact that Ramey's own information officer at the Fort Worth Army Air Field by the name of Major Charles A. Cashon had been ordered to take two pictures of Marcel holding a piece of the *alleged* Roswell debris. And a third and final photograph would be taken of another military officer who was brought in to identify the wreckage in front of Mr. Johnson and Ramey.

Warrant Officer Irving Newton, a weather officer on duty at the time, was ordered by Ramey to attend the media photo shoot with Mr. Johnson and answer some questions. Newton was briefed by a colonel just prior to entering Ramey's office where it is alleged, according to Newton's own words in original testimony, he was told "that an object had been found by a major in Roswell and that the general had decided that it was really a weather balloon and wanted him to identify as such".

As soon as Newton walked into the office, there stood General Ramey and Mr. Johnson apparently waiting for him. Ramey allowed Mr. Johnson to ask Newton what he thought the material lying on the floor was. Newton took a brief moment to look at the tangle of metallic foil and balsa wood frame on the floor and concluded without hesitation to Mr. Johnson that it was a weather balloon. In fact, he didn't even need to pick up pieces and examine them. He knew from the moment he looked at the debris what it was.

First photograph taken by Major Charles Cashon showing Marjet Jesse Marcel with the substitute wreckage. (UTA AR406-6 #2026 10000258.tif)

For Newton to be able to state so quickly what the object was, this is quite remarkable considering Marcel and his military colleagues at Roswell AAF, including Blanchard himself, could not make an immediate conclusion along this line despite their vast experience of man-made aerial objects, or the numerous attempts to grab hold of the materials and look at them up close and personal, or even the testing done on the materials. Extraordinary. Does this mean Mr. Newton is a genius in recognizing pieces of an object at first glance? Well, if he was, he wasn't the only genius. According to Newton's testimony, it would appear Ramey was also able to identify the remains as a weather balloon. As Newton stated:

"There's no doubt that what I was given were parts of a balloon. I was later told that the major from Roswell had identified the stuff as a flying saucer but that the general had been suspicious of this identification from the beginning..."[40]

In a later affidavit for the Air Force, Newton said:

"I was convinced at the time that this was a balloon with a [kite] and remain convinced ... There were figures on the sticks lavender or pink in color, appeared to be weather faded markings with no rhyme or reason."

So why were Marcel and his colleagues at the Roswell AAF not able to identify it as such? Perhaps Newton or Ramey should have been put in charge of the investigation and recovery operation instead of any of the men at Roswell AAF.

After the interview, Mr. Johnson did not realize that another photograph was taken, this time by Major Cashon of Warrant Officer Newton squatting and holding metal pieces from the weather balloon in both hands to help make the "balloon theory" look even more plausible to the media. All three additional photographs from Cashon were slipped in to form Mr. Johnson's complete set for the U.S. press to publish to the world.

As Mr. Johnson said:

"A FWAAF PIO photographer may have taken the Newton photo to accompany the 'cover up' story but I do not know any details. Now I do not think there was a 'grand scheme' by the AAF to 'set me up' with the photo. I have no reason to believe other than that at that time they simply did not know what they had."[41]

He recalled a couple of other interesting things about the interview with Ramey:

"It remains curious to me after 49 years that (1) Gen. Ramey spread the wreckage out on the carpet in his office;

63

[and] (2) That he (Gen. Ramey) put on his class 'A' uniform complete with jacket and frame cap for the photos taken in his own office while Col. Dubose, Maj. Marcel and Warrant Officer Newton all were photographed in their class 'B' uniforms—with no hat, tie or jacket."[42]

Then Mr. Johnson added:

"I still believe the General thought at the time of the photo session that he was living an important day in history."[43]

Mutual UFO Network (MUFON) gets involved

After the Australian research group SUNRISE briefly published preliminary details on the history of the Roswell case for online public discussion and gave Mr. Johnson an opportunity to present his views, SUNRISE received an email on February 7, 2006 from the U.S. Eastern Regional Director of MUFON and former USAF officer, Major George A. Filer[44], titled *Filer's Files #06-2006*. The email gave brief information about an interview allegedly conducted by Neil Morris with Mr. Johnson claiming a more prosaic set of events had occurred in Ramey's office. In fact, Mr. Johnson now claims he actually spoke to and helped Marcel to spread out the original Roswell wreckage on the floor of the office.

SUNRISE immediately noticed a discrepancy.

As MUFON stated:

"Marcel was met at the flight line by the second highest ranking officer at Carswell Army Air Base Colonel Thomas DuBose. Together they drove to General Roger Ramey's office with the debris, and waiting for them was James Bond Johnson a Fort Worth Star Newspaper photographer who took these photos.

Johnson who is still alive told me [Neil Morris], when the officers arrived the package was still wrapped and they unwrapped the metallic burnt debris and spread them out on the General's rug. He remembered the debris smelled burnt like it had been in a crash.

NO SAUCER—Brig Gen. Roger M. Ramey, commanding general of the 8th Air Force here, looks beneath the pile of crumpled tinfoil, broken sticks and ragged rubber which set off an international report Tuesday that a flying saucer had been found in New Mexico. Ramey identified it as the remnants of a weather balloon and attachments.

This is believed to be the first photograph taken by James Bond Johnson, showing Brigadier-General Roger Ramey with the 'substitute' Roswell debris (Original photograph missing from UTA collection).

General Ramey arrived holding a message in his hand. Six photos are taken of General Ramey, Colonel DuBose, and Major Jessie Marcel with the debris. Four of the 4 x 5 negatives are now stored in the Archives of the University of Texas."

The discrepancy can be seen from the official interviews conducted with Marcel, where he never mentioned, even once, seeing or being assisted by Mr. Johnson in unpacking the wreckage and placing it on the carpet before pictures were taken using Mr. Johnson's camera.

Why the difference?

Mr. Johnson had a personal web site and email address[45] for the public to contact him. SUNRISE took the opportunity to notify him on February 13, 2006 of the latest Australian research into the Roswell case and asked for his comments regarding the history chapter of this book. The chapter he saw contained the version of events in General Ramey's office according to Marcel from his own affidavit. In his email reply, Mr. Johnson focused on only one thing: no other reporters were present in General Ramey's office when his photos were taken. On February 14, 2006, Mr. Johnson said:

"I was in General Ramey's office late in the afternoon of July 8, 1947, and took the photos of the debris displayed along with Ramey, COL DuBose, MAJ Marcel and Warrant Officer Newton. My hat is in one of the photos, too! On a chair in the background. As far as has been determined after more than a half century I was the ONLY media representative to be allowed in Ramey's office that day....

I know that there have been some rumors circulated as to a 'press conference' but if such was to have been held it is inconceivable that the Fort Worth Star-Telegram, which at that time was the South's largest newspaper, would not have been included. I was at that time the military reporter for the Star-Telegram and we were NEVER once ignored by Fort Worth Army Air Field, later Carswell Air Force Base."

Third photograph from James Bond Johnson. Brigadier-General Roger Ramey and Colonel Thomas J. DuBose with Roswell debris. (Courtesy of Fort Worth Star-Telegram Photograph Collection, Special Collections, University of Texas Arlington Library, UTA No. AR406-6 #2026 10000207.tif)

Third and final photo taken by Major Charles Cashon shows Warrant Officer Irving Newton with the substitute wreckage. This photo first appeared in the LOOK Magazine, 1967. (Not in UTA collection)

It is clear Mr. Johnston makes no mention of Marcel being present at the same time as he was. Is this because the original story as told by Marcel was incorrect?

As Marcel said on more than one occasion in interviews with investigators, he was ordered to accompany part of the original material to Fort Worth AAF. Upon arriving, he recalls leaving the box containing a sample of the original materials (as provided by William Brazel) on Ramey's desk. Then, another officer came in and told Marcel that Ramey was on the phone in another room and would arrive soon. Ramey returned to his office and asked Marcel to follow him into another room where a large wall map could be seen. Ramey asked Marcel to show where the wreckage was found. Marcel pointed out the location.

Fourth photogaph from James Bond Johnson. Brigadier-General Roger Ramey and Colonel Thomas J. DuBose with substitute Roswell debris. (Original photograph missing from UTA collection)

Marcel re-entered Ramey's office at his request and noticed the box had disappeared and was replaced by the remains of a weather balloon by someone else. Ramey was quick to say to Marcel, "Don't say anything. I'll take care of it."[46]

To his surprise, Marcel was ordered to pose in two pictures with the substituted wreckage, even though he knew without a doubt that this was *not* the original material. In order to make sure he could not answer questions from Mr. Johnson (or any other reporters—he didn't know precisely how many would arrive), Marcel was ordered to leave the room, leaving Ramey and the others to do the talking. Apparently it was so important to have Marcel away from the media that according to Carey & Schmitt's book *Witness to Roswell*:

> "Marcel was kept in seclusion for the next 24 hours."[47]

From this moment on, Marcel knew a cover-up had begun.[48]

To verify if Marcel's version of events is true, SUNRISE again contacted Mr. Johnson on February 15, 2006 requesting further clarification on this issue. He said he needed a little extra time to answer our questions on February 16, 2006. Unfortunately we never received a reply.

On February 15, 2006, the U.S. Eastern Regional Director of MUFON sent SUNRISE an email[49] with a subscription to pay and join MUFON.

On April 5, 2006, another email from Major Filer stated that Mr. Johnson had died on March 25, 2006:

> "I'm sorry to report the passing of James Bond Johnson on March 25, 2006, in Long Beach, California due to cancer."

In the same email, MUFON reiterated the version of events in Ramey's office according to Mr. Johnson but with further details:

> "A special B-29 flight flew Marcel to Fort Worth and he was directly to the General's office carrying meat wrapper paper packages which held the debris. Johnson told me he worked for *Fort Worth Star Telegram* and was assigned on the afternoon of July 8, 1947, to take photos of General Ramey and two other officers in the general's 8th Air Force

Second photograph taken by Major Charles Cashon showing Major Jesse A. Marcel with the substituted 'weather balloon' Roswell debris. Courtesy of Fort Worth Star-Telegram Photograph Collection, Special Collections, University of Texas Arlington Library.(UTA AR406-6 #2026 10000208.tif)

HQ offices. Bond told me, 'He was ushered into the General's office and the debris was still wrapped, so he helped unwrap the debris and displayed it so photos could be taken.'

Both Johnson and Marcel felt the debris material was unusual and smelled like it had been burned in a crash. The general was away from his office but expected to return shortly. Johnson first assisted with the debris unpacking and then 'posed' Marcel with what later was described by the major as some of the 'less impressive pieces' of the wreckage. When General Ramey returned to his office holding a piece of paper, Johnson photographed him alone examining the pieces of wreckage and then he was joined in other shots by Colonel Thomas J. DuBose. Johnson felt he photographed very unusual debris, which was not a weather balloon since they do not burn when landing."

And now we can no longer verify Marcel's version of events with Mr. Johnson.

Lieutenant Walter Haut supports a cover up

At any rate, Marcel was quick to reject the weather balloon explanation. He stated unequivocally and without retraction of his statements to investigators that a switch of the original debris was made for the remains of a weather balloon to help cover-up the real story.

When Lieutenant Haut later heard about the weather balloon story from Marcel, he said:

> "A balloon may have crashed, but it certainly had nothing to do with the downed saucer. What most people don't realize is that back then, you didn't ask questions—you did whatever your superior told you. Today, there'd be a lot of questions, or even a Congressional hearing, but it was a different era then."[50]

Questions from a skeptical media

As General Ramey ordered Marcel and DuBose to make statements to the press in support of the weather balloon explanation, an inconsistency started to creep into the military's statements. Later Ramey had to rectify the problem personally during another radio interview.

According to Associated Press (AP), Marcel was attributed as saying that he'd found balloon debris "scattered over a square mile". Some reporters found this inconsistent, given the small amount of balloon debris shown in Ramey's office. Ramey later spoke to reporters and was quoted by the *Washington Post*, United Press, Associated Press and others as saying it was a "box kite" covered with foil and if reconstructed it would have been "about 25 feet in diameter".

Ramey ordered one of his intelligence officers, Major Kirton, to make statements. He too was inconsistent regarding the type and size of weather balloon, and where it was being transported.

Major Kirton stated the balloon was "20 feet" to the Dallas FBI office and Reuters news agency; it had attached radar reflectors to the FBI and *Dallas Morning News;* and the *Dallas Morning News* was told the flight to Wright Field for identification was canceled because the object was clearly identifiable as a weather balloon. But ABC News contacted Wright Field and were told they disagreed with the weather balloon explanation and were awaiting the arrival of the debris for detailed analysis.[51]

Top military brass acting like it was not a weather balloon

As the media and public were being spoon fed with the USAF's new official explanation, behind the scenes we learn of another top U.S. military official who had to be stirred into action by the event near Roswell. We see this in the high ranking military official, General Nathan Farragut Twining (1897-1982), in charge of the Air Force Matèriel Command based at Wright–Patterson AAF in Dayton, Ohio, USA.

On July 7, 1947, Twining suddenly changed his travel plans to attend a matter of great importance in New Mexico. The official letter[52] mentioning the sudden change in plans was written to Mr. Julius Earl Schaefer (1893-1974) of the Wichita Division of Boeing Airplane

Company. It showed that Twining had canceled his scheduled meeting for July 10, 1947 with the Boeing company because, in his words, of "a very important and sudden matter that developed here."

The U.S. Air Force later officially explained the visit by Twining as a "routine inspection".

An urgent "routine inspection" of a weather balloon? Must be a very special weather balloon to get Twining off his feet and visit New Mexico in such an urgent manner. Or, as the USAF suggested in their 1994 *Roswell Report*, it was to attend a Bomb Commanders' Course on July 8, 1947:

> "An example of activity sometimes cited by pro-UFO writers to illustrate the point that something unusual was going on was the travel of Lt. General Nathan Twining, Commander of the Air Materiel Command, to New Mexico in July, 1947. Actually, records were located indicating that Twining went to the Bomb Commanders' Course on July 8, along with a number of other general officers, and requested orders to do so a month before, on June 5, 1947."[53]

This must have probably been one of the most important Bomb Commander's course ever attended by the General, and would appear to have been organised at the very last minute for the General to suddenly change his travel plans.

However, a secret memo[54] dated July 15, 1947 and written by General Twining, released to the public under Freedom of Information (FoI), seems to reveal the true reason for the sudden urgent and rather important visit and possibly the type of object that was recovered. On page 2, it states:

> 1. As ordered by Presidential Directive, dated July 9, 1947, a preliminary investigation of a recovered "Flying Disc" and remains of a possible second disc, was conducted by the senior staff of this command [Air Material Command at Wright Patterson AFB]....
>
> 2. It is the collective view of this investigative body, that the aircraft recovered by the Army and Air Force units near

Victorio Peak and Socorro, New Mexico, are not of U.S. manufacture for the following reasons:

(a) The circular, disc-shaped 'planform' design does not resemble any design currently under development by this command nor of any Navy project.

(b) The lack of any external propulsion system, power plant, intake, exhaust either for propeller or jet propulsion, warrants this view.

(c) The inability of the German scientists from Fort Bliss and White Sands Proving Ground to make a positive identification of a secret German V weapon out of these discs. Though the possibility that the Russians have managed to develop such a craft, remains. The lack of any markings, ID numbers or instructions in Cyrillic, has placed serious doubt in the minds of many, that the objects recovered are not of Russian manufacture either.

The document is highly revealing in the sense that it discusses the initial examination of the wingless disc-shaped craft. It reveals interesting insights into the construction, and likely propulsion system, stating "...the craft itself comprises the propulsion system", "...the reactor [or engine] to function as a heat exchanger and permitting the storage of energy into a substance for later use", "...storage battery", "...no moving parts discernable within the power room", "...this motivation of a electrical potential is believed to be the primary power to the reactor [engine]", "...air outside the craft would be ionised", "...[the] crew compartments were hermetically sealed via a solidification process", "...[the] craft components appear to be molded and pressed into a perfect fit", "...no weld marks, rivets or soldered joints".

Assuming this remarkable document is genuine, it suggests that by July 15, 1947 it came to the attention of the USAF (probably by the Army) of a second crash site where a disc was found and that this disc had unusual construction techniques and a propulsion system not recognised by any military or scientific establishment at the time. The information had presumably been passed on to General Twining after he made an unplanned visit to Roswell AAF in New Mexico.

General Nathan F. Twining (1897–1982)

Also within the document is mentioned "remains of a possible second disc". Is this a mistake? Or did the military discover a second disc?

It's feasible to suggest that the first crash location was presumed to be the second disc at this early stage of the military's investigation given the amount of debris found scattered over a large area despite not revealing any clear evidence of its shape. The fact that the document mentions nothing about the internal structure of the second disc to give a sense of comparison through discussion of similarities and/or differences suggests the military had assumed a "two discs" theory (as they were probably not told by the rancher about the "odd explosion" during the lightning episode), with the second disc being sufficiently disintegrated over the rancher's property to the point where it wasn't worth discussing its internal structure according to General Twining. Certainly, if we look more closely at the document, we do see the words "...remains of a possible second disc".

In addition, 200 kilometers (for a *single* badly damaged disc) to stay in the air before crashing would be unheard of for normal military or civilian aircraft. It would seem natural to assume the possibility of two discs.

Disregarding for the moment the number of discs, it is beginning to look like the U.S. Army and Air Force found at least one disc. We are led to believe it was a symmetrical, wingless "aircraft...not of U.S. manufacture", requiring a high electrical potential or voltage to move the object, ionization of the air to perhaps reduce air pressure near the surface, with no sharp metal points, and no identifying marks. This is one very special type of weather balloon designed to move and protect itself using obscure electromagnetic principles. Or maybe it isn't meant to be a weather balloon? The fact that the document describes the disc as a "craft" suggests it could carry instruments and possibly people inside.

Does this mean the military have found a new type of high-speed aircraft as well as the remains of pilots, if any, during their recovery? If so, what type of craft is it? And who are the pilots? Is McBoyle's earlier statement of "little green men" related in some way to these pilots, which we presume had been recovered during the clean-up operation at the first crash site?

Or there is always the possibility that military life was getting to General Twining in his later years to the point where he wanted to write the first chapter for a new science fiction novel he had been thinking about. Depending on how scientifically accurate he wanted to make it sound in his novel through the use of interesting electromagnetic concepts and manufacturing techniques, he decided he would do it in a secret memo. A rather strange place to express his heavily restrained creativity.

Or if the document is meant to be a fake, it is a particularly good one considering the amount of interesting electromagnetic concepts being revealed by the disc.

More evidence supporting a "disc" with "victims"

In 2000-01, intriguing new information emerged suggesting Marcel's claim of a cover-up as well as the possible recovery of at least one "disc" and "bodies" could be true. It involves a secret memo carried in General Ramey's hands in a rather blasé fashion during the interview. It can be seen in four of Mr. Johnson's photographs; only to find that in one of the photographs he had inadvertently held it towards Mr. Johnson's camera.

It began in 1991, when a lawyer by day and dedicated Roswell researcher in his free time (and already a best-selling author in this area of interest) Donald Schmitt sent the best available copy he had of the photograph containing Ramey's top secret memo to a respected NASA research scientist, Dr. Richard Haines. Without the benefit of digital enhancements, Dr. Haines used a microscope to analyze the message. He reported to Mr. Schmitt in his letter dated February 13, 1991 that he could observe words containing a certain number of letters in each one, but not enough letters could be recognised to help him uncover familiar words or at least ones that could create meaningful sentences in his opinion.

In 1994, as part of the USAF official attempt to investigate the Roswell incident, the photograph was sent to an unnamed organization for more detailed analysis. The result of the analysis can be seen in the following quote taken from the *Executive Summary: Report of Air Force Research Regarding the 'Roswell Incident'*:

"It was noted that in the two photos of Ramey he had a piece of paper in his hand. In one, it was folded over so that nothing could be seen. In the second, however, there appears to be text printed on the paper. In an attempt to read this text to determine if it could shed any further light on locating documents relating to this matter, the photo was sent to a national level organization for digitizing and subsequent photo interpretation and analysis.... This organization reported on July 20, 1994, that even after digitizing, the photos were of insufficient quality to visualize ... details sought for analysis"[55]

However, more sophisticated digital technology became available in 1998, allowing the retired James Bond Johnson to assemble a team of specialists, including Ron Regehr, a space and satellite engineer, to inspect the original negative of the photograph he took of General Ramey. This time the section of the photograph showing General Ramey's secret memo was properly enlarged and digitized to at least 600dpi using special camera equipment. He then digitally enhanced the image.

On this occasion, Johnson and his team could make out certain words and construct basic sentences. The following is what the team managed to extract from the memo:

"AS THE ... 4 HRS THE VICTIMS OF THE ... AT FORT WORTH, TEX ... THE 'CRASH' STORY ... FOR 0984 ACKNOWLEDGES ... EMERGENCY POWERS ARE NEEDED SITE TWO SW OF MAGDALENA, NMEX ... SAFE TALK ... FOR MEANING OF STORY AND MISSION ... WEATHER BALLOONS SENT ON THE ... AND LAND ... rOVER CREWS ... [SIGNED] ... TEMPLE."

Realizing the potentially explosive nature of the memo and its contents, other researchers have also attempted a similar enlargement and digital enhancement of the memo to help corroborate on Johnson's latest findings. Among the researchers was a former USAF serviceman with top-secret crypto clearance named Thomas J. Carey who joined MUFON (Mutual UFO Network) State Section Director for Southeastern Pennsylvania (1986-2002) before specializing in the

79

Roswell case. He later joined with lawyer and author Don Schmitt in 1998—these two becoming the two principal Roswell investigators in the U.S.

In 2000, Tom Carey, Kevin Randle and Dr. Donald R. Burleson gave their own interpretations of the memo. The results suggest they were inconclusive in the sense that there appeared to be different interpretations of a certain number of words shown in the memo, which could lead to different sentence constructions.

As Burleson stated in his article "Looking Up" published in *Vision*, a monthly magazine from the *Roswell Daily Record*:

> "A number of attempts have been made to read the Ramey letter. Quite frankly, most of these attempts are amateurish, and even some ufologists have concluded that there is nothing in the Ramey image that advances the case for the Roswell incident. They are mistaken."[56]

Then in late 2000, Mr. David Rudiak got involved in the Roswell case. An optometrist by day, he had skills with a computer that allowed him to include one other additional tool to his work: an electronic dictionary.

According to the work of David Rudiak who applied his own digital enlargement and enhancement techniques to the memo and combined with his exceptional eyesight and ability to recognise obscure letters at a distance, his efforts using an electronic dictionary were rewarded with enough reliable and readable words allowing him to hone in on the most likely sentences being displayed in the memo. It is from these key words and from the most reliable sentences that we learn how the military had continued to support the idea of a "disc", but the word "victims" was added in association with the "disc" as if there were casualties.

Furthermore, Rudiak has maintained since 2001 that the memo does claim the disc is a "new find", as if the USAF hadn't manufactured the object.

The memo also indicates the use of "weather balloons" and "demo Rawin crews", and keeps these terms sufficiently separate from the other more controversial key words and positioned toward the end of the document as if suggesting a more prosaic man-made military explanation would be used for the "disc" and "victims" should the media ask questions.

Before we get into the details of the message, let us explain the steps used by Mr. Rudiak (and others) to make the memo more readable.

Firstly, using magnification and clever digital enhancement techniques, Mr. Rudiak was able to discern the following image[57]:

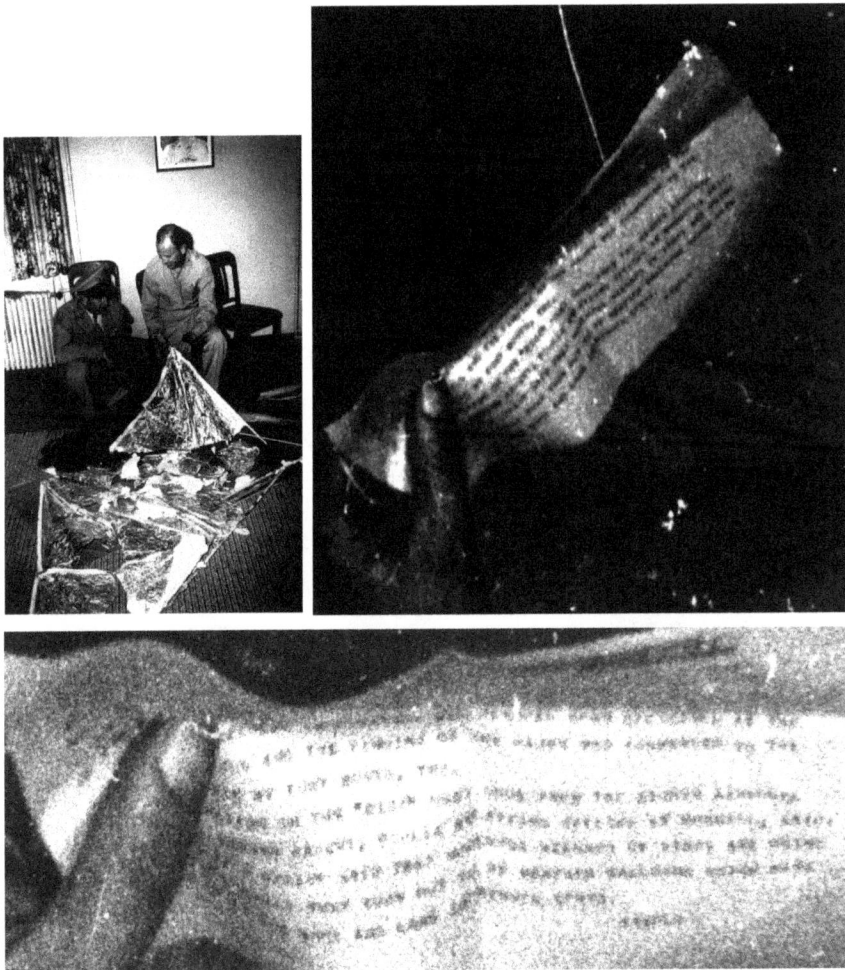

As an example, here are a couple of words extracted directly from the enlarged image:

With digital enhancement of the image, this is how Mr. Rudiak was able to see the above two words:

Secondly, by counting the number of characters in each word, Mr. Rudiak was able to narrow down the list of possible words using an electronic dictionary:

VICTIMS   "DISC" or *DISC*
REMAINS   "RISK" or *RISK* etc
FINDING etc

Thirdly, by looking at the individual characters in each word, one can discern familiar shapes for which the available list of possible words obtained from the electronic dictionary (or a Crossword Puzzle Helper tool) can be further refined:

VI _ _ I _ S

VILLIS   VIOLINS
VIBRIOS   VITRIOS
VIROIDS   VILNIUS

And fourthly, the word or words that best fit the pattern of letters are checked by reading the sentence to see if it makes sense. Bearing in mind the memo is not meant to be a piece of masterful poetry or to use the most sophisticated words in the English language, the choice of simple words and basic sentence construction will often narrow down the list of words to just one or two in most cases; otherwise it is left blank.

The word most likely to fit in the context of the memo:
VICTIMS

Perhaps the only tricky part is whether the USAF used unfamiliar acronyms within the memo. Luckily for Mr. Rudiak, he has not found too many of those.

After going through this laborious analysis, it is now possible to make out certain readable English words[58]. Shown below is Mr. Rudiak's best interpretation:

(3) ..."RANCH" AND THE VICTIMS OF THE WRECK YOU FORWARDED TO THE

(4) ...TEAM AT FORTH WORTH, TEX.

(5) AVIATORS IN THE "DISC" THEY WILL SHIP FOR A1-8TH ARMY AMHC

(6) BY B29-ST OR C47. WRIGHT AF ASSESS AIRFOIL AT ROSWELL. ASSURE
ASSIST FLY-OUT

(7) THAT CIC/TEAM SAID THIS MISSTATE MEANING OF STORY AND THINK

(8) LATE TODAY NEXT SENT OUT PR OF WEATHER BALLOONS SENT ON THE

(9) BETTER IF THEY ADD LAND DEMO RAWIN CREWS.
[SIGNED] ... TEMPLE

Mr. Rudiak also indicated some success in flattening the fold at the top sufficiently to make out additional words. He believes the words he can see in the top paragraph are:

(1) FWAAF ACKNOWLEDGES THAT A "DISC" IS NEXT NEW FIND WEST OF

(2) THE CORDON. AT LOCATION WAS A WRECK NEAR OPERATION AT THE
IN ADDITION   "POD"

Ignoring the fold at the top, it is still easy to make out certain discerning and rather important words such as "disc" (line 5), "weather balloons" (line 8) and "land dem[o] Rawin crews" (line 9). In addition, we also see from the previous paragraph (line 3) the word "victims" is included for some reason by the USAF. More importantly, the position of all these key words within the memo provide further insight into what was happening within the USAF at the time.

Of course, to the skeptic, all this may be seen as a matter of interpretation and depends on whether the person interpreting the words are UFO-biased and, therefore, likely to *choose* words to suit a particular story. To support this claim, skeptics have noted the work conducted by James Houran.

In 2002, James Houran of Adelaide University conducted an experiment where he asked a statistically small sample of 176 people to make their own interpretation of a standard digitization and magnification image of the memo presented by Kevin D. Randles without performing digital enhancements or supplying an electronic dictionary to the participants. At the end of the experiment, Houran found the memo can have a variety of interpretations even though certain words do appear to consistently come up as the same for most people with clear eyesight and who made a reasonable effort to interpret the words. He also notes that the words chosen are dependent on whether the participants were told whether the memo refers to the Roswell incident or not.

As Houran stated:

> "The surprisingly high agreement between our participants and previous investigators on specific words in identical locations in the Ramey memo suggests that some of the document is indeed legible, even without computer enhancement. However, the meaning or context of those words remains ambiguous because the degree of interpretation of the document is strongly influenced by suggestion effects and the interpreter's cognitive style."

The problem with this experiment, however, is that the way participants are asked to interpret the memo is not dissimilar to sitting inside a car on a cold winter's morning with the windscreen fogged up, and you are asked to interpret the words displayed on an unfamiliar road sign outside the car. You can either get a bunch of people to sit in the car and make their own interpretations, or you can use the technology of a cloth to wipe away some of the moisture built-up on the windscreen to see the words more clearly. In other words, the experiment needs to be done again, and this time apply the latest digital enhancements that present-day technology can muster, together with information on exactly how many letters are in each word, and include any letters considered discernible. Then, with an electronic dictionary showing the list of best fit words, let the participants choose the words and ensure that the sentences make reasonable sense. Only then can we ask the question: What percentage of participants agree that the word "disc", "victims", "weather balloons" and "Rawin demo crews" appeared in the text? If more than 50 percent of the participants agree, it is on the basis of probability that we can say those words are likely to be the ones contained in the original text.

Even without proper digital enhancement and an electronic dictionary, we can already see from the survey a consistency in several key words in the memo.

Where digital enhancement technology was applied by researchers Tom Carey, Neil Morris, David Rudiak, Don Burleson and John Kirby from a standard scan made by the University of Texas Arlington Libraries of the original negative, the consistency is more striking. For example, the word "DISC" or "DISK" is mentioned in 5 out of the 7 interpretations presented by the researchers for line 5. The word "VICTIMS" is consistently noted in 7 out of the 8 interpretations made for line 3. The words "WEATHER BALLOONS" has been recognised in all 7 interpretations of the text. The one that might be a little more contentious are the words "LAND DEMO RAWIN CREWS". Certainly the word "CREW" appears in 4 out of the 7 interpretations made of the text.

In the survey where no digital enhancements or electronic dictionary is applied, the results vary considerably, but the key words do appear to crop up consistently.

For example, Houran says for the group of participants who were told the memo refers to an atomic bomb (total of 58), only 5 recognised the word "WEATHER BALLOONS" compared to only 2 for "BALLOONS" in another 59 participants who were not told of the context of the memo (i.e. the blind condition). As for "DISC" (it is more likely written as *DISK*) and "VICTIMS", we don't know (presumably they were not asked). On the other hand, out of the 59 participants who were primed to think the memo referred to the Roswell incident, 20 recognised "WEATHER BALLOONS". That's a significant difference. Houran interprets this as due to the way a biased mind primed to a certain condition for understanding the context of the memo can change the way participants spend time on interpreting the words (depending on how interesting the document is) which in turn affects the choice of words selected to give the sentences some meaning.

Or more likely the participants were simply not making any reasonable effort to analyze the text and instead preferred to rush the job because they were not told of the memo's importance. What if participants were told that the best interpretation was a matter of life and death? Would the results be the same as before, or different and closer to the results of the researchers, or those primed to think it related to the Roswell incident?

In that case, the experiment needs to be done again. And this time give the participants the benefit of the latest digital capture technology from Mr. Johnson's original negative together with the best digital enhancements to bring out the number of letters and their shapes as clearly as possible, and with an electronic dictionary showing the best available words that most likely fit the letters contained in each word in the memo. Without telling them the nature of the memo, it should be interesting to see how many of these participants will recognise the same keywords.

Until the experiment is done again, we must rely on digital enhancement techniques and an electronic dictionary from other researchers to get a reasonable consensus on the keywords. It is here where we find important keywords, and from it produce a picture different from the one being officially given to the media and the public by the USAF.

Assuming Mr. Rudiak's interpretation of a "new find" is correct, in which case the recovery of "weather balloons" cannot be a "new find" (unless they are highly advanced Russian-made weather balloons made of new materials and an electromagnetic propulsion system), we find that the USAF still maintained the view that it was a "disc" in its top-secret paper and considered it a "new find" as if the military still had no idea what had crashed or who had made it, despite stating the contrary to media officials.

In other words, the subject of this memo is the "disk", not the "weather balloons". The former word is positioned nearer to the beginning of the memo. Weather balloons are only added right at the end as a suggestion by the Pentagon to cover up the real story in case the inquisitive media learned more details about what was discovered at the rancher's property.

Sure, one could vigorously argue the possibility that the inclusion of the word "disc" might be in reference to the news release issued by Walter Haut. However, there is one tiny problem: the fact that the news release never mentioned "victims". The failure to do so would effectively discount this theory. It means the memo must strongly be pointing to a genuine discovery of a "disc" (perhaps a man-made disc-shaped object) exactly in keeping with the news release and thus cannot be incorrect, and it gives greater credence to General Twining's supposed original creative writing piece of a memo dated July 15, 1947, which describes the inside of the disc.

Too controversial? In that case, let us look at this situation more logically.

Why do we have "weather balloons" in the memo? Could this be referring to the "disc" in terms of its shape? The only problem with making this link is the way the memo states "weather balloons" in the plural.

To give an example, the photo on the right was used by the USAF to explain the Roswell object. It shows several weather balloons holding up the reflector. This would effectively support "weather balloons" in Ramey's memo. Even so, a bunch of independently floating balloons and a series of reflectors attached underneath will not make it look more like a "disc". Quite the contrary. It would look like a jumbled mess. If the USAF still wants to call these multiple balloons and/or reflectors a "disc", the Ramey memo does not support it. The memo clearly states a "disc" in the singular sense.

How can this be so? Or could it be that we are referring to just one weather balloon?

The photo on the next page shows a reflector attached to a single balloon, taken on July 10, 1947 by the *Fort-Worth Star-Telegram* photographer at Fort Worth AAF. This was part of a demonstration by the USAF to show what had probably crashed near Roswell, except the object is not revealing itself to look anything like a disc. The balloon at

Fort Worth AAF airmen demonstrating a radar device and balloon as the official explanation for the Roswell debris. Courtesy of Fort Worth Star-Telegram Photograph Collection, Special Collections, University of Texas Arlington Library.

close range looks near-spherical in shape, and the reflector is definitely not a disc. There is simply no resemblance. Yet someone in the USAF still wanted to call it a disc.

Here is an example. The following teletype sent by the USAF to United Press at 10:30 PM on July 10, 1947 attempts to link the word "disc" to "weather balloons":

"EXR44S

ALAMOGORDO, NEW MEXICO---AN ARMY AIR FORCE MAJOR HAS SUGGESTED THAT THE MYSTERIOUS FLYING DISKS MIGHT BE REFLECTORS OFF SOUNDING BALLOONS USED IN THE V-2 ROCKET TESTS.

MAJOR JAMES R. PRITCHARD, PUBLIC RELATIONS OFFICER AT THE ALAMOGORDO AIR FIELD---JUST EAST OF WHITE SANDS, NEW MEXICO, PROVING GROUNDS---DEMONSTRATED THE BALLOONS TO THE PRESS.

HE SAID THE BALLOONS--WHICH CARRY A SILVER REFLECTOR WHICH COULD BE A DISK--- HAD BEEN RELEASED AT ALAMOGORDO FOR THE PAST 15 MONTHS AT FREQUENT INTERVALS---AND WHICH HAVE LANDED THROUGHOUT THE ROCKY MOUNTAIN REGION.

THE MAJOR ADDS THAT DISK IN THE PACIFIC NORTHWEST COULD BE THE RESULT OF SIMILAR ACTIVITY BY THE NAVY.

S01030P7/10."[59]

What does this mean?

It looks like either the balloons themselves have been combined together to give the impression it is disc-shaped, or the reflector itself is disc-shaped.

Well, in terms of the amount of metallic foil found at the first crash site, the photograph showing a single balloon and simple reflector is insufficient to cover up to a mile in length and 200 to 400 feet wide in debris according to Marcel. This means there must have been a series of radar reflectors; and to lift the weight of these reflectors, there must be multiple balloons. This would effectively support "weather balloons" in Ramey's memo and help to explain the amount of foil found at the crash site.

In that case, if the balloons were stuck together with glue, the following diagrams show how balloons can be arranged to look vaguely disc-shaped from afar:

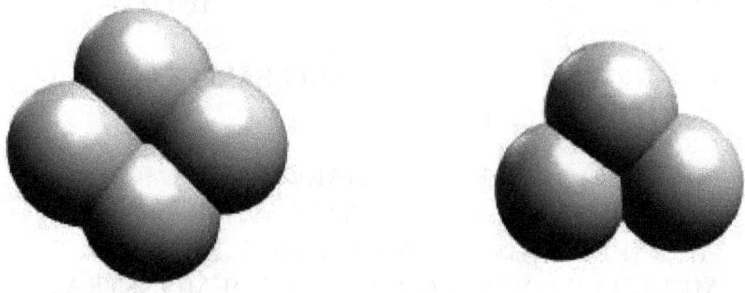

Top View

Side View (most likely to be interpreted as a "disc" when seen far away)

Or, if we wanted to push the boundaries of reality, it is possible for two balloons stuck together to be slightly disc-shaped from a great distance.

In the case of just one balloon, something has to flatten it to reveal a disc. However, at close range, and presumably torn up on the ground, it would be extraordinarily hard to continue calling a single balloon (or multiple balloons) a "disk". Yet that is exactly the problem the USAF had. In fact, the military still had trouble describing the balloons in any other shape at close range during the recovery operation except to call them a "disc" in Ramey's memo.

Why use the word "disc"?

It is the close-range observation of the object's remains on the ground that is the problem when trying to understand how the USAF can describe it as a "disc". Even if we assume the object was undamaged, three or four spherical balloons arranged like a triangle or square as shown in the pictures above will not look like a "disc" at close range. The USAF would have described them as triangular-shaped, or square-shaped. Even six balloons—four to make a square and two above and below the centre of the square to give it a slightly more "disc"-shaped appearance from all angles from a distance—would still be described at close range as a pyramid-shaped object. Any more balloons and we can describe it as two cones stuck together at the base. So why not call it a "double cone-shaped" object. Or why not a "dumbbell-shaped" object in the case of just two balloons stuck together?

Close-range observations of the downed object on the rancher's property cannot support the shape as anything remotely like a "disc" according to the memo.

However, according to an FBI teletype[60] dated July 8, 1947:

> "Telephonically advised this office that an object purporting to be a flying disc was recovered near Roswell, New Mexico, this date. The disc is hexagonal in shape and was suspended from a balloon by a cable..."

Amazing! It is now a hexagonal object. So why not call it as such? All the details regarding the shape should have been provided after the object was recovered.

In fact, on carefully examining this quote, we see that the shape is not in the number of balloons stuck together but in the object attached underneath. The hexagonal shape is in the reflector part of the weather balloons, and the number of balloons lifting the metallic object is only *one*.

From the amount of wreckage found at the first crash site, this must make it a very large balloon carrying a very big "hexagonal" reflector. Unfortunately this does not conform to the memo where it clearly states "weather balloons" in the plural as if more than one balloon were present. Neither does it remotely support a "disc" for a "hexagonal" reflector.

In fact, there continues to be considerable confusion among military observers in the number of balloons found and the shape (either a hexagon or disc), let alone whether this object and the balloon or balloons it is attached to is a weather balloon or something else. We see evidence of this in the same FBI teletype:

> "[Major Curtan] further advised that the object found resembles a high-altitude weather balloon with a radar reflector, but that telephonic conversation between their office and Wright field had not borne out this belief."

We are led to believe that neither the Wright–Patterson staff who analyzed the object and its materials nor the USAF staff at the Roswell AAF that recovered the object or at the Fort-Worth AAF where pieces were brought in at the request of Ramey can agree on anything in relation to the object. If these experts in observational powers and scientific analysis cannot figure out how many balloons and what shape the entire object should be, how likely is it that they can determine whether it is a weather balloon? For all we know, the USAF could have found pieces to support the world's first evidence of a flying pig and we still would not know what it is from the way these experts were behaving.

However, all experts, no matter who they are, must agree on something, especially at close range. When you inspect an object at close range with your hands, you have to get a reasonable and consistent idea of how the materials look and how they behave when certain tests are performed on them, what type of object it was given the type of materials used and its construction, and how many balloons there are if it is a weather balloon. And given the fact that the FBI memo claims the object was "hexagonal", the shape should have been known too.

Indeed, at close range, and from the evidence presented so far from the FBI memo, the USAF should have identified the shape and type of object that crashed as quickly as Irving Newton was able to identify a scrunched-up weather balloon that was physically flattened out on the floor of Ramey's office. They could have just said in the memo that it was a weather balloon with a "hexagonal" reflector and left it at that for the reader.

Simple. Why complicate the matter?

Unfortunately the experts can't agree. From the way Ramey's memo is written, the USAF is not making any direct association between "disc" and "hexagonal" reflector, or "disc" and "weather balloons" to help give the object a more precise and definitive "man-made" description.

The only thing that the military can agree on is the fact that the word "disc" continues to be maintained in General Ramey's and General Twining's memos and is consistent with the use of the term "disc" in the Roswell AAF news release.

However, the "disc" cannot be a "weather balloon".

You see, the word "disc" in Ramey's memo is separated from the other key words including "weather balloons", and left hanging on its own with no other words next to it or in very close proximity to give it a more precise meaning. That is, we don't see a "weather balloon disc" or "disc-shaped weather balloon", or "a hexagonal-shaped reflector looking like a disc edge-on". The word "disc" remains without the immediate context of something with which we would be familiar. Even more telling is the decision to put quotation (or asterisk) marks around the word as if the object is indeed unfamiliar or unusual in some sense. With no markings to say what model number or make this object is, the best the military could say was that the object was a disc. It is from this basic observation that we can begin to understand the way the memo was written. In other words, the memo tells us that the military had no better explanation at the time (or even by July 15, 1947 when Twining wrote his secret memo on the internal structure of the disc) to explain the stricken object even when observed at close range and given a preliminary examination. If this is untrue, why use "disc" at all? The word "balloon" would have sufficed to explain the entire object for what it was. Not so according to the memos. Instead, the only description the military could give for this mysterious flying object that dropped pieces from the air on a rancher's property is to continue using the word "disc". In addition, from the way the memo was written with the intriguing words "new find", it is a "disc" not known to the military.

Therefore, we must interpret this interesting part of the memo held by Ramey as stating that the USAF will use weather balloons as part of the official cover-up should the real story of a "disc" get out to the media. Even if this is untrue, why cover up multiple radar reflectors (or a hexagonal metallic reflector) attached to one or more weather balloons, assuming we can allow the USAF to interpret this object as a disc?

This suggests that the cover-up involves something new about the disc. Perhaps another type of man-made flying object. If so, what was so special about it? In other words, was it new type of object with unusual manufacturing techniques not familiar to the military at the time? Or was this "disc" manufactured by someone not from the USAF?

Or perhaps both?

So what is the "new find" in relation to this "disc"?

As we have stated earlier, a weather balloon cannot be the new find. Weather balloons are flown in the air all the time—nothing new here. Nor is the shape of a "disc" enough to describe it as a "new find". It is certainly not shocking enough to cause General Twining to change his travel plans to visit New Mexico to look at a disc-shaped weather balloon. The cover-up must clearly relate to another aspect of the object. It suggests we are dealing with a new type of technology, which could include the materials making up the object. In other words, we are not dealing with a weather balloon of any sort. Perhaps it is a new type of military aircraft, or missile? Either that, or the object was manufactured by someone not in the U.S. military (possibly another nation). Whatever this object is, we know it cannot be a weather balloon.

Supporting this possibility is the way the USAF insisted it was a "weather balloon", but ABC News claimed sources at Wright–Patterson told them they were waiting for the object to arrive for analysis. What would be the point of the analysis if the USAF already knew the shape, origin and character of the object in question, which we presume must be a weather balloon? Are we to assume Wright-Patterson experts had not seen a weather balloon before? Certainly not. It would only reaffirm what the USAF already knew. Unless, of course, the object was so badly damaged that the analysis was required, except that would imply the shape of the object was beyond recognition just as the rancher had observed. So how did they know the object was a disc? Without lightweight gases to fill the balloon (or multiple balloons, whichever is closer to the truth) and with the reflector part looking significantly damaged as seen in Ramey's office, there is no way of telling the shape just by looking at it. It would require detailed analysis by a scientific laboratory to determine the shape, and that would take time.

Yet the USAF already assumed a "disc", "hexagon" or whatever shape it was, and wanted to describe it as a "weather balloon" before the analysis could be carried out.

The only way this is possible is to say that either the military already knew the shape of the object the instant they arrived at the second location during the recovery phase, or were informed of the disc near Socorro by the U.S. Army. Or they found something else on the rancher's property telling them it had to be a "disc", perhaps from the "victims" they had recovered, and/or even the unusual types of materials used to construct the "disc".

Furthermore, if the USAF wanted to assume it was a disc just on the basis of arriving at the scene of the crash site and looking at the materials, then clearly the materials cannot be flimsy. They had to be incredibly tough to retain the shape of the object. Otherwise something in the materials or whatever else they had found told them to consider the object a "disc".

Either way, it is starting to look like this "disc" is something unique and special.

Are we dealing with a new type of highly advanced Russian disc-shaped aircraft? Or was the discovery more to do with the "victims" recovered on the rancher's property?

Perhaps the appearance of these "victims" was such that it made military officials in the USAF seriously think it was one of the flying discs mentioned in the newspapers and they did not need to elaborate further on the "disc" as there was nothing in our arsenal of man-made objects to give it context or familiarity to help explain what it was. Better to call the remains a "disc" because of how the "victims" looked and leave it at that. It makes sense according to the way the memo was written. If true, this would point to a potentially disturbing extraterrestrial explanation; otherwise, the USAF would not be keeping the identity of these "victims", the materials used, or the type of disc it had recovered, a secret to this day.

While on the subject of recovered bodies, the word "victims" is highly suggestive of two or more people injured or killed. Certainly it does not suggest mannequins or dummies as the USAF would later use in its final report on the incident in 1997 (to be discussed in Chapter 4). Sure, we do see virtually at the end of the memo the words "demo Rawin crew", which was in reference to a set of dummies. Yet the two terms are not directly associated. In other words, we do not see "demo Rawin crew victims". In fact, it would not make sense, considering dummies do not have feelings. When they hit the ground, they are not victims. They remain dummies in every sense of the word.

This means the same argument used before for the "disc" and "weather balloons" can be applied to the "victims" and "demo Rawin crew".

Furthermore, the use of the term "victims" strongly suggests something that was, at one point, alive. It suggests there were living entities sitting inside the object as it flew in the air and that these had either died or were injured at the moment the object exploded or hit the ground. Perhaps we are dealing with a canary or some other animal? Not likely. It would have to be more human. It is the only way to explain the USAF's sudden need to track down Brazel and harshly interrogate him, even to the point of threatening him with his life should he speak out about anything he might know. It would be strange to behave this way over dummies, a bunch of canaries, or even monkeys sitting inside an ordinary weather balloon. The odd behavior of the military toward at least one civilian simply reinforces the view that we are dealing with something real, biological and more human-like in the "victims" found.

Indeed, for the USAF to maintain secrecy to this very day over who or what was killed would have to make the "victims" more human-like rather than familiar and ordinary experimental animals or plastic dummies.

In that case, who was injured or killed?

Certainly none of the civilian and military witnesses gave a hint of being injured, or having known other people to be injured or killed in association with the Roswell materials. To this day, investigators have never found a record in the local hospital at Roswell or elsewhere of any military or civilian victims associated with this incident. Were there unknown human victims? Even so, would the USAF not have worked out who they were by this time? Why do they remain anonymous? Surely the USAF would know the names by now. If not, what is wrong with telling us whether the victims were humans, at the very least? Surely no harm in telling us this?

This is highly unusual considering it is the job of the U.S. military to identify the names of those who died in a crash on U.S. soil, even if they are not part of their own military establishment or country. Even enemies of the U.S. get more respect than was given to these alleged victims, assuming there was a way to identify the victims. In addition, not one family member in the world has come forth to explain a loved one going missing in the New Mexico desert in early July 1947. Either these victims have no family on Earth, or they were not human victims at all. Perhaps we are dealing with a number of highly intelligent and expensively trained monkeys designed to fly this mysterious object. This would explain why the USAF will not reveal any names. Yet the USAF refuses to say they recovered "monkeys". The word "victims" remains the only and best description the USAF could give under the circumstances.

Does this mean the USAF had loved their experimental animals so much to the point where the military had to call them "victims"? Unfortunately, the USAF is not willing to say the victims were well-loved and highly trained experimental monkeys or some other animal and be done with it. In that case, it simply does not make sense to associate "victims" with experimental animals. The highly enduring nature of the secrecy behind these "victims" tells us we have to be dealing with "humans", and certainly not "dummies" or something else. And apparently highly sensitive individuals too considering the USAF still wants to keep a tight lip over their identity. So who are these people?

Perhaps we could be dealing with thin, short and highly intelligent Japanese prisoners of war with large heads who were forced to participate in the secret military experiment and the USAF wanted to keep their names secret. It is possible. But even the most hated enemies at the time would be given some respect by mentioning their names after they had died. What would cause these Japanese men to be treated any differently?

There is clearly something odd about these victims, and the USAF needs to explain what it is.

Two discs, or one disc?

Further analysis of Ramey's memo also indicates something else: there appears to be another location separate from the main operation where the USAF had been clearing debris from the Foster Ranch. Could this second location be where the victims and possibly the disc itself were found? Clearly, it can't be the first wreckage site, as the rancher never made statements to the effect of finding a disc and/or noticing victims. There was not even the tiniest smell indicating the possibility of bodies lying around and decaying under the desert sun for days as he picked up larger pieces to store in the shed at the back of the Hines House. The bodies had to be lying at some distance away from the initial wreckage site, and likewise the disc itself.

Seriously, how could bodies get transported to such a distance and not get noticed by the rancher? The flying object clearly must have remained in the air.

Well, we have already heard the story of Barnett about the victims and a disc. He claims the object landed 200 kilometers west of the initial crash site. That is quite a long way off from the initial site, suggesting the object was still under some power and traveling at a high speed before crashing to the ground. Or the crash site was much closer than originally thought.

Or are we dealing with two discs?

Unfortunately, Twining's memo reveals the internal structure for one disc. No other disc was uncovered and analyzed for comparison purposes. Unless the USAF assumed a second disc had disintegrated over the first site and, therefore, could not find anything useful from this second disc, it suggests we are dealing with one disc, which dropped debris over one site and came to grief together with the bodies at another site.

However, at present there is a dispute over the exact location of this second site.

New witnesses have come forth claiming the location of the disc to be much closer to the initial crash site than originally thought, perhaps 40 miles away or less. Then again, it could also be true that the U.S. Army, the ones who allegedly recovered the disc and victims in the Plains of San Agustin over 200 kilometers away, had mentioned the discovery to the USAF where the connection was made.

Or was there a third location where potentially more bodies and some more debris were found?

The interesting thing is that these new witnesses are claiming a virtually intact (i.e., with very little damage) bright silver disc was found 40 miles from the initial debris field. And its direction from the initial site is toward Roswell. Hence, the reason for calling this the Roswell case, and not the Socorro case.

Below is an example of a quote from an anonymous member of the U.S. military claiming that a family member was part of the U.S. military involved in the recovery of the bright-silver disc and bodies near Roswell:

"My grandfather was a member of the Retrieval Team, sent to the crash site, just after the incident was reported. He died in 1974, but not before he had sat down with some of us, and talked about the incident.

I am currently serving in the military, and hold a [high] Security Clearance, and do NOT wish to 'go public', and risk losing my career and commission.

Nonetheless, I would like to briefly tell you what my own grandfather told me about Roswell. In fact, I enclose for your safekeeping samples that were in the possession of my grandfather until he died, and which I have had since his own estate was settled. As I understand it, they came from the UFO debris, and were among a large batch subsequently sent to Wright–Patterson AFB in Ohio from New Mexico.

My grandfather was able to appropriate them, and stated that the metallic samples, are PURE EXTRACT ALUMINIUM…

As my granddad stated, the Team arrived at the crash site just after the AAF/USAF reported the ground zero location. They found two dead occupants, hurled free of the Disc.

Two surviving occupants, were found within the Disc, and it was apparent, One, it's left leg was broken. There was a

99

minimal radiation contamination, and it was quickly dispersed with a water/solvent wash, and soon the occupant was dispatched for medical assistance and isolation. The bodies were sent to the Wright–Patterson AFB, for dispersal. The debris was also loaded onto three trucks which finished the on-load just before the sunset."[61]

This is pushing us towards the *two-discs* theory.

Were there two discs involved in the incident? Twining's memo makes it clear that the military could only analyze one disc. It means the disc was either located much closer to the initial crash site about 40 miles away, or it was as far away as 200 kilometers to the west.

Which is it?

Perhaps there is a way to figure this out.

Notice how the disc allegedly recovered 40 miles southeast from the initial crash site and closer to Roswell is described as a bright silver object reminiscent of aluminium. Whether this disc is part of the *two-discs* theory, or is the only disc responsible for the initial debris site, we immediately discover a discrepancy in the alleged new observation.

Firstly, Barnett, the man who was in the Socorro region observing a disc and bodies, claims a "dirty stainless steel" disc was observed. The word "dirty" is significant, because it means the metal was darker in color. He is suggesting a smooth dark-grey metal in the outer skin component of the object. Considering the rancher and other firsthand witnesses also made it clear the Roswell metallic foil at the initial wreckage site was smooth and dark-grey in color, it would have to be a lucky guess on the part of Barnett to have chosen the right color for the metal he allegedly saw.

And secondly, why did the military personnel at Roswell AAF, who had inspected a lot more of the debris from the first crash site as brought in by Marcel and Cavitt, not notice pieces of a bright silver metal among the materials, something to at least buck the "dark grey" trend?

If a single bright silver disc was involved, and had been brought down by a lightning strike, or if the disc had a run-in with another disc, we should reasonably expect it to sustain a similar amount of damage, including loss of materials, just as had been the case for the super-tough materials used in the construction of the dark-grey disc. In that case, the witnesses would have seen bright silver metallic foil mixed in with the dark-grey one, or they would have seen only a bright silver metallic foil if this was meant to be a single disc incident crashing to Earth some 40 miles away.

Instead, the original witnesses at Roswell AAF and the rancher and his family members and friends are claiming only one type of metallic foil—the dark-grey variety.

On the basis of probability, it is more reasonable to state that a single disc was found and had been located in the area stated by Barnett. Any talk of another site closer to the first may be referring to additional bodies that could have fallen out of the stricken object as it tried to race away from the thunderstorm. New witnesses claiming to have seen a disc much closer to the initial site, without indications of a dark-grey metal seen in conjunction with the object itself, these testimonies are to be considered less reliable and could be an example of a wild goose chase (either people wanting to get in on the act for fame and/or money, or someone is trying to keep scientists off the track).

Perhaps it is somewhat like MUFON's story about what happened in Ramey's office.

Roswell debris sent to Wright–Patterson AAF for analysis

As media interest in the incident subsided, the original Roswell wreckage was placed inside numerous wooden boxes (varying in size from four to five feet to smaller boxes of less than 2 feet) by the recovery team and delivered to Roswell AAF in trucks. There was absolutely no time for anyone to remove some of the dirt from the wreckage. All the boxes were numbered and had the words "Wright Field" marked on them. These boxes were allegedly loaded onto several Boeing B-29 Bombers and flown to Fort Worth AAF (now Carswell AFB) in Texas, and eventually to Wright–Patterson AAF in Dayton, Ohio (site of the Army Air Force's scientific and technical laboratories) for analysis.

Confirmation that Wright–Patterson AAF was the place where the wreckage was examined may be found in the FBI teletype message dated July 8, 1947 sent to J. Edgar Hoover, who was then director of the FBI office in Cincinnati.[62]

Further confirmation can also be found in the statement made by Thomas Jefferson DuBose to investigators. In an interview with Billy Cox, DuBose, who retired from the USAF in 1959, confirmed the destination of the Roswell debris was at Wright-Field:

"Later, after the whole thing was over, I asked Clements what happened to it [the Roswell debris], and he said he sent it out to Wright Field so they could analyze it..."[63]

The quote from Dubose appeared in Timothy Good's highly acclaimed 1997 book, *Beyond Top Secret*. A more complete quote is shown here[64].

Thomas Jefferson DuBose was Chief of Staff to Major General Roger Ramey at Fort Worth Army Air Field during the Roswell incident. A colonel at the time, he retired from the Air Force in 1959 with the rank of brigadier general. In an interview with Billy Cox (whom I know to be reliable), DuBose confirmed that a 'containment strategy' was ordered by Major General Clements McMullen, Deputy Commander, Strategic Air Command. 'Knowing General McMullen,' said DuBose, '[the cover-up] was an effort to get it off the front pages, to keep people from thinking about it. I couldn't blame him for that.'

On the evening of 6 July 1947, after a stopover in Fort Worth and by order of McMullen, the debris was flown to Washington, according to DuBose. '[Some of] this stuff, this junk, this whatever you want to call it, came in a mail pouch,' he recalled:

I didn't look at it, I wasn't supposed to. McMullen told me to send it to him immediately, and for me not to say anything about it to anyone, to forget about it, and that was an order. I sealed it personally with a lead seal and handcuffed it to the wrist of [Colonel] Al Clark, which is a rather unusual step, and he delivered it to McMullen. Later, after the whole thing was over, I asked Clements what happened to it, and he said he sent it out to Wright Field so they could analyze it . . .

Also, the official radio broadcasts at the time made a point about mentioning Wright Field as the location for analysis of the debris.

General Exon suggests titanium in the composition of the Roswell metallic foil

Brigadier-General Arthur E. Exon (1916–2005) has supported DuBose's statement of a cover-up and location for where all the debris ended up for analysis. He was the former commander of Wright–Patterson AAF in the early 1950s.

In an interview held in 1990 (before he began suffering from dementia), Exon claimed that many top military and intelligence officers, including U.S. President Harry S. Truman, debated in top level meetings over the materials and what to do with them. He was sure everyone agreed the materials were not man-made. Someone else, not from Earth, had made the materials and the object itself.

Missler and Eastman wrote in their 1997 book:

World War II aerial photograph of Wright-Patterson AAF.

"Retired General Arthur E. Exon was stationed at Wright Field in Dayton, Ohio, as a lieutenant colonel in July of 1947 during the time the wreckage from Roswell was brought in. In a 1990 interview, General Exon said of the testing that was done, *The overall consensus was that the pieces were from [outer] space.*"[65]

From a scientific viewpoint, this would have been a remarkable claim. But, given the way the U.S. military behaved and were looking like they had no idea what happened until they were informed by a local rancher, it is a possibility a scientist cannot ignore.

For a more complete quote, Exon gave the following to investigators:

> "...They knew they had something new in their hands. The metal and material was unknown to anyone I talked to. Whatever they found, I never heard what the results were. A couple of guys thought it might be Russian, but the overall consensus was that the pieces were from space. ...Roswell was the recovery of a craft from space."[66]

Exon also hinted at the possibility that the Roswell metallic foil could contain titanium to help explain the blowtorch result and other high toughness tests, but he could not be absolutely sure. As he stated to investigators:

> "I don't know, at that time, if it [the Roswell metallic foil] was titanium or some other metal...or if it was something they knew about and the processing was something different."[67]

If the foil had contained titanium, a study of the history of titanium should reveal a status change in the metal (at the very least) for the U.S. military. And if so, would this correspond to a period just before, or just after 1947?

Should evidence be found to support this status change in titanium, it would mean Exon was probably right in his claims on the type of metal used.

Did two U.S. Presidents know what happened?

Little more would be known about the Roswell case except for a rumor and further quotes from Exon of two U.S. Presidents having direct knowledge of what happened. One of them was Harry S. Truman, the other, Dwight D. Eisenhower.

According to the rumor, U.S. President Dwight D. Eisenhower (1890–1969) allegedly visited Edward AAF on February 20, 1954 to view the remains of so-called "alien bodies" recovered from Roswell not long after the following statement from the President was made:

> "I do not think that it would be correct to say they come
> from another planet, that is, from one single planet, as
> General Twining said."[68]

It was a time when Eisenhower's whereabouts on this particular Saturday night in 1954 were unusually unclear. Official records indicate that an unexpected trip to the dentist was made, but it seems no one has been able to properly verify the claim. Neither the dentist's office could confirm his presence there, nor can the dentist's wife recall an important man like the President ever requiring the services of her husband's dentistry skills on that Saturday night.

In the case of Harry S. Truman (1884-1972), several people who had worked for him noticed a rather intense and secretive interest in UFOs as if he was trying to determine just how much of a threat to national security UFOs posed to his country. Of these people, only Exon, the military general who worked at Wright–Patterson AFB, is prepared to go further claiming Truman was briefed on the UFO crash, which was the instigator for his interest in UFOs.

In a telephone interview with investigator Kevin Randles on May 19, 1990, Exon claimed that a number of top military and intelligence officers debated with Truman over what to do with the Roswell materials. In other words, should the materials be released to the world? Or would it be better to keep them secret in the interest of protecting national security while gathering more information on the UFO situation?

If such a briefing did take place, one can understand why there was a debate. Truman would probably have been told about the preliminary analysis of the "disc" and "victims". This should have told him that

105

there were no weapons visible anywhere, suggesting to him that UFOs were no threat to national security. But to be absolutely sure, and to probably appease the top military officials, he decided not to release these materials to the public. Instead he reorganised all the separate intelligence organizations held in various departments within his country. He then persuaded the reluctant Rear Admiral Roscoe Henry Hillenkoetter to be Director of Central Intelligence (DCI), and the CIA was formed in September 1947 where it was given powers by the President to centralize all intelligence information. Part of this intelligence gathering involved the collecting of UFO reports on a covert basis.

So, on the one hand, Truman officially claimed no interest in flying saucers, but on the other, as General Robert B. Landry, the USAF Aide to the President, said:

> "I was directed to report quarterly to the President after consulting with Central Intelligence people, as to whether or not any UFO incidents received by them could be considered as having any strategic threatening implications at all. The report was to be made orally by me unless it was considered by intelligence to be so serious or alarming as to warrant a more detailed report in writing. During the four and one-half years in office there, all reports were made orally. Nothing of substance considered credible or threatening to the country was ever received from intelligence."[69]

Then, one day, Truman allegedly dropped a bomb by revealing his possible insights from the Roswell briefing (unknowingly) to his listener.

In 1949, Truman commissioned a report into the mysterious Foo Fighters that menaced military aircraft on both sides during World War II. General Jimmy Doolittle, who headed the study, concluded that both German and Allied forces had seen them. All the witnesses claimed the objects were not secret German or Allied weapons; and the objects were not caused by natural phenomena.

When Doolittle told Truman of his results and suggested the objects were "most likely of extraterrestrial origin", Truman replied in a message he received April 4, 1950 saying:

Harry S. Truman (1884-1972)

"I can assure you that flying saucers, given that they exist, are not constructed by any power on earth."[70]

Does this mean Truman did know something about the Roswell materials that the public doesn't?

CHAPTER 3

The Witnesses Speak

WEATHER BALLOONS or not, something did fall from the sky on the night of July 2-3, 1947. And it would appear the USAF was in some bother trying to figure out what it was that crashed. In the end, it seemed the USAF needed to cover up the situation by claiming a weather balloon as the cause, but later added the existence of victims just to complicate the situation further. To this day, the USAF has never revealed the names of these victims (assuming they were humans).

For a scientist looking at this controversial case, determining the nature of the object (let alone the names of the victims) is dependent on whether the materials were flimsy and contained aluminium as the principal element of the metallic foil, or the materials were very tough, able to return to its original shape, and may have contained a high temperature metal such as titanium to help withstand the effects of a blowtorch.

Either way, a scientist could assume the case is nothing more than a military experiment of some sort, even if the weather balloon explanation is dubious.

But can we make this assumption?

Not according to the witnesses. Talk of something exotic—namely an alien spaceship crashing to Earth (that is, a symmetrical flying saucer with odd-looking bodies)—remains paramount in their minds. Even a retired high-ranking general who worked at Wright–Patterson AFB has suggested the only reasonable explanation for the Roswell case is this exotic explanation after scientists at the base had analyzed the Roswell materials for many years.

Is this true?

Leaving aside this explanation, one thing does seem clear among almost all the firsthand witnesses—the materials found were not from a weather balloon. We say *almost* everyone because there is one firsthand witness, and the only one, who has decided to go against the consensus. His name is Captain Sheridan Cavitt, the guy who accompanied Marcel to the first crash site.

Cavitt follows the USAF position

When Cavitt was tracked down by investigators—after the USAF had already spoken with him in May 1994 as part of their investigations for the *Roswell Report* released later that year—he said nothing unusual was found at the crash site. He thought the whole incident was caused by a downed weather balloon.[1]

Clearly contradicting his close friend and work colleague, Marcel commented to investigators in 1979:

> "I was pretty well acquainted with everything in the air at that time, both ours and foreign. I was also acquainted with virtually every type of weather-observation or radar-tracking device being used by either the civilians or the military. It was definitely not a weather or tracking device, nor was it any sort of plane or missile....It was something I had not seen before, or since, for that matter. I didn't know what it was, but it certainly wasn't anything built by us and it most certainly wasn't any weather balloon."

Here we have two intelligent, experienced officers sent by Blanchard to the same place at the same time to pick up the same materials while both men had presumably discussed in an amicable and rational way the results of their early examination of the materials, and not once did Cavitt show dissent or express a general disagreement to Marcel's findings or his other colleagues at the Roswell base during a meeting by

110

claiming that the whole situation was the result of a downed weather balloon. All we find from the collected testimony of the available military officers who were present at the time is an agreement that there was no explanation for these materials or where they came from. But, incredibly, Cavitt kept quiet all this time and then manages to suddenly change his mind so late in his life just after being approached by the USAF. How odd?

From a scientific viewpoint, if all observational data came from just these two people, scientists would ignore *both* accounts. However, when there are *many* more witnesses who have seen the materials and support Marcel's position, a scientist must study the consensus view and ignore the one or two anomalies. In the case of Cavitt, let us ignore his testimony.

With this in mind, let us review the available quotes from witnesses showing what it is they found, and what's so unusual about the materials that made them think it was not a weather balloon, or some other man-made flying object.

Balsa wood-like materials

Quotes from Loretta Proctor

1: "In July 1947, my neighbor William W. 'Mac' Brazel came to my ranch and showed my husband and me a piece of material he said came from a large pile of debris on the property he managed. The piece he brought was brown in color, similar to plastic. He and my husband tried to cut and burn the object, but they weren't successful. It was extremely light in weight."[2]

2: "The piece he [Mac Brazel] brought looked like a kind of tan, light brown plastic. It was very lightweight, like balsa wood. It wasn't a large piece, maybe about four inches long, maybe just a little larger than a pencil. We cut on it with a knife and would hold a match on it, and it wouldn't burn. We knew it wasn't wood. It was smooth like plastic, it didn't have a real sharp corners, kind of like a dowel stick. Kind of dark tan. It didn't have any grain, just smooth. I hadn't seen anything like it."[3]

111

3: "...he [Mac Brazel] did bring a little sliver of a wood-looking stuff up, but you couldn't burn it or you couldn't cut it or anything. I guess it was just a sliver of it, about the size of a pencil and about three to four inches long. I would say it was kind of brownish-tan but you know that's been quite a long time. It looked like plastic, of course there wasn't any plastic then, but that was kind of what it looked like."[4]

Quotes from William Brazel, Jr. (Bill)

1: "[There were also] some wooden-like particles like balsa wood in weight, but a bit darker in color and much harder.... It was pliable but wouldn't break. Weighed nothing, but you couldn't scratch it with your fingernail.... [There was no writing or markings on the pieces I had] but Dad did say one time that there were what he called *figures* on some of the pieces he found. He often referred to the petroglyphs the ancient Indians drew on the rocks around here as *figures*, too, and I think that's what he meant to compare them with."[5]

2: "Some of it was like balsa wood: real light and kind of neutral color, more of a tan. To the best of my memory, there wasn't any grain in it. Couldn't break it—it'd flex a little. I couldn't whittle it with my pocket knife."[6]

3: "There were several different types of stuff....it sure was light in weight. It weighed almost nothing. There was some wooden-like particles I picked up. These were like balsa wood in weight, but a bit darker in color and much harder. You know the thing about wood is that the harder it gets, the heavier it is. Mahogany, for example is quite heavy. This stuff, on the other hand, weighed nothing, yet you couldn't scratch it with your fingernail like ordinary balsa, and you couldn't break it either. It was pliable, but wouldn't break."[7]

1: "There was all kinds of stuff—Small beams about three-eighths or a half-inch square with some sort of hieroglyphics on them that nobody could decipher. These looked something like balsa wood, and were of about the same weight, except that they were not wood at all. They were very hard, although flexible, and would not burn."[8]

2: "A lot of it had a lot of little members [beams] with symbols that we had to call them hieroglyphics because I could not interpret them, they could not be read, they were just symbols, something that meant something and they were not all the same. The members that this was painted on—by the way, those symbols were pink and purple, lavender was actually what it was—could not be burned."[9]

Quotes from Dr. Jesse Marcel, Jr.

1: "Many of the remnants, including I-beam pieces that were present, had strange hieroglyphic typewriting symbols across the inner surfaces, pink and purple, except that I don't think there were any animal figures present as there are in true Egyptian hieroglyphics."[10]

2: "The figures were composed of curved geometric shapes. It had no resemblance to Russian, Japanese or any other foreign language. It resembled hieroglyphics, but it had no animal-like characters."[11]

Dark parchment-like plastic paper or cloth

Quotes from Major Jesse Marcel

1: "Then there was a kind of parchment, brown and very tough..."[12]

2: "There was a great deal of an unusual parchment-like substance which was brown in color and extremely strong... One thing that impressed me about the debris was the fact that a lot of it looked like parchment. It had little numbers with symbols that we had to call hieroglyphics because I could not understand them. They could not be read, they were just like symbols, something that meant something, and they were not all the same, but the same general pattern, I would say. They were pink and purple. They looked like they were painted on. These little numbers could not be broken, could not be burned. I even took my cigarette lighter and tried to burn the material we found that resembled parchment and balsa, but it would not burn —wouldn't even smoke."[13]

Quotes from Jesse Marcel, Jr.

1: "[There was] a quantity of black plastic material which looked organic in nature.... There were...bits of black, brittle residue that looked like plastic that had either melted or burned."[14]

2: "[There was] a brittle, brownish-black plastic-like material, like Bakelite"[15]

3: "Some of the debris was not metallic but more like pieces of black plastic fragments thicker than the metallic skin"[16]

Quote from Bessie Brazel Schreiber

"There was what appeared to be pieces of heavily waxed paper.... Some of these pieces had something like numbers and lettering on them, but there were no words that we were able to make out....It looked like numbers mostly, as least I assumed them to be numbers. They were written out like you would write numbers in columns to do an addition problem, but they didn't look like the numbers we use at all. What gave me the idea they were numbers, I guess, was the way they were all ranged out in columns."[17]

"Whatever he found it was all in pieces and some of it had some kind of unusual writing on it—Mac said it was like the kind of stuff you find all over Japanese and Chinese firecrackers; not really writing, just wiggles and such. Of course, he couldn't read it and neither could anybody else as far as I heard..."[18]

The metallic foil

Quotes from Major Jesse Marcel

In an interview with investigators, Major Jesse Marcel gave the following description for the foil:

"There was a great...number of small pieces of a metal like tinfoil, except that it wasn't tinfoil...the pieces of metal that we brought back were so thin, just like the tinfoil in a pack of cigarettes. I didn't pay too much attention to that at first, until one of the boys came to me and said: 'You know that metal that was in there? I tried to bend the stuff and it won't bend. I even tried it with a sledgehammer. You can't make a dent on it.'...This particular piece of metal was about two feet long and maybe a foot wide. It was so light it weighed practically nothing, that was true of all the material that was brought up, it weighed practically nothing..."[19]

Could this foil be flexed back and forth? Marcel said:

"...Now by bend, I mean crease. It was possible to flex this stuff back and forth, even to wrinkle it, but you could not put a crease in it that would stay, nor could you dent it at all. I would almost have to describe it as a metal with plastic properties."

In a television interview in 1979, Marcel gave more specific information about the mysterious metallic foil's physical behavior:

"...Then there was...many bits of metallic foil, that looked like, but was not, aluminium, for no matter how often one crumpled it, it regained its original shape again. Besides that, they [and all the rest of the materials] were indestructible, even with a sledgehammer."[20]

Marcel is not alone. Other original witnesses to the scene of the first crash site or who had seen the debris up close (and weren't influenced first by the USAF) confirm the *dark-grey* appearance of the metallic foil and its unusual *shape-memory* property.

Quotes from Bill Brazel

One such witness, Mr. Bill Brazel, recalls the appearance and behavior of the metallic foil after finding a few pieces on his dad's ranch "after a good rain"[21]:

"There were...several bits of a metal-like substance, something on the order of tinfoil except that this stuff wouldn't tear and was actually a bit darker in color than tinfoil—more like lead foil, except very thin and extremely lightweight. The odd thing about this foil was that you could wrinkle it and lay it back down and it immediately resumed its original shape. It was quite pliable, yet you couldn't crease or bend it like ordinary metal. It was almost more like a plastic of some sort, except that it was definitely metallic in nature."[22]

When pressed by investigators for more details, Bill said:

"...a little piece of—it wasn't tinfoil, it wasn't lead foil—a piece about the size of my finger....The only reason I noticed the tinfoil (I'm gonna call it tinfoil), was that I picked this stuff up and put it in my chaps pocket. Might be two or three days or a week before I took it out and put it in a cigar box. I happened to notice when I put that piece of foil in that box, and the damn thing just started unfolding and just flattened out. Then I got to playing with it. I'd fold it, crease it, lay it down and it'd unfold. It was kind of weird. I couldn't tear it. Hell, tin foil or lead foil is easy but I couldn't tear it. I didn't take pliers or anything. I just used my fingers. I didn't try to cut it with my knife. The color was consistent through the pieces I found. It was a dull color. It was about the gauge of lead foil. Thicker than tin foil. It was pliable. Real pliable. I would bend it over and crease it and if you straighten back up, there wouldn't be a crinkle in it. Nothing. It would flatten out and it was just as smooth as ever. Not a crinkle or anything in it. [It didn't make a sound.] ...As best as I can remember, it was smooth."[23]

Was this the only material he found? No. Bill claims he also found bits and pieces of a wooden-like beam, a brown parchment or plastic paper-like substance, and a plastic wire like a fishing line which, we are led to believe by his own signed affidavit, he couldn't cut or break:

"There was some thread-like material. It looked like silk and there were several pieces of it. It was not large enough to call string, but yet not so small as sewing thread either.... Whatever it was, it too was a very strong material. You could take it in two hands and try to snap it, but it wouldn't snap at all. Nor did it have strands or fibers like silk thread would have. This was more like a wire—all one piece or substance."[24]

Corroboration for Bill's claim of the shape recovery behavior of the dark-grey metallic foil came from Ms Sally Strickland-Tadolini, the daughter of Marian Strickland, one of Brazel's neighbors in 1947. According to Hasemann and Mantle in *Beyond Roswell*, Sally, who was nine years old at the time, recalls:

"What Bill showed us was a piece of what...was something like aluminium foil...a dull metallic greyish silver....

Bill passed it around and we all felt of it. I did a lot of sewing, so the feel made a great impression on me. It felt like no fabric I have touched before or since....[W]hen I crumpled it in my hands the feel was like that you notice when you crumple a leather glove in your hand. When it was released it sprang back into its original shape, quickly flattening out with no wrinkles. I did this several times, as did the others."[25]

However, during the summer of 1949, four USAF officials confiscated the foil and other materials which Bill had found, after he told someone about it in a bar in Corona, New Mexico. Bill recalls:

"I was in Corona, in the bar, the pool hall, sort of the meeting place, domino parlor.... That's where everybody got together. Everybody was asking...they'd seen the papers (this was about a month after the crash) and I said, 'Oh, I picked up a few little bits and pieces and fragments'. So, what are they? 'I dunno.'

Then lo and behold, here comes the military (out to the ranch, a day or two later). I'm almost positive that the officer in charge, his name was Armstrong, [was] a real nice guy. He had a [black] sergeant with him that was real nice. I think there was two other enlisted men. They said, 'We understand your father found this weather balloon.' I said, 'Well yeah.' 'And we understand you found some bits and pieces.' I said, 'Yeah, I've got a cigar box that's got a few of them in there, down at the saddle shed.'

And this (I think he was a captain), and he said, 'Well, we would like to take it with us.' I said, 'Well...' And he smiled and he said, 'Your father turned the rest of it over to us, and you know he's under an oath not to tell. Well,' he said, 'we came after those bits and pieces.' And I kind of smiled and said, 'OK, you can have the stuff, I have no use for it at all.'

He said, 'Well, have you examined it?' And I said, 'Well, enough to know that I don't know what the hell it is.' And he said, 'We would rather you didn't talk very much about it.'"[26]

There are more quotes from witnesses who claimed to have seen, or heard from others about, the "shape-memory" ability of the dark-grey metallic foil.

Quote from Marian Strickland

"The time that he [Mac Brazel] brought the sample of what he had picked up, he was at the corral. My daughter and two sons and husband were at the corral, and they saw it. My daughter says that it could be crumpled up and straighten right back out."[27]

Quote from Loretta Proctor

1: "He said the stuff that looked kind of like aluminium foil, he said you'd crumple it up and then it would straighten out, it wouldn't stay creased, it would just open out.....He said he couldn't cut it or anything."[28]

2: "He was telling us about more of the other material that was so lightweight and that was crinkled up and then would fold out."[29]

Phyllis Wilcox McGuire
(Daughter of Roswell Sheriff George Wilcox)

"When I read in the Roswell paper about the Flying Saucer being found, I went into his [George Wilcox] office to ask about it... I asked my father if he thought the information about the saucer was true. He said: 'I don't know why Brazel ... would come all the way in here if there wasn't something to it.' He said Brazel had brought in some of the material to show, and that it looked like tinfoil, (a material like aluminium foil), but when you wadded this material up it would come right back to its original shape. He felt it was an important finding..."[30]

Civilian witnesses are not the only ones to notice the shape-memory nature of the metallic foil.

Sergeant Robert E. Smith

As a member of the First Air Transport Unit, Smith's involvement was to load three planes at Roswell AAF with the crates containing the mysterious debris. According to an affidavit dated October 10, 1991:

"We started out in the morning [July 9], say about eight or nine o'clock. Seems like it lasted up to nearly four o'clock with a break for lunch...

We were taken to the hangar to load crates. There was a lot of farm dirt on the hangar floor. We loaded it [the crates] on flatbeds and dollies. Each crate had to be checked as to width and height. We had to know which crates went on which plane....We weren't suppose to know their destination, but we were told they were headed north."[31]

Smith was shown a piece of the metallic foil picked up near Roswell. As he said:

"We were talking about what was in the crates and so forth and he (another of the NCOs) said, 'Oh do you remember the story about the UFO? Or rather the flying saucer?' That was what we called them back then. We thought he was joking, but he let us feel a piece and he stuck it back into his pocket. Afterwards we got to talking a little bit more about it and he said he'd been out there helping clean this up. He didn't think taking a little piece like that would matter.

"It was just a little piece of metal or foil or whatever it was. Just small enough to be slipped into a pocket. I think he just picked it up for a souvenir."

Then he noticed something unusual about the physical behavior of the metallic foil:

"It was foil-like, but it was a little stiffer than foil that we have now. In fact, being a sheet-metal man, it kind of intrigued me, being that you could crumple it and it would flatten back out again without any wrinkles showing up in it. Of course we didn't get to look at it too close because it was supposed to be *top secret*. He just popped it out there real quick and let us feel it and so forth while everybody was doing something else."[32]

Smith was asked by investigators to elaborate on what he saw:

"All I saw was a little piece of material. You could crumple it up, let it come out. You couldn't crease it. The piece of debris I saw was two to three inches square. It was jagged. When you crumpled it up, it then laid back out. And when it did, it kind of crackled, making a sound like cellophane, and it crackled when it was let out. There were no creases."[33]

Quotes from Brigadier General Arthur E. Exon

Another military person who supports Smith's observation is retired Brigadier General Arthur E. Exon. After speaking to some people during the course of his work over many years at Wright Field, he said:

"We heard the material was coming to Wright Field. [Testing was done in the various labs.] Everything from chemical analysis, stress tests, compression tests, flexing. It was brought into our material evaluation labs. I don't know how it arrived, but the boys who tested it said it was very unusual."[34]

Exon then focused his recollection on statements made by others on the metallic foil:

"...[The metallic foil] couldn't be easily ripped.... You could wad it up, you could change the shape, but it was still there and...there were other parts of it that were very thin but awfully strong and couldn't be dented with heavy hammers and stuff like that...which at the time were causing some people some concern...[some] say it was a shape of some kind, you could grab this end and bend it, but it would come right back. It was flexible to a degree. It had them pretty puzzled.

I think the full range of testing was possible. Everything from chemical analysis, and resist chemicals, stress tests, compression tests, flexing.... I don't know, at that time, if it was titanium or some other metal...or if it was something they knew about and the processing was something different."[35]

Brigadier General Arthur E. Exon (1916-2005)

Other interesting information gleaned from Exon include the following:

"There was another location where...apparently the main body of the spacecraft was...where they did say there were bodies.... They were all found, apparently, outside the craft itself but were in fairly good condition. In other words, they weren't broken up a lot.

That's my information [that the bodies went to Wright Field], but one of them went to the mortuary outfit ... I think at that time it was in Denver, but the strongest information was that they were brought to Wright-Pat."[36]

Despite the bodies and some materials being moved from Wright Field to another location on a B-29 Bomber at a later date, Exon felt sure some of the materials might still remain housed at Wright–Patterson . He suspects reports about the incident and what the USAF learned about the materials and bodies, including some photographs of bodies and of the autopsies, the crash site, and the debris would still be there, probably in the Foreign Technology Building, if there is a way to get in.

Quotes from Major Sergeant Lewis Rickett

Major Sergeant Lewis Rickett was stationed at the Roswell AAF and was privy to examining the materials at the emergency meeting on July 8, 1947. He recalls:

"[The foil] was very strong and very light. You could bend it but couldn't crease it. As far as I know, no one ever figured out what it was made of...."[37]

Rickett was walking the debris field with Cavitt when he found a large enough piece of the mysterious grey foil about two feet square. He crouched down to pick it up and began to bend it:

"...[Cavitt] says, do what we couldn't do. And, go ahead, touch it! ...the best that I could recall ... what in the hell is that stuff made out of... it can't be plastic. I said, don't feel like plastic... just flat feels like metal...There was a slightly curved piece of metal, real light. It was about six inches by twelve or fourteen inches. Very light. I crouched down and

123

tried to snap it... It didn't feel like plastic and I never saw a piece of metal this thin that you couldn't break."[38]

Quotes from June Crain

Alright then. If the Roswell materials had arrived at Wright–Patterson AFB for analysis, was there anyone else at the base other than General Exon who could vouch for seeing the metal and its behavior?

By fortuitous chance we do have another witness.

June Crain (1924-1998) worked at Wright–Patterson AFB between 1942 and 1952. She has testified to having played with the original Roswell metallic foil herself sometime in 1951 or 52.

In 1990, June released part of her remarkable story to UFO researcher Kevin Randle under a pseudonym. Those details would be published in Randle's book, *The Truth about the UFO Crash at Roswell.* Feeling more confident of her newfound position of power and anonymity, she went in 1993 to the new modern Ocean Shores Library where she was a major financial contributor to its construction and supported the aims of providing open public access to education. She knew there was a lecture on UFOs being held there.

James E. Clarkson, a detective sergeant from the Aberdeen Police Department with an interest in UFOs, stated during the lecture that the U.S. Government knew more about UFOs than it would publicly admit. At the end of the lecture, Crain approached Clarkson and stated she agreed with his view. When asked how she knew he was correct, June responded, "Because I worked there."

Curiosity got to the better of him, so Clarkson asked for more details. Crain was reluctant, saying she might get arrested by the government if she said anything. Crain left the library.

That would have been the end of the story until 4 years later, in 1997, when Clarkson learned that Crain had made enquiries about him at the police station. When she realized he could be trusted, she decided to tell her story.

In an interview held on June 27, 1997, Clarkson learned Crain was, for a long time, torn between loyalty to her country in keeping an oath and a desire to see the public told the truth about UFOs. In the end, she decided it was in the best interest of the public to know what was going on. Clarkson recalls from the interview:

"June was angry because of what she perceived as a great hypocrisy, that on the one hand the existence of UFOs is officially denied, and yet in classified laboratories where she worked, she overheard scientists and engineers discussing artifacts and bodies from recovered, crashed UFOs. She believed that the public deserves to be told the truth."

She began the interview stating she was a civilian employee who worked for the U.S. Government at Wright–Patterson AFB from 1942 to 1952.

During her time there, Crain showed her loyalty to the country and dedication to her job as she worked hard to earn a living—so much so, in fact, that her persistence and hard work gained the trust and respect of those she worked for. The papers revealed to Clarkson and later verified by the National Archives showed Crain had, as Clarkson puts it, a "steady series of promotions leading to her last position as a Clerk-Stenographer in 1951."

In the late 1940s and early 50s she heard various interesting conversations about some sensitive work, especially relating to a so-called crashed disc and bodies. Before she left her job, she was shown a piece of metal from a military officer.

As Crain said:

"He [Lieutenant Rose or Captain Wheeler] walked in.... He threw it on my desk, and it was a piece, well it was a piece [about the size of a business card?] ...Yeah, about that size, and a half of this. ...it was bent like this. And he says, 'June, you're good. Tear that thing apart, break that up.' And I took it and I bent it and I twisted it and I laid it back down, and it went...right back to the same shape. I got back to my desk and he said, 'Cut it. Cut it. Try cutting it.' ...I got my scissors out and I snipped at it, and you know there was no way I could even cut that piece of metal. And it was as light as a feather. I had it in my hand and I couldn't—I would say that it didn't weigh as much as these two cards—it wasn't that heavy...It was so light but strong, and it was [fairly thick], but it had no weight at all, it was like a feather. And so strong. It was sort of a greyish, gun-metal type of color, and you could see that on the inside that there was a different...coating on the outside of it. Both sides were the same and the insides seemed to have a sort of a lead-colored, light lead-colored center to it."

Clarkson asks:

"Were the edges even or like part of something else?"

Crain replied:

"It was even, all even and I said 'What is it?' He said, 'It's a piece of a space ship.' ...He said, 'I just came back from New Mexico and I brought it back with me.'"

Being a loyal citizen to her country, she kept her experience of the metal and stories she heard of analyzes done on a disc and bodies at Wright–Patterson AFB a secret for a long time—until 1990.

When she heard about the USAF explanation of the Roswell case after the release of the Roswell Report, she decided she would try to set the record straight herself.

Quotes from Major Ellis Boldra

More testimonial evidence would emerge of the mysterious dark-grey shape-memory metal through Kevin Randle and Don Schmitt's 1994 book, *The Truth About the UFO Crash at Roswell*.

According to this book, the son and friends of Major Ellis Boldra can testify on oath that they heard Boldra say to them he found a piece of the metallic foil found in the crash debris near Roswell. It was kept in a safe at the Roswell AFB engineering department in 1952. It is claimed he tested the foil personally. Boldra was able to show the foil could be crumpled, but quickly returned to its original shape. It was thin, and incredibly strong and lightweight. He subjected the foil to an acetylene torch, but it wouldn't melt. He didn't notice the foil glow under the intense heat, suggesting the metal was quickly dissipating the heat. On removing the flame, the foil became cool to the touch within seconds. He tried to cut it with a variety of tools. All attempts were unsuccessful. Boldra claimed he couldn't identify what type of metal it was.

Quotes from John Kromschroeder

John Kromschroeder is a dentist and a retired military officer. He was a friend of Pappy Henderson, a World War II bomber pilot. In 1977, Henderson told him about work transporting a crashed disc and bodies found near Roswell to Wright Field AAF for analysis. About a year later, Henderson felt comfortable enough to show him a piece of the metal he had taken from the collection of wreckage. This is the piece that came from the safe at Roswell AAF removed by Boldra.

Kromschroeder confidently stated the following during an interview held in 1990:

> "I gave it a good, thorough looking-at and decided it was an alloy we are not familiar with. Grey, lustrous metal resembling aluminium, lighter in weight and much stiffer. [We couldn't] bend it. Edges sharp and jagged."[39]

Kromschroeder also claimed Henderson had told him the piece of metal was part of the material lining the interior of the craft. When energized, it produced an even and effective soft-light illumination.

The patterns we can identify about the metallic foil

Focusing on the mysterious metallic foil for the remainder of this book, a scientist can clearly identify at least six consistent patterns across the range of statement made by the witnesses.

Observation 1
The foil was dark-grey in color

Witnesses observed a dark-grey metallic foil. The color of the foil was highly reminiscent of someone placing a thin film of black grease or dirt evenly over the surface of a piece of stainless steel or some other silvery colored metal to make it look darker. The closest thing in appearance would be lead foil, but lead foil is not lightweight, nor does it behave in the way this metallic foil does.

As Timothy Good wrote in his 1997 book, *Beyond Top Secret*:

> "The disc [allegedly observed on the Plains of San Agustin on the morning after the odd explosion near Roswell]

seemed to be made of a metal that looked like dirty stainless steel."[40]

And in Stanton Friedman and Don Berliner's 1992 book, *Crash at Corona*, Bill Brazel is quoted as saying:

"The color was in between tinfoil and lead foil."[41]

Observation 2
The foil was extremely tough to break and would not melt under high heat

It is claimed the Roswell metallic foil could not be melted by a blowtorch, let alone rip it by hand or cut it with scissors or other cutting implement. This important observation is one of the factors that convince witnesses that whatever flying object it was that had dropped these super-tough materials to the ground, it was not a weather balloon of any sort, or any other familiar man-made object.

Observation 3
The foil could return to its original shape

The foil's shape-memory effect, together with its super-tough nature, are considered the two biggest factors in swaying witnesses to believe in its unconventional nature. For these witnesses to observe a metal returning to its original shape was something unheard of, let alone something so thin that it wouldn't tear or suffer some form of metal fatigue.

Timothy Good said:

1: "Several witnesses to the Roswell, New Mexico, incident of July 1947 described some of the recovered metal foil as being impervious to bending or folding, in that it always returned to its original shape."[42]

2: "Major Marcel, it should be noted, was familiar with balloon debris and was convinced that the material [the metallic foil] he handled was unfamiliar, in that it was impossible to dent or burn, and that, no matter what was done to it, the foil-like metal always returned to its original shape."[43]

Observation 4
The foil was extremely lightweight

Virtually all the firsthand witnesses who examined the foil said they were impressed by its extreme low-weight. One witness commented that a good sizeable piece of this foil felt as light as a feather when held in the hand.

Observation 5
The foil was an alloy of some sort

Observers described the mysterious dark-grey, metallic foil as a metal rather than a plastic or polymer composite. It was a solid piece of metal that behaved like a highly flexible plastic (i.e., it returned to its original shape). For example, CIC agent Lewis S. Rickett said, while analyzing the metallic foil:

> "For God's sake! What in the hell is this stuff made out of? It can't be plastic. Don't feel like plastic, but it just flat feels like metal."[44]

Indeed, military witnesses—who had more time to closely examine and test the foil with various equipment, and were trained to understand differences between metals and plastics—were much more certain the foil was a metal.

Civilian witnesses, on the other hand, had a harder time finding the right words to describe this curious material. Was it a metal or a plastic? To circumvent this dilemma, they typically chose to focus on what they knew was definitely true, which was the metallic look of the foil. In other words, it looks like a metal, but it could be something else. If one were to push them to give a slightly more accurate description for this material, they often likened it to the foil found in a packet of cigarettes. This would be the closest thing to describing it as a metal as well as indicating how thin it was.

In terms of the military descriptions to support the view that the foil was a metal, Major Jesse Marcel said:

> ...But something that is more astounding is that the piece of metal that we brought back was so thin, just like the tinfoil in a pack of cigarette paper. I didn't pay too much attention to that at first, until one of the GIs came to me

and said, "You know the metal that was in there? I tried to bend that stuff and it won't bend. I even tried it with a sledge hammer. You can't make a dent on it."

CIC Officer Bill Ricket who was present at Roswell AAF to test and gather information about the foil, commented:

We collected a boxful of samples of this material. As I recall, there were some metal samples here, too, of that same sort of thin foil stuff.

First Lieutenant Robert Shirkey said:

A call came in to have a B-29 ready to go as soon as possible....Blanchard waved to somebody, and approximately five people came in the front door, down the hallway, and onto the ramp to climb into the airplane, carrying parts of the crashed flying saucer. I got a very short glimpse, asked Blanchard to turn sideways so [I] could see too. Saw them carrying pieces of metal. They had one piece that was eighteen by twenty-four inches, brushed stainless steel in color.

In the words of Sergeant Robert Smith:

Afterwards we got to talking a little bit more about it and he said he'd been out there helping clean this up. He didn't think taking a little piece like that would matter. It was just a little piece of metal, or foil, or whatever it was. Just small enough to be slipped into a pocket. I think he just picked it up for a souvenir. It was foil-like, but it was stiffer than foil that we have now. In fact, being a sheet metal man, it kind of intrigued me, being that you could crumple it and it would flatten back out again without any wrinkles showing up in it.

June Crain worked at Wright-Patterson AFB between 1952 and 1953, and was the sole lady permitted to play with a sample of the original Roswell foil. She stated:

"Cut it. Trying cutting it. ...I got my scissors out and I snipped at it., and you know there was no way I could even cut that piece of metal. And it was as light as a feather....I ve never seen [anything like it since]. I always look at things, metal things and I still have that curiosity, because it still bothers me and I have yet to see anything that would have those properties and looks like that.

Pappy Henderson went one step further, claiming the foil was an alloy (a material made of a metal and one or more other elements, but usually different metals) when he said:

"I gave it a good, thorough looking-at and decided it was an alloy we are not familiar with. Gray, lustrous metal resembling aluminum, lighter in weight and much stiffer. [We couldn't] bend it. Edges sharp and jagged."

Of course, we also have the statement from Brigadier General Arthur E. Exon who was prepared to go on the record that the foil probably contained titanium.

Combine this with the testing done by military personnel at the Roswell AAF using a blowtorch and this "metal" view (or more accurately an "alloy" since no metal in the periodic table displays a shape-memory response after bending it to a significant degree by hand) would be the most obvious choice. Even polymers (i.e., non-metals) in modern times would struggle to withstand the high heat from a blowtorch (see Chapter 9 for the closest known polymers to match the properties of the Roswell foil).

Together with its great hardness to withstand cutting and piercing, great strength to resist ripping by hand, incredibly lightweight, and as thin as, if not thinner than, a sheet of paper, this Roswell foil had all the hallmarks of the most advanced aerospace material ever conceived by humans in 1947.

Observation 6
The foil is "top-secret"

And finally, the metallic foil was treated by the U.S. military just like the rest of the Roswell debris (including the victims), as top-secret material, and remains so to this day.

Is there a metal like it?

We are naturally left with the question: What is this super-secret, dark-grey, thin and super-tough shape-memory metal? Is there anything in science to explain what it is?

CHAPTER 4

The U.S. Government Speaks

SO HOW does the USAF explain the witnesses' alleged claims about the materials they had seen? In particular, is there an explanation for the shape-memory foil found at the Roswell crash site, which we are confident was found in large quantities and, therefore, cannot be ignored?

What we do know from the records is that upon hearing about the rancher's discovery and having collected and tested the materials, virtually every single Roswell AAF personnel who observed and examined the Roswell materials were baffled by what they had found. Not even Cavitt would pipe-up with his own opposing view on the unusual nature of the materials. Of course, the option was there for anyone in the military base to claim a weather balloon if he wanted to. However, instead the commanding officer of the Roswell AAF decided to send a news release to media outlets in Roswell, and eventually the rest of the world, confirming that a disc had been recovered as the best explanation.

This clearly confirms the Roswell debris was not comprised of the usual run-of-the-mill man-made materials people have come to expect. Not even to this day has the scientific community heard of another object that has crashed to earth containing these same super-tough materials of which the metallic foil component had the ability to return to its original shape.

This is a rare, and possibly one off, situation.

So here we have a news release claiming a disc was recovered.

Finally, it may have taken a while, but after a barrage of calls from the media hit Roswell AAF, the USAF changed their story. They now claimed that the materials were nothing more than the remains of a non-secret Rawin weather balloon made of balsa wood, aluminium foil and plastic. If this was true, Marcel should have identified it immediately. He was an expert in practically all aerial objects. He should have known. Indeed, after seeing the materials carefully laid out on the floor of Ramey's office, he did identify it as such without hesitation. Yet he could not identify the materials before he had arrived in the office (and supported by his colleagues at the Roswell air base). This is why he said that the original materials recovered had been substituted and a cover up had begun.

It is possible this balloon explanation was intended to avoid potential panic among the public caused by the unfortunate choice of the term "disc" in the news release (given the intense interest American citizens had over the flying saucer situation in their own nation), because it really was a weather balloon or some other man-made object. Even so, the decision by top military officials in the Pentagon at Washington, D.C. to explain what it was took considerable time since the first call from Blanchard the previous day, but they finally had the answer within two hours of the official news release but only when the calls from the media started arriving en masse. When the explanation did finally arrive, apparently the top Pentagon officials, who had not seen the materials up close, accepted the weather balloon explanation as so obvious that all the military personnel on the ground who had been able to see and touch the original Roswell materials were made to look inept.

These top officials must have known more than anyone else in the military.

In that case, one would presume that every military personnel at Roswell AAF would have had to be put through a refresher training course, as they had clearly failed to properly identify this type of weather balloon after the Roswell incident. As far as can be ascertained from history, this did not happen. The military simply got back to business as usual, meaning there could be another misidentification of the same weather balloon should another incident occur in the future. How odd?

It is almost as if the Pentagon officials were certain such an incident will not happen again and that the object that crashed near Roswell was definitely not a weather balloon but something else. A secret military aircraft? A secret missile under test? Or was it something else entirely unexpected?

Even if it was possible to explain the materials as having come from a weather balloon, the official explanation given by the USAF was not sufficient to solve the controversy. The biggest problem in all of this was, why were the U.S. military and government officials still acting in a secret and evasive manner whenever questions were asked about the incident?

This was the question foremost on the minds of a number of inquisitive people, including the former New Mexico Congressman Steven H. Schiff (1947-1998).

In an attempt to resolve the controversy, Schiff sent a letter to Defense Secretary Les Aspen, asking him to provide a personal briefing on the Roswell case. Aspen forwarded the letter to the Pentagon for an official response. A letter from an Air Force lieutenant colonel arrived advising that he should refer to the files of Project Blue Book—the official USAF study into UFOs between 1952 and 1969—which is kept in the National Archives in Washington. Not happy with the response (as Project Blue Book had never officially investigated the Roswell case), Schiff approached Aspen once more on May 10, 1993. Again, Schiff was issued the same response.

Schiff said:

"I was getting pretty upset at all the running around...

Generally, I'm a skeptic on UFOs and alien beings, but there are indications from the runaround that I got that whatever it was, it wasn't a balloon. Apparently, it's another government cover-up."[1]

Annoyed with the rambunctious runaround from the Pentagon, Schiff submitted a request to the U.S. General Accountability Office (GAO) in Washington D.C. in October 1993. The GAO agreed, and on

February 9, 1994 announced that it would conduct an audit on all available U.S. government documents relating to the Roswell crash of 1947.

Steven H. Schiff (1947-1998)

The then Secretary of Defense, William J. Perry, received the GAO letter on February 15, 1994, The request for information was written to be broad in scope so as not to infer a link to the Roswell event. As the letter stated:

136

"In response to a congressional request, the General Accounting Office is initiating a review of DoD's policies and procedures for acquiring, classifying, retaining, and disposing of official government documents dealing with weather balloon, aircraft, and similar crash incidence. The review will involve testing whether DoD, the military services, specialized defense agencies, and others such as the National Archives, have systematically followed the proper procedures to ensure government accountability over such records."

Except that the GAO and Mr. Schiff revealed a little too much information to the military to the point where it got a whiff of what was happening.

The letter from Perry was sent to the DoD Inspector General who in turn wrote a memo dated February 23, 1994 for distribution to "the Secretaries of the Services and other affected parties"[2]. In it, we see for the first time since the GAO letter was sent the words "Roswell Incident" and "UFOs" being mentioned as if somehow relevant for the other military officials to know about. Also considered important for the USAF was the name of the person who had officially made the request. The name was gathered by the USAF during the in-briefing process with the GAO, and later checked the Defense records where a link was made between Mr. Schiff's letters and the Department of Defense (DoD) Legislative Liaison Office for information relating to the Roswell incident.

As the Inspector General explained the purpose of the GAO investigation:

"Representative Schiff requested the GAO review two issues of concern (1) the DoD records management procedures for crash incidents involving weather balloons and unknown aircraft, such as UFOs and foreign aircraft, and (2) the facts regarding the reported crash of an UFO in 1949 [sic, 1947] at Roswell, New Mexico. Since the UFO story appeared in an episode of the television program 'Unsolved Mysteries', Representative Schiff has received many requests for an investigation into the alleged 'DoD cover-up'. Apparently, reports on the incident were attributed to a weather balloon crash."[3]

As top officials decided what to do next, other USAF officials working below the top brass in the Pentagon were apparently caught off guard by the GAO situation. In a letter he wrote to John H. Gibbons, Ph.D., Assistant to the President for Science and Technology in Washington dated February 17, 1994, Scott Jones said:

"From what I have been told, a military Pentagon spokesman was less than civil in a verbal response to the GAO investigator when asked about Roswell. The alleged response was *Go shit in your hat*."[4]

Later, the USAF offered a slightly more civil response in a letter to Senator Patty Murray, dated September 2, 1953, when Lieutenant Colonel Thomas W. Shubert of the USAF Congressional Inquiry Division of the Office of Legislative Liaison stated:

"The Air Force possesses no records regarding this [Roswell] incident."

Upon learning about the sudden change in response, Scott revealed some possible reasons for the Pentagon spokesman's poor choice of words:

"It could reflect that he was having a bad day, or perhaps a personal opinion about UFO phenomena; or it could be a recent example of unwillingness to discuss the subject."

Whatever the reason, the change in public response was quickly matched by a unwavering determination on the part of the USAF to make it appear as if it was serious about getting at the truth. Despite claiming not being able to find records, within a matter of days, the USAF counterintelligence branch decided to conduct its own investigations into the controversial matter. This time the USAF would check its records as if anticipating the possibility that they might indeed exist.

Seriously, this was a remarkable turnaround from an organization which had, for a long time, been seemingly convinced that there were no records on the Roswell event to be given to the public. Now the organization believed it might find something. Extraordinary. So what had the USAF done in the past when people asked about the event? Twiddling their thumbs and whistling in the wind pretending nothing did ever happen? Of course something must have happened. The newspaper reports, a Roswell AAF press release, and the witnesses' testimony are too much information to ignore. Could the USAF not have prepared at least a first-class public information kit containing a copy of the original Roswell AAF press release, all available 1947 newspaper clippings of the Roswell event, and a brief official USAF report to explain why they thought it was a weather balloon? That's the least they could have done considering that the salaries come from the taxpayers. Still, the military in doing nothing when asked by the public had simply added to the controversy, to the point where now the USAF had been forced to provide an explanation for the continuing secrecy to the GAO.

But if they had made an effort and found *nothing* in their records, why did they again publicly state that they would recheck their records in the hope they might find something? Was the USAF not confident of their own ability to search the records? It makes one wonder just how successful the USAF would be to find information this time around. And if they would be successful, what the USAF would learn. Or is it possible that the USAF was expecting to find new information that would quell the mounting public interest in the incident? Either the USAF was lazy, incompetent, or they didn't want to explain what had happened.

It was clear. The USAF had to do something; and it had to be done quickly, before the GAO could discover anything potentially damaging to the military.

One of the first things that had to be done in controlling this escalating and potentially explosive situation was to somehow keep quiet any military personnel who was being a little too outspoken about aspects of the Roswell case. Brigadier-General Arthur E. Exon was one to receive special treatment from unknown person(s) who wanted to keep him quiet. By the time several high-level congressional staff members interviewed Exon as part of the GAO investigation of the Roswell incident, Exon suddenly felt he had to be extremely guarded in his talks to the point where the staff members had to write in their report:

"General Exon is afraid. He was afraid he was being monitored at that point. He was probably afraid his whole house was bugged."[5]

With the biggest security risk already in a state of paranoia and forced to keep quiet, it was a matter for the USAF to find something in its records to explain the materials and the type of object that crashed near Roswell in July 1947.

Colonel Richard L. Weaver, who headed the team of USAF investigators, allegedly interviewed all the remaining first-hand witnesses, searched for documents, and followed leads that eventually led them to interview Prof. Charles B. Moore (1920-2010), a renowned researcher on atmospheric physics from New York University (NYU).

It was learnt that Moore had worked on a formerly top-secret research project code named Mogul at the Alamogordo AAF in June and July of 1947. Moore told the investigators that the project was designed to monitor Soviet nuclear tests by listening for sound waves trapped in the layer between the stratosphere and troposphere using acoustic equipment attached to clusters of high-altitude weather balloons.

Moore stated that the balloons were made of polyethylene and the reflector consisted of balsa wood coated with "Elmer's-type" glue to give it strength, and covered with aluminium foil reinforced with tape containing "pinkish-purple abstract flower-like designs" that looked like "hieroglyphics". A train of these reflectors was often strung together. On hearing this, the USAF investigators concluded in their 23-page

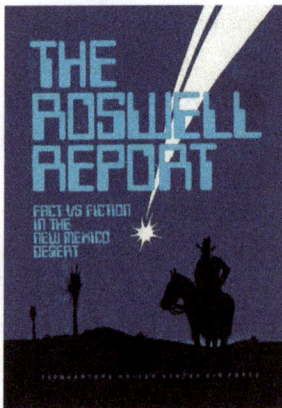

executive summary titled, *Report of Air Force Research Regarding the 'Roswell Incident'* July 1994, issued on September 8, 1994[6]:

"...there was no indication in official records from the [1947] period that there was heightened military operational or security activity which could have been generated if this was, in fact, the first recovery of materials and/or persons from another world."[7]

Prof. Charles B. Moore holding the metallic reflector component of a Mogul balloon.

In other words, there was nothing in the evidence presented by Dr. Moore, and from the available and official archive records searched by the USAF to suggest the UFO crash could be construed as extraterrestrial in nature. Given how similar in appearance the materials found by witnesses at the time were to those that made up the balloons of Project Mogul—namely metal, plastic and something resembling wood, and all very lightweight—the USAF publicly announced that the crash was probably from one of the Project Mogul balloons. As the USAF stated in its report:

"Project MOGUL balloon train components can be compared with the debris recovered from the Foster ranch and shown at Fort Worth Army Airfield with Maj. Jesse Marcel. Crashed saucer theorists allege that the debris depicted with Major Marcel is not the original debris collected from the Foster ranch. A switch is alleged to have taken place after the material arrived from Roswell AAF. However, detailed analysis and interviews with individuals who viewed and handled the debris, verify it to be completely consistent with the materials launched by Project MOGUL and subsequently recovered at the Foster ranch."

The conclusions of the GAO, released in July 1994 (and later published as a book in 1995), together with nearly 1,000 pages of reprinted ancillary documents, confirmed the USAF results. Well, this was because the USAF had quickly provided the GAO with what they *claimed* was everything they knew about the Roswell incident.

Back to square one.

Even when the GAO attempted to find the logs of outgoing phone calls from Roswell AFB for July 1947, they discovered that the USAF had destroyed them nearly 50 years ago. The only thing the GAO could say about this was that the logs were destroyed *without* proper authorization. With nothing else to go by, the GAO could do nothing except slap the wrist of the USAF for the poor decision made with the phone logs.[8]

Or could it be that the USAF wanted the phone logs destroyed to hide important evidence relating to the Roswell case? We can only speculate.

In the end, Schiff had no choice but to accept the USAF's latest conclusion unless, of course, he could find new evidence to seriously question it. But unfortunately his life came to an abrupt end a few years later, presumably of natural causes.

Mr. James Bond Johnson, the photographer at Ramey's office, had his own views about the investigation:

"I was not impressed with the attitude of the Air Force Intelligence officer who interviewed me in 1994 in connection with the GAO investigation in that he seemed to be just going through the motions and not too concerned with finding the true facts—his mind seemed already made up!"

When civilian investigators pursued the matter with the USAF concerning the recovery of so-called bodies from the 1947 debris, which had not been mentioned in their report, a USAF spokesman stated that these were probably anthropomorphic test dummies flown in some high-altitude balloons at the time of the incident.

Weaver explains:

"What quite likely happened is that people who saw these dummies mistook them for aliens."

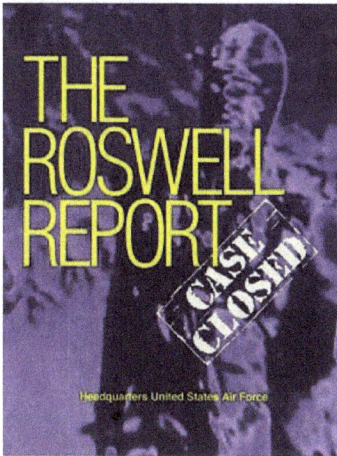

In July 1997, the USAF report was updated to include the "dummies-for-aliens" theory as the official explanation.[9]

The USAF held a press conference on Tuesday June 24, 1997 in Washington D.C. to endorse the new official conclusion, and explain to reporters the results of their study and the follow-up report. With so much riding on this press conference to end all civilian and scientific speculation on the Roswell controversy, the USAF gave a spectacular presentation, including a video-tape showing weather balloons flown in Project Mogul and other similar experiments at the time.

Colonel John Haynes, the USAF spokesman for the case, said it was an attempt to "set the record straight" and to present the latest USAF report on the Roswell incident published in the following month.

As Colonel Haynes said:

"Today we are releasing the final report to address questions about alleged bodies associated with the Roswell story.

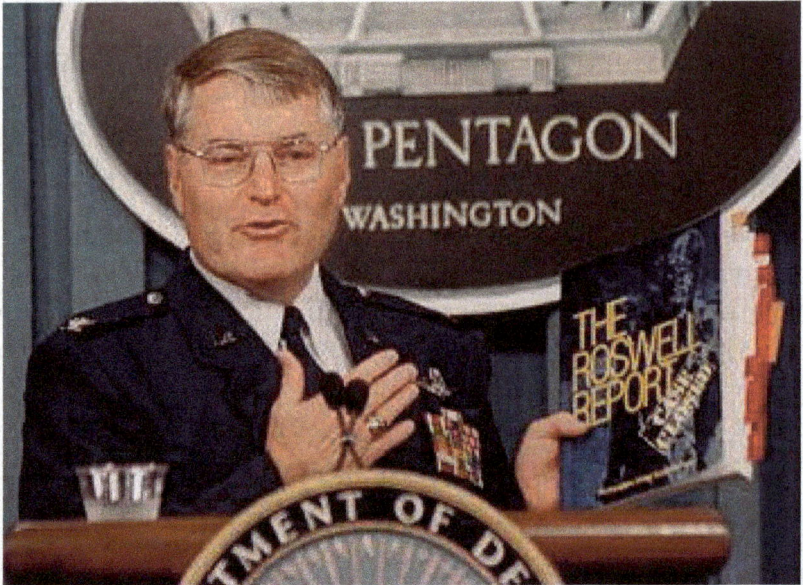

Colonel John Haynes

This report has four main conclusions:

1. Air Force activities which occurred over a period of many years have been consolidated and are now represented to have occurred in two or three days in July 1947.

2. Bodies observed in the New Mexico desert were probably test dummies that were carried aloft by U.S. Air Force high altitude balloons for scientific research.

3. The unusual military activities in the New Mexico desert were high altitude research balloon launch and recovery operations. The reports of military units that always seem to arrive shortly after a crash were actually accurate descriptions of Air Force personnel engaged in the dummy recovery operations.

4. Claims of bodies at the Roswell Army Air Field hospital were most likely a combination of two separate incidents. One was a 1956 KC-96 aircraft accident in which eleven Air Force members lost their lives, or a 1959 manned balloon mishap in which two Air Force pilots were injured."

In fact, so sure was the USAF of having the right explanation for what happened that Colonel Haynes said at the beginning of his speech to the media:

> "We're confident that once the report is out and digested by the public that this will be the final word on the Roswell incident."

Somehow, he wasn't reading the right script, because, on further examination of the Roswell report and realizing very little had changed, researchers began to wonder whether the report could set the record straight. More specifically, was the USAF looking at the right record of the Roswell incident?

SUMMARY OF FLIGHTS 1–6

Flight no.	Date	Launch Site	Configuration	Landing Site
1	4/3/47	Bethlehem, PA	See NYU *Tech. Report No. 1*, Table VII	Sandy Hook, NJ
2	4/18/47	Bethlehem, PA	See Appendix NYU *Special Report No. 1*	Unknown
3	5/29/47	Alamogordo, NM	Same as flight no. 2*	Unknown
4	6/4/47	Alamogordo, NM	Same as flight no. 2*	Unknown
5	6/5/47	Alamogordo, NM	See NYU *Tech. Report No. 1*, Table VII	East of Roswell, NM
6	6/7/47	Alamogordo, NM	See NYU *Tech. Report No. 1*, Table VII	South of Highrolls, NM

Source: USAF Roswell Report

For example, when the report was released in September 1994 and again in 1997, researchers learned that Moore had suggested that the likely candidate for having created the Roswell debris was the undocumented flight path of NYU Balloon Flight #4. The date of launch for this balloon was on June 4, 1947 with a flight duration of less than ten hours[10]. USAF supporters argue that this balloon, which was apparently never recovered, could have landed on the Foster ranch, since the date of launch had weather conditions that would have propelled the balloons northeast. Yet, somehow, William Brazel did not notice anything unusual on the property while checking on his sheep for nearly a month prior to the incident.

The only other balloons released in early July 1947 were the cluster neoprene and polyethylene balloons of Flight #7 and #8, but the observed path of these fragile balloons did not reach the Foster ranch.

Even if one could assume the balloon of Flight #4 had been responsible for the debris over the Foster ranch, there is no evidence mentioned in the USAF investigation and subsequent reports in 1994, and revised in 1997, to suggest that they had doubts about this and other balloons of Project Mogul as a possible explanation when witnesses made it clear that the dark-grey foil exhibited a shape-memory property as well as other tough materials. Clearly such materials had not been used in Project Mogul. The USAF had simply assumed that the witnesses had not seen anything unusual. Or, if the USAF did take into account the strange shape-memory behavior of the foil, it didn't suggest the possibility that Project Mogul could be testing new secret materials, or that another secret military experiment could have been involved. Instead it was better not to say anything. We can now understand why. The USAF's statement in its report of a supposedly "detailed analysis and interviews with individuals who viewed and handled the debris" involved nothing more than a personal interview of only *two* witnesses—Major Irving Newton and Captain Sheridan Cavitt. The only other people approached by the USAF were those involved in Project Mogul. Apart from Cavitt who appeared to have changed his story rather conveniently and coincidentally after being approached by the USAF before anyone else did, everyone else relied on by the USAF in interviews had to assume the debris was from a weather balloon.

As for civilian witnesses who saw the original debris, some quotes were obtained, but these were restricted principally to those provided by William Brazel as published in the *Roswell Daily Record* titled, "Harassed Rancher who Located 'Saucer' sorry he told about it", dated July 9, 1947. However, we know this was the time when Brazel was harshly interrogated by the USAF, and later he was accompanied by a USAF security officer to the local newspaper, in order to ensure he said the right thing.

Any other quotes obtained for the report from witnesses such as Loretta Proctor, Bessie Brazel Schreiber, Sally Strickland Tadolini, Jesse A. Marcel, Jr., were not gathered and, therefore, would not reveal anything unusual in the Roswell metallic foil for the USAF in order for the military to question its own Project Mogul explanation. Instead, the emphasis was on showing the aluminium-like nature of the metallic material. The same situation was also applied to the plastic-like and wood-like materials found near Roswell by stating that these appeared to be ordinary man-made substances. The super-tough nature of these materials were not on the radar of the USAF investigators as being in any way unusual.

True, the USAF did make some reasonable effort to indicate in the report that it was aware of William L. Moore and Charles Berlitz's 1980 book on the Roswell incident and understood that the book contained original quotes from other firsthand witnesses including Major Jesse A. Marcel (who was present at the debris field), and even made a point in the report of noting "exotic metals" were mentioned by witnesses. However, the USAF saw nothing in the metallic foil descriptions to remind them of the shape-memory effect of known alloys in the scientific literature let alone any secret studies the military may have done in this area at around the time of the Roswell incident. As the USAF stated:

> "There are also now several major variations of the 'Roswell story'. For example, it was originally reported that there was only recovery of debris from one site. This has since grown from a minimal amount of debris recovered from a small area to airplane loads of debris from multiple huge 'debris fields'. Likewise, the relatively simple description of sticks, paper, tape and tinfoil has since grown to exotic metals with hieroglyphics and fiber optic-like materials."[11]

In essence, the USAF had effectively ruled out any likelihood of the military having carried out a secret study into shape memory metals at the time of the incident, assuming the foil was from a military experiment.

Another concern is the size of the object responsible for the debris. No single Project Mogul balloon would have produced foil to cover "three-quarters of a mile" by 200 to 400 feet across, not to mention the alleged dark-grey disc found further west.

And if that is not enough, not even the shape of the Project Mogul balloons is in direct agreement with the official news release, the secret memo held in the hands of Ramey in his office, or the memo written by General Twining on July 15, 1947. The original object has to be disc-shaped, or of symmetrical metallic design. A torn up Project Mogul balloon on the ground or an intact version flying through the air could not be described as disc-shaped.

Furthermore, Dr. Moore never mentioned using dummies in Project Mogul to the best of his recollection. This idea of "dummies" was something the USAF decided to add at the last minute in an attempt to deal with the more difficult questions relating to so-called "victims" as allegedly observed and heard by some witnesses and supported later by the analysis of Ramey's top secret memo. If dummies had been used in secret for the first time in July 1947 (pushing back the official use of these plastic mannequins by nearly 10 years), the USAF could not give details of the exact flight numbers where these dummies had allegedly been flown. Nor could the USAF explain why a number of dummies were without parachutes when they were recovered, according to the witnesses.

All in all, this was a remarkable effort, considering the USAF was trying to conduct an independent investigation of its own affairs surrounding the Roswell incident at a cost to the U.S. taxpayer said to be in the hundreds of thousands of dollars.

CHAPTER 5

An Australian Researcher Speaks

THERE COMES a point in this research where a scientist must face two distinct views on what was found.

The first comes from the witnesses themselves, a news release, and two secret USAF memos. Here we find astonishing claims of a dark-grey, metallic, disc-shaped object that is 9 meters in diameter, and made of unusually tough materials. In the case of General Twining's memo, it also reveals unusual fabrication techniques to not only handle these materials, but also to support some kind of unknown electromagnetic technology.

Furthermore, there was talk of at least half a dozen diminutive victims in skin-tight metallic suits found next to the disc. Closer to the first wreckage site northwest of Roswell, military witnesses alleged more bodies were found, as if they had fallen out of the stricken object as it continued its fatal flight across the desert at high speeds before crashing to the ground.

Leaving aside the issue of what type of electromagnetic technology we might be dealing with and whether bodies were found, if we focus our attention solely on the metallic foil, we find remarkably consistent claims made by the original witnesses. In particular, the foil was described as:

1. Dark grey in color;
2. Extremely lightweight;
3. Very hard, and strong enough to withstand cutting, scratching and denting with scissors and sledgehammers;
4. Highly resistant to high temperatures from a blowtorch; and
5. Having a strong shape-memory effect at room (or early-morning desert) temperature.

Also, the pieces of this foil showed absolutely no signs of welding, riveting, or nuts and bolts to suggest multiple sheets had been assembled to form the outer hull of this flying object. As far as Marcel could ascertain after joining together a few pieces on his kitchen floor, and observed in the more complete form by Barney Barnett, the foil was very smooth, and it covered a large area of the object. According to Mr. Barnett, this dark-grey foil appeared to have covered the entire skin of the flying disc. Coupled with the claim that little or no wreckage was scattered around the object upon impact with the ground, even with presumed high force, this suggests the materials were indeed tough. This is further supported by witnesses at the first crash site where more of the dark-grey metallic foil, along with other materials, were found.

On the other hand, the USAF—whose people we understand had cleaned up the initial wreckage site (and were probably informed of the main crash site by the U.S. Army) and should have performed the most thorough study of the wreckage after it was sent to Wright–Patterson AFB for analysis—claimed the object was an ordinary flimsy weather balloon made of aluminium, plastic tape, and a balsa-wood frame of some reasonable size, allegedly carrying an unspecified number of dummies.

There is quite a difference in the two opposing views. Just on the foil alone we notice the following discrepancies:

1. The USAF had claimed the foil was a flimsy and bright silvery-colored metal made of aluminium (or possibly tinfoil if the

military would attempt to account for the foil's unusually dark color);

2. The witnesses had claimed the foil was a metal, dark-grey in color, which was very tough to melt and break, and maintained its original shape.

This is a rather curious discrepancy, to say the least.

Why?

In the previous chapter, we learned how the USAF relied on three people. One was Cavitt. However, we can understand why his story changed compared to everyone else at Roswell AAF. The USAF said something to Cavitt to convince him to support its position on the Roswell case. Other than Cavitt, the other person we find is the military officer named Newton. The only thing is, he had no choice but to support the weather balloon explanation after he was brought in to view a substitute version of the original wreckage in Ramey's office. Even the *principal* military officer, Marcel, who had handled the original wreckage, knew it was a weather balloon when he saw the *substituted* wreckage. No disagreement. Unfortunately for Marcel, he was taken out of the office just before the arrival of James Bond Johnson, the official photographer from the *Fort Worth Star-Telegram*. Finally, the third person interviewed, this time a scientist, was involved in the secret weather balloon experiments of Project Mogul. When the scientist spoke to USAF investigators, he mentioned nothing about a shape-memory foil. It was simply aluminium foil. Somehow, we get the impression this is not exactly a *balanced* report. Something is just not right. As such, we must consider the likelihood that the original firsthand witnesses may have something to contribute to this scientific discussion.

For a start, we cannot dispute the number of witnesses making the claim of a shape-memory property in the dark-grey metallic foil. They are far too numerous and consistent to ignore. It is the kind of observation the USAF should have examined in detail if they had spoken to other witnesses, gathered all the known official quotes from witnesses obtained from previous investigators, and provided a satisfactory explanation of what this metallic foil was, which they haven't. With this in mind, let us take a closer look at this observation and see what we can learn.

151

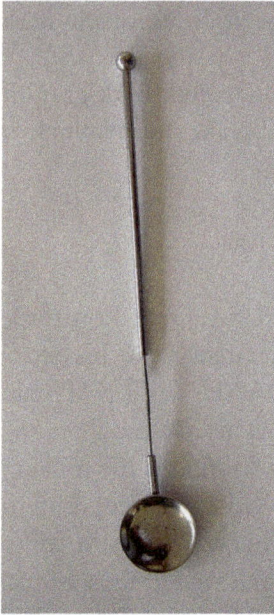

Perhaps a study into this type of foil may reveal the real reason for the cover-up of "victims" in the Roswell crash and the type of object that was found.

Revealing the dark-grey shape-memory alloy (SMA)

In the early 1980s, an Australian business called Dick Smith Electronics sold an unusual toy to the general public known as the "Miracle Spoon". The manufacturer of the spoon describes it as a *miracle* because of the way the super-alloy behaves—it is capable of returning to its original shape.

The alloy attached to the stem of the spoon happens to be dark-grey in color when compared to the handle for holding the spoon and the end of the spoon containing the shallow round bowl for stirring contents in a cup.

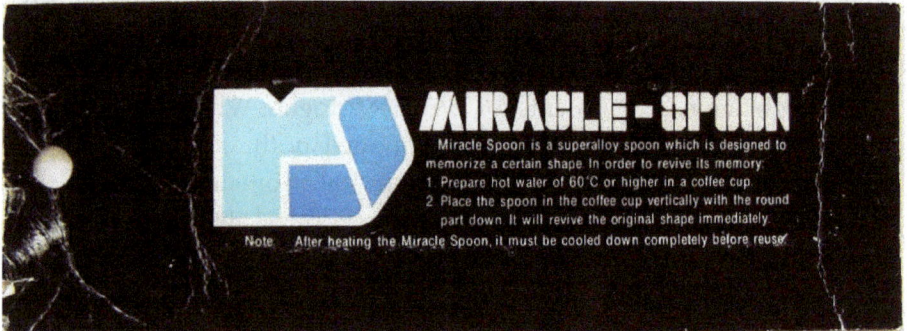

Original product package of the shape-memory spoon.

To activate the shape-memory effect, all you do is bend, twist and turn the end of the spoon in different directions when it is cool in order to distort the alloy. When the stem of the spoon is dunked into a cup of hot water, the alloy immediately resumes its original shape with great force, thus permitting the user to stir the contents in the cup.

Not long after, in the March/April 1984 Australian edition of the *Omega Science Digest*, an article was published discussing this dark-grey alloy as having the unmistakable ability to return to its original shape when heated. The article is titled "Miracle Metal (The Metal That Science Can't Explain)—There is Absolutely Nothing like it".

Remarkable.

This distinctly dark-grey metal behaves and looks very much like the witnesses had claimed for the Roswell foil. Coincidence or not, does this mean we will have to take the witnesses' claims a little more seriously? Or should we assume the USAF is correct in their present claims of *aluminium* foil as the best explanation for the Roswell metallic foil? Maybe it was a SMA, except it contained a reasonable amount of aluminium and the military had somehow forgotten about it? Well, it will have to be dark-grey in color, not the bright silver color we know aluminium to be. Or was it a completely different type of SMA? And how many dark-grey SMAs are there to consider?

The only way to be certain about this was to check through the available metal journals and chemical abstracts from the ANU to find out more about the metal's physical and chemical properties, manufacturing requirements, and history.

After conducting this research, what we learn has to be considered rather remarkable. Apparently, we learn that the USAF was indeed officially studying this intriguing dark-grey metal, among several other shape-memory metals, roughly 6 to 12 months after the witnesses had observed a metallic foil that looked and behaved just like this metal. Yet, the USAF chose not to discuss the work in either the 1994-95 edition or the final 1997 edition of the *Roswell Report: Case Closed*.

More disturbingly, one of the elements in this metal and others of this type not only experienced a sudden change in its status after 1947 in the U.S., but it required advanced technology to reach a minimum of 99.995 percent purity in order to reveal this shape-memory property. Unfortunately, the technology didn't exist by July of 1947 or any time during 1947, and certainly not to the quantity required to cover the amount of Roswell metallic foil debris that had been found. So, instead, the USAF had to ask for scientific assistance from the world-renowned Battelle Memorial Institute after 1947 to study this type of metal, among several others, and search for ways to improve the purity of the metal (for some unexplained reason) so it can study certain types of SMAs.

Let us take a closer look at the available scientific literature for information relating to SMAs and their history, and see where it will take us.

We will begin by determining whether the super-alloy in the Miracle Spoon can have the same physical characteristics as the Roswell foil.

154

CHAPTER 6

Nitinol: The World's Most Powerful Memory Metal

THE METAL responsible for the Miracle Spoon's amazing behavior is called *nitinol (TiNi, NiTi, Ni-Ti, Ti-Ni* or *nickel-titanium alloys)*, composed of nickel (Ni) and titanium (Ti). This is a naturally dark-grey shape-memory alloy (SMA), and the only distinctly dark-greyish SMA scientists know of (and certainly in the titanium SMA family).

This chapter will present the scientifically-established facts concerning the physical properties of nitinol and what scientists know about the shape-memory effect. For a quick summary of the facts relevant to the Roswell case, please see the *Summary* section at the end of this chapter.

Nitinol's classic composition range

Also known as a muscle, super, martensitic, martensitic memory (or marmem), martensite-to-austenite, mechanical memory, shape-memory, or smart alloy, nitinol's remarkable shape recovery property reaches maximum effect when the composition contains nearly equal quantities of nickel and titanium. The "shape-memory" composition range is between 47 and 53 atomic % (or 53.5 to 56.5 weight %) nickel, with the greatest interest for scientists and engineers in the range of 50 to 51 atomic % (or 55 to 55.5 weight %) for nickel covering a transformation temperature range of 0°C to 100°C.[1]

Two examples of nitinol. Picture on the left (manufactured by Adv-Ti Titanium Industry (Group) Co. Ltd in China) is in the natural form with the surface polished. Despite the dark-grey nature of the alloy, notice how it reflects light. Picture on the right (from http://www.stemfinity.com/Nitinol-Memory-Wire-1ft-Pk-30) is said to be made of the same alloy, except the surface has been anodised to a dark grey, almost black color with very little light reflected off its surface.

Nitinol's main claim to fame and its principal uses

When a highly pure sample of nitinol is activated (or energized) above what is known as the *martensite transformation temperature*, or when an electrical current passes through it, it can return to its original shape with great force and speed. This is because the atoms in the alloy reach a state where they are able to move from one crystalline structure into another with relative ease. While in this energized state, scientists describe the alloy as being as strong as steel, but it is over 40 percent lighter than steel. It is highly resistant to corrosion, is non-magnetic, and has a melting point above 1200°C, making it difficult to melt using a blowtorch.

It is the kind of alloy respected by the engineering world as a useful aerospace material for building spacecraft components. As the 1969 edition of *Hackh's Chemical Dictionary* stated on page 456:

> "**nitinol** Generic name for nickel-titanium intermetallic compounds used in spacecraft construction, i.e., Ti-Ni. They have high ductility and impact resistance at low temperatures."

156

As for the alloy's resistance to corrosion and inert properties within the human body, the material has also found its way into medicine, through uses such as tooth replacement, stents to widen blood vessels, and more.

Nitinol's crystalline structures

Scientists have identified three crystalline structures in the low-temperature phase (below the transformation temperature), and at least two crystalline structures in the high-temperature phase (above the transformation temperature).

When nitinol is allowed to solidify inside an oven for the first time to form the chemical bonds needed to hold together the nickel and titanium atoms, the first crystalline structure to be formed is called a cubic body-centred unit cell. In crystallography, this is known as B2, or CsCl-type (because the CsCl crystalline structure is the same as NiTi other than the fact that it uses different atoms). The cubic B2 structure is considered the simplest repeating unit cell in the high-temperature state.

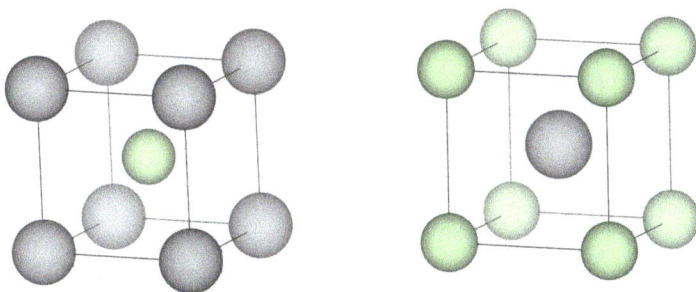

The cubic B2 body-centred unit cell. Either of these two diagrams can be used to represent this unit cell. See Appendix H for details of how the two diagrams relate.

The dimension of this highly compact cubic unit cell based on single-crystal x-ray diffraction method is somewhere between 2.960Å and 3.019Å.[2] It is generally on the upper end of this scale for a freshly made and relatively warm sample of NiTi coming straight out of the oven (for these crystalline diagrams, we have used 3.015Å).[3] With regular cold-working of the alloy and heating and cooling above and below the transformation temperature, the lattice dimensions decrease slightly.

As for understanding what happens in the low-temperature state, this B2 cubic structure is a little too simplistic to use. Scientists have chosen to add a few more nickel and titanium atoms to form the unit cell called a tetragonal, with height and width of 3.015Å and length is 4.264Å.

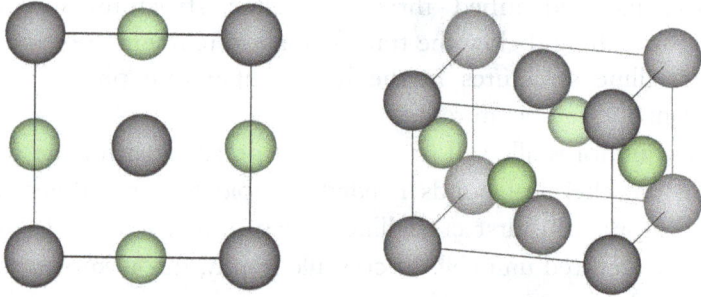

Tetragonal face-centred unit cell (top view on left and 3D view on right).

Now as the crystalline structure cools to below the transformation temperature, the dimensions of the tetragonal unit cell changes to become what scientists call *orthorhombic*. In crystallography, this is described as B19.

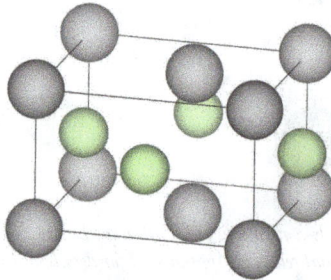

Orthorhombic B19 unit cell

In mathematical terms, *orthorhombic* simply means that all three sides are of unequal length and all the angles remain at 90°. In the case of NiTi, the difference compared to the original high-temperature tetragonal unit cell structure is in the way the length and width increases, and the height shrinks. The increase in length and width is due primarily to two face-centred nickel atoms being displaced in what we will call the forward direction (i.e., to the right in the previous diagram). This in turn pushes the two formerly face-centred titanium atoms above and below in the same direction. As a result of the two nickel atoms no longer being in their expected face-centred positions, the height of the unit cell shrinks to 2.956Å and width increases to about 4.189Å. The length along the direction where the two nickel atoms have moved forward is marginally increased to 4.686Å. Well, to be precise, it is a little bit longer because one of the nickel atoms is outside the unit cell. Therefore, the length is closer to around 5.623Å. As a result of this stretching of the unit cell in at least a couple of directions (i.e., width and length), scientists refer to this by the technical term, *martensite*.

However, things are never quite as they seem. We see this most commonly in a freshly made sample of NiTi taken straight out of the oven and allowed to cool once (or where there has been evidence of cold-working of the alloy to bend it in one direction). When scientists applied the x-ray diffraction method[4] to look at the crystalline structure of NiTi, they discovered a distortion of the orthorhombic B19 unit cell whereby not only are the edges of unequal lengths, but one angle is greater than 90° (the other two remain at right angles), causing the cell to slant as shown in the diagram below:

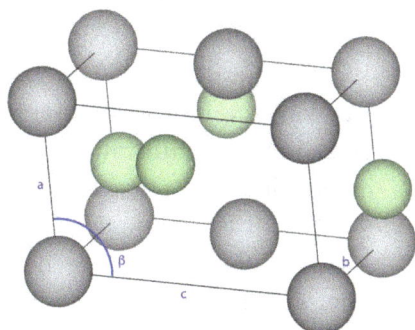

Monoclinic B19' unit cell

Perhaps as a consequence of this slanting in one direction, the eight titanium atoms acting as the "cage" (or corners) of the unit cell seem to push the remaining atoms in the formerly face-centred positions upwards (or downwards according to the crystalline diagrams provided

159

by Vishnu et al.). In crystallography, scientists classify this structure as B19' (note the use of the apostrophe to distinguish it from the orthorhombic form).

There is also a mathematical word to describe the shape of this non-symmetric unit cell with its characteristic slanting in one direction: *monoclinic*.

The maximum dimensions of this monoclinic B19' martensite unit cell so far measured by the scientists is length=4.686Å, width=4.189Å, height=2.956Å, and β=98.26°.[5] However, as the alloy goes through a number of thermal cycles above and below the transformation temperature and without any bending or twisting to induce stress (or strain) into the alloy, an increasing number of these monoclinic B19' structures convert to the orthorhombic B19 crystalline form.[6] Otherwise, as soon as the alloy is bent in a particular direction a few times, the B19' structure is usually more predominant in the low-temperature state.

Just to add a little more complexity to the picture, Vishnu and Strachan (2010) reported of a new phase (or crystalline structure) called B19" (note the additional apostrophe). This is not quite the lowest energy state and the most stable form. Also included in the range of structures for NiTi in the low-temperature state is what scientists call the Body-Centred Orthorhombic (BCO) structure. The main differences observed in these two additional structures compared to the B19' structure is the wider angle of the slanting, to 103.20° for the B19" structure, and 107° for the BCO structure. And while the atoms move in a similar way as with the B19' structure, both B19" and BCO have the atoms displaced more significantly. Furthermore, the width and height shrinks marginally, but the length increases to around 4.819Å for B19" and 4.936Å for BCO.

Despite the complexity of the monoclinic B19' unit cell structure and the presence of other crystalline structures in the low-temperature phase, the one unmistakable feature that distinguishes the tetragonal (or B2 cubic) from the B19' (as well as B19" and BCO) structure is definitely in the way the nickel and titanium atoms are spaced apart in the low-temperature state (mostly in one direction along the length) while still retaining the fundamental crystalline structures needed to be called a solid substance.

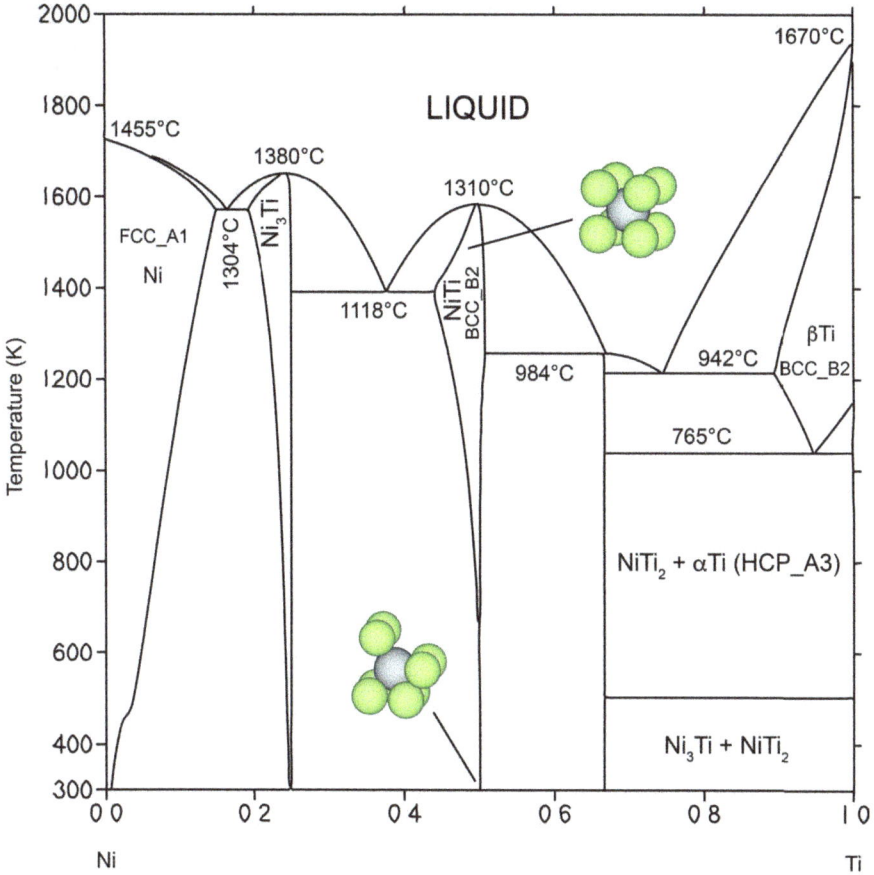

NiTi Phase Diagram

Whichever structure gets formed in the low-temperature phase, metallurgists call all these structures the *martensite*[7] (or *martensitic*) phase.

Phase 2: Low temperature stress-induced structure

As a result of the greater separation between atoms, the martensite phase is characterised by the way the alloy is relatively soft[8] and pliable (especially for a freshly made sample coming straight out of an oven, and before it hardens over time through continual cold-working of the alloy[9]) and so can be easily distorted[10] by physically bending the alloy. When the structure is bent in this way, some scientists call this phase the stress-induced martensite phase. It can also be called the *twinning* of a SMA.

The effect of this stressing on the alloy to the crystalline structure is to push a number of these B19 orthorhombic unit cell structures into a distorted monoclinic form, or exaggerate the existing monoclinic structures by forcing the angle that is over 90° to become wider. Or, depending on the direction of the stress, it could also reduce the existing angle greater than 90° back to 90°, and even push the crystalline structure into a different direction where a different angle can exceed 90°.

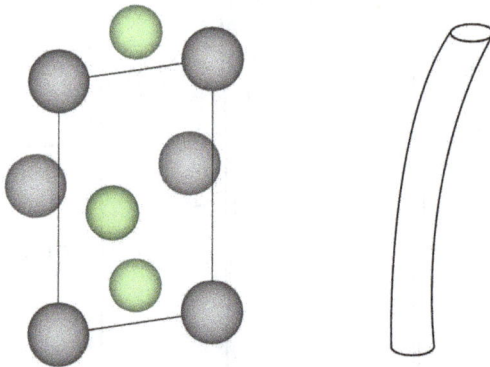

A side-on view of how the monoclinic B19' unit cell is likely to relate to the stress-induced phase (showing the direction of a bent wire),

Phase 3: High-temperature austenite structure

When the alloy is heated above the transformation temperature, or if electricity (i.e., the "flow of electrons") passes through it, the crystalline structure rapidly undergoes a de-twinning (or reorientation of the martensite, which is just another way of saying the atoms move) to a highly symmetrical and compact body-centred cubic (or face-centred tetragonal), with all atoms returning almost precisely to their original lattice positions.

As the atoms move into this compact form, it causes the alloy to return to its original shape (hence the term *shape-memory*) with tremendous force, incredible hardness, and remarkable speed (described as virtually instantaneous depending on purity). While it is in this energized state—called the *austenite*[11] (or *austenitic*) phase—nitinol is described as being as strong as steel and difficult to bend, cut, pierce, and scratch its surface.

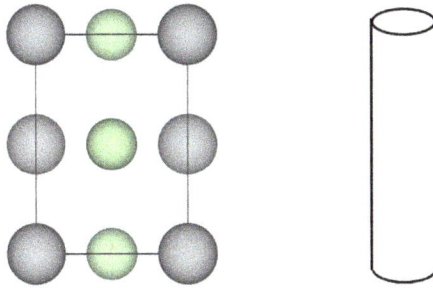

Phase 4: The R-phase

Just when we thought we have seen the last of the phases in NiTi, we came across another apparently distinct crystalline structure that precedes the martensite structure, called the *R-phase*. This was first observed by some scientists in the 1970s and later confirmed by H. C. Ling and Roy Kaplow. Basically, as NiTi is cooled, the high-temperature cubic B2 austenite structure changes first into a "transitional" crystalline structure best described as a "rhombohedral distortion of the cubic austenite phase"[12]. Hence the use of the letter "R" in "R-phase". With further cooling, this R-phase eventually gives way to the monoclinic B19' structure. As Holec et al. stated in their article, "Ab Initio Study of Point Defects in NiTi-based Alloys":

> "The martensitic transformation converts the high temperature parent B2-austenite phase (CsCl-type ordered cubic lattice) into the 'soft' martensite R-phase (P3 trigonal lattice), and further into the low temperature martensite B19'-phase (P2$_1$/m monoclinic lattice)."[13]

Not all the austenite unit cells get into the R-phase. It depends on various factors, such as:

1. Training the unit cells

    Cold-working the alloy (i.e., by bending it) and thermal cycling helps to "train" enough of the austenite unit cells to transition more easily into the R-phase.[14] As Brachet et al. stated in 1997:

    > "For binary TiNi alloys, cold-working and partial recovery heat treatment is necessary to obtain such as [i.e. R-phase] properties"[15]

163

2. Lowering the temperature

Once enough unit cells have been "trained", a lowering of the temperature sees an increasing number of the austenite unit cells get into the R-phase.

3. Adding ternary elements

Adding small amounts of a ternary element to the NiTi alloy system without destroying the shape-memory effect can determine whether the R-phase exists or not. As Duerig & Bhattacharya can confirm:

> "Certain ternary additions, such as Fe, Co, and Cr suppress M [Martensite] resulting in alloys that form R [R-phase] upon cooling, but do not form M without the application of a stress."[16]

This is in agreement with Brachet et al.:

> "Fe addition decreases the martensitic transformation temperature (Ms) promoting the occurrence of intermediate 'premartensitic' R phase."[17]

In other words, a freshly made sample of NiTi taken straight out of the oven is said to have very little, if any, R-phase no matter what the temperature is. However, apply the thermal cycling above and below the transformation temperature a few times, or any form of cold-working of the alloy in the low-temperature state, and the R-phase develops.

Think of the R-phase as an intermediary crystalline structure to help the atoms to transition more easily into the cooler martensite B19' structure and vice versa. This is why you will often see in the scientific literature the phrase "premartensitic transformation" in connection with the R-phase to suggest it is somewhere between the austenite and martensite structures. For example, Dieter Stoeckel of the Californian company Nitinol Devices & Components, Inc. located in Fremont, mentioned on page 3 of his article "The Shape Memory Effect – Phenomenon, Alloys and Applications":

> "...Ni-Ti alloys exhibiting a premartensitic transformation (commonly called R-phase)."

At the same time, the R-phase seems to serve another purpose. As mentioned before, regular cold-working of NiTi in the low-temperature martensite phase does develop the R-phase. So once the B19' structure gets heated, it goes into the R-phase and eventually settles into the B2 structure. Similarly the reverse process occurs as well. There is a reason for this. It concerns the formation of another interesting property of the alloy. More specifically, the presence of the R-phase helps the alloy to remember a second "shape" in the cool state. In other words, NiTi is a two-way SMA. Basically, the more monoclinic B19' structures there are in the martensite phase after going through the R-phase, the more easily the alloy attains this second shape when it is cooled from the austenite phase.

The link between the R-phase and the two-way shape-memory effect (TWSME) is supported by Urbina et al. (2010) when they stated:

> "...to obtain the best results (that is, a substantial two-way memory strain together with a minimum of final plastic strain and easy TWSME activation temperatures), the NiTi wire should be thermally cycled at zero stress prior to training until the R-phase is totally developed and trained by thermal cycling under constant load."[18]

Or to put it another way, NiTi needs the R-phase to be present and ready to accept new changes in direction in the bending performed on the alloy (i.e., the training part). Once the "training" commences through the application of a load on the alloy and maintaining it during the regular thermal cycling, the alloy is more easily able to "learn" a second shape in the cool state, and to get to that shape more easily and quickly once the temperature is reduced. The training seems merely to help re-align many of the R-phase unit cells to slant in a particular direction and so make it easier to attain this second shape in the cool state. That is, as soon as the load is removed, the alloy keeps to the new shape in the cool state.

The R-phase was originally observed during the 1970s, but it took Ling and Kaplow's landmark paper in 1981 to settle the debate and confirm its existence.[19]

Phase 5: The annealing phase

But we are not quite finished yet.

Some metallurgists have decided it is important to define a fifth phase known as the *annealing phase*. This means that nitinol can be hardened with continual cold-working—such as bending, twisting or rolling—while it is in the low-temperature structure, making it more difficult to bend or put a dent in it. This interesting discovery of nitinol was patented[20] by U.S. scientists William J. Buehler and G. Rozner in 1967.

What this means at the crystalline level is that the atoms are no longer separating as much in the low-temperature phase as when they did in a freshly made sample of the alloy. The atoms are, in a sense, "learning" to settle into a more compact form, described mostly as orthorhombic B19 with the slightest tendency to elongate into the martensite, when it is cooled. Perhaps with regular cold-working, this low-temperature structure will eventually become the tetragonal, which is the B2 cubic structure of the high-temperature phase. However, because of NiTi's two-way shape-memory response, we can expect the tetragonal shape not to be perfect. A slight slant in a number of unit cells will exist and in a particular direction (mainly in response to an applied strain to deform the alloy that is predominantly in that slanted direction). At any rate, the atoms will over time move into the best position they can to ensure not only that the shape-memory response appears in both the hot and cool phases, but also that the force exerted by the alloy to get into the second "memorised" shape while in this cool state gets stronger as it gets harder. It means the alloy will have an uncanny ability to get better at attaining the two different shapes over time thanks to this inherent tendency of the alloy toward becoming increasingly hard in the martensite.

For even greater hardness (and so to make it extremely difficult to bend in a material of sufficient thickness), make sure it is heated well above the transformation temperature. Then you will have an alloy in the austenite phase that behaves like armour. It is at this point that the chemical bonds holding the atoms together are extremely short and strong (because the crystalline structure is at its most compact form), making it more difficult for projectiles or sharp objects to penetrate, pierce, or cut the alloy.

Scientists have yet to test exactly how thin a sheet of hardened nitinol can get to resist cutting or piercing from, say, the fine tip of a pair of scissors.

Simplifying the crystalline structures

For the purposes of this book, let us not look too hard at the precise crystalline structures for NiTi. All the scientists are essentially saying is that:

1. There are expanded "low symmetry" crystalline structures (B19, B19', B19", and BCO) called *martensite*, and a compact "high symmetry" crystalline structure (B2, and to a certain extent the R-phase as well) called *austenite*.

2. The expanded crystalline structures of the martensite make NiTi softer and more pliable as needed to bend it into different shapes.

3. The compact crystalline structure of the austenite makes the alloy harder and more difficult to bend and always retain its original shape.

4. When the alloy transforms from the austenite to the martensite, a great deal of heat is emitted. Likewise, bending the alloy in the martensite phase also emits heat.

5. When reversing the transformation process, heat must be applied to force the martensite crystalline structures back to the austenite crystalline structure. As it does so, this energy is absorbed and used in a way that helps to bring the atoms together into a highly compact form and with its original shape.[21]

6. The more NiTi is cold-worked, it increasingly gets into a more compact crystalline structure, meaning the alloy will get harder and tougher within the two main shapes it has "memorised".

Or, leaving aside the R-phase and other fine crystalline structures between B2 and whatever the final martensite structure should be, think of NiTi as a material that can transform between two main crystalline structures. The difference in the two crystalline structures can be best summed up in the words of T. W. Duerig and K. Bhatttacharya in their May 16, 2015 article, "The Influence of the R-Phase on the Superelastic Behavior of NiTi":

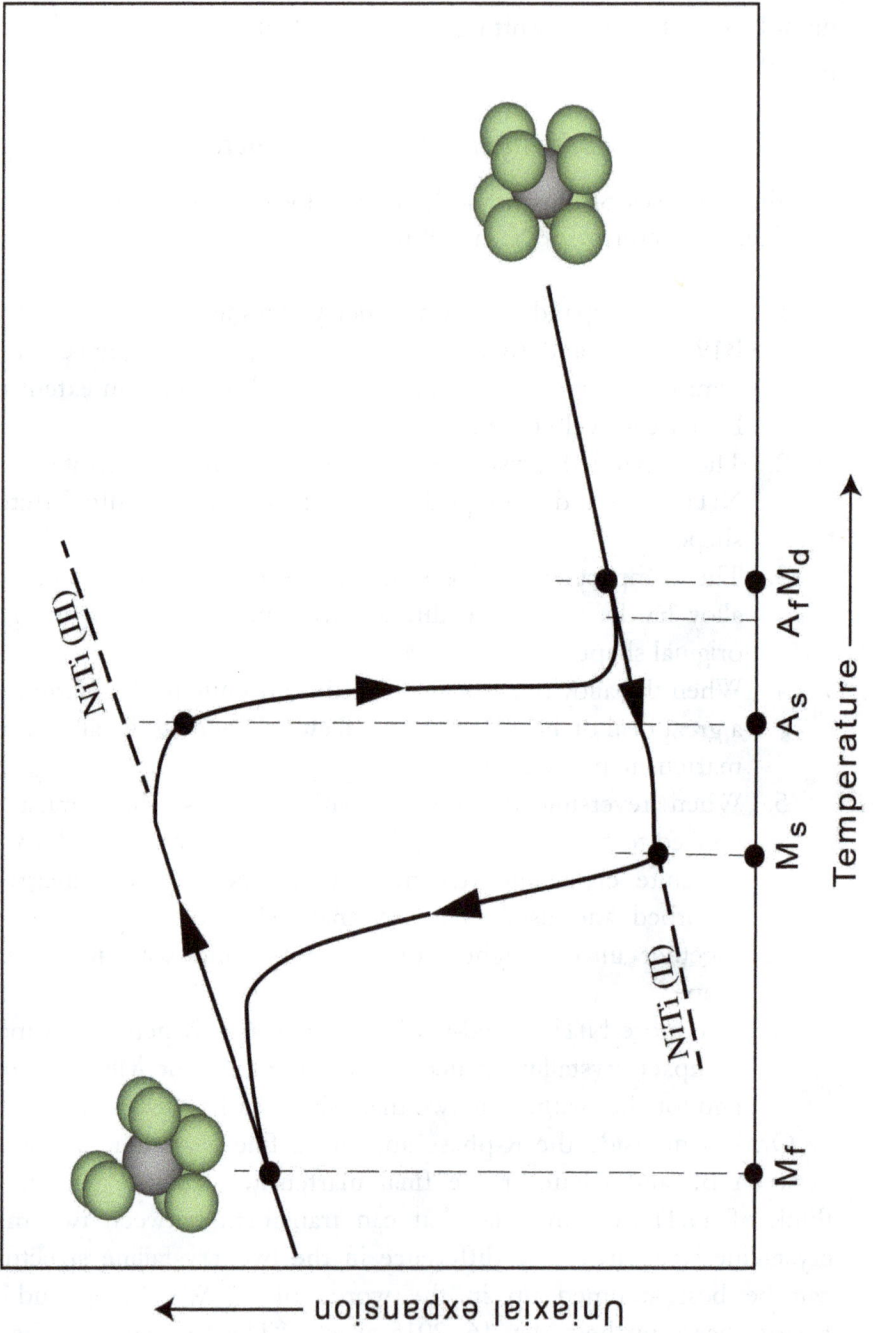

Graph of Uniaxial Expansion for NiTi.

"The M-phase [martensite phase] can be imagined as a monoclinic distortion of the cubic austenitic phase..."[22]

Do the atoms of nitinol move?

When NiTi performs its famous shape-memory effect during the transformation from the martensite to the austenite and back again, there is always a movement of the atoms at the crystalline level as confirmed through x-ray diffraction.

Bill Hammack

Bill Hammack, a professor in the Department of Chemical Engineering at the University of Illinois at Urbana-Champaign, confirms this in his Engineering and Life Radio Program that he delivered to the public in 2003:

> "It [nitinol] works because the atoms in the metal rearrange into a new phase. Usually changing phase is something dramatic: an ice cube melts, changing phase from a solid to a liquid, or we put water to boil on the stove and it changes phase from a liquid to a vapor. Less well-known are that such changes occur within a solid: solid-to-solid phase changes involve rearrangement of the position of the atoms. Nitinol remembers its shape because above a certain

temperature it returns to a rigid arrangement of the atoms."[23]

As further evidence, scientists have measured the uniaxial dimensional change (or expansion behavior) of the alloy with increasing temperature as shown in the graph on the previous page.

The results from the graph show that as the temperature of nitinol increases, the atoms of nitinol naturally move apart slightly as one would expect in this circumstance. Then suddenly the unexpected occurs at position A_s when the temperature somehow, and in a rather dramatic fashion, causes the atoms to move in the opposite direction, resulting in a more compact crystalline structure. This is the moment when the alloy exhibits its shape-memory effect. When the atoms come together into this form, the entire alloy not only remembers its shape, but it also reduces in size. After the temperature exceeds position M_d, the increasing temperature sees the compact B2 structure expand normally.

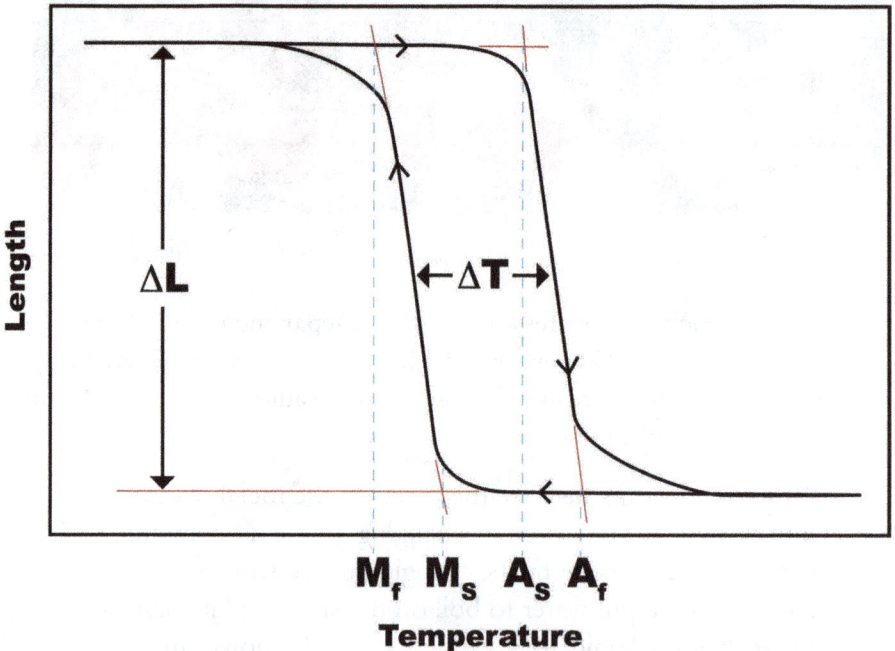

To calculate the width of the thermal hysteresis (ΔT), draw the red lines as shown here and where they intersect the temperature axis, calculate $\Delta T = A_f - M_s$

Pure nitinol results in extremely efficient shape recoverability effect

In addition to this, studies have conclusively shown that nitinol, with very low levels of impurities (especially oxygen and carbon), is capable of transforming heat and electrical energy into mechanical energy (i.e. the process of returning to its original shape) in a highly efficient manner. The fewer impurities there are in nitinol, the more pronounced the shape-memory effect is. When it does so, the heat differential (also known as the thermal hysteresis ΔT in the graph of uniaxial expansion, which is the difference between temperatures M_s and A_f) needed to complete the martensite-to-austenite transformation phase is extremely small with virtually all the energy quickly absorbed by the crystalline bonds.[24]

Nitinol never wears down

Although there is a limit to the level of physical stress that can be applied to nitinol (above which it remains permanently deformed, or may break) its shape-memory behavior can be produced repeatedly within these strain limits without showing signs of metal fatigue[25]. In fact, it gets better over time because of its increasing hardness through repeated mechanical movement of the alloy, suggesting perhaps that the chemical bonds holding the metal atoms together are getting shorter and/or developing some other characteristic to help increase bond strength. Nitinol is like a muscle; the more you use it, the stronger it gets.

Eliminate crystalline microdefects

Moreover, maximum shape-memory recovery effect occurs when there are absolutely no impurities and microdefects in nitinol's atomic crystalline structure and where the composition is very near to Ni-Ti$_{49at.\%}$, with the shape-memory effect occurring at room temperature.[26]

As Nespoli et al. stated in "New Developments on Mini/Micro Shape Memory Actuators", page 33:

"Lattice defects, dislocations and nano-scaled precipitates... cause significant changes in transformation temperature and in the capability of the material to recover a deformation."[27]

Why does nitinol move?

This is an interesting question for the scientists. What we do know so far is that practically all metals and alloys are good conductors of heat and electricity. This is due to the free movement of electrons flowing through the crystalline structure, usually without any displacement of its atoms. Nitinol, on the other hand, behaves differently from any ordinary metal or alloy in that the atoms do move, and quite significantly, when electrons are flowing. Even when subjecting a piece of nitinol in the Miracle Spoon to an electrical current from a 9-volt battery[28] (more than enough energy to energize the alloy), one can immediately and unmistakable observe the shape-memory effect taking place between the negative and positive wires supplying the electrical current to nitinol.

Another curious observation is the way nitinol must have a crystalline structure and be near-to-perfect in the crystalline shape for this shape-memory effect to occur with *extreme* efficiency. In fact, if you try to break up the crystalline structure and randomise the position of the atoms of nitinol in some way into what is known as the *amorphous state*, the electrons stop flowing and there is absolutely no shape-memory response no matter how much heat or electrical current is applied.

The term *amorphous* simply means a random positioning of the atoms, thereby breaking any chemical bonds that may exist between the atoms. A classic example of an amorphous state is when a metal or alloy is melted at high temperatures into a liquid. What is not commonly known is that the same metals and alloys can be rendered into the amorphous state while remaining in the solid phase. One technique scientists use to render an alloy into the amorphous state is rapidly cooling to the point where there is little time for the atoms to get into a crystalline structure.

Another technique discovered by Hirotaro Mori and Hiroshi Fujita can be seen in U.S. Patent No. 4,564,395. In this invention, an amorphous metal can be formed by the simple process of "irradiating a metal with an electron beam having an energy large enough to damage the metal thereby introducing a lattice defect into the metal and controlling the concentration of the introduced lattice defect"[29]. The method is used mainly on the surface of an intermetallic compound such as NiTi, Zr_2Al, CuTi, FeZr etc (the chief examples mentioned in this patent).

Why render a metal or alloy into the amorphous state? Apart from understanding how certain alloys work, such as to gain a greater insight into the inner workings of the shape-memory effect of nitinol, it also has another unexpected benefit. That is, the amorphous phase can significantly lower the temperature of metals and alloys for fashioning into any shape and so reduce the costs of manufacturing sophisticated metal products.

You see, amorphous metals and alloys behave very much like glass, such as exhibiting a dramatic increase in hardness in the cool amorphous state (and hence a superior strength-to-weight ratio), and can often be as brittle as glass when stressed (in other examples, the brittleness is non-existent and so allowing an extremely hard and highly durable material to be produced[30]). But when an amorphous metal or alloy is heated slightly it will become soft and pliable like hot glass, allowing anyone to create sophisticated shapes with relative ease. This is why some U.S. companies have decided to name this type of material *LiquidMetal*.

Once the amorphous metal is moulded into shape, it becomes a simple matter of applying other techniques to help re-crystallise the metal atoms and establish the lattice framework for holding the atoms in a tight-packed structure in order to reveal the properties of the atoms to their full potential. Those techniques include annealing (i.e. raising the metal to a higher temperature and letting it cool slowly), stressing (i.e. cold-working the metal or alloy), or applying lasers (which has the benefit of selecting specific areas where you want the crystallization to occur).

In the case of nitinol, we know it does render into the amorphous state through rapid cooling. With regards to its shape-memory effect, scientists have shown that crystallization of the alloy is absolutely critical for the effect to work. If the alloy is in the amorphous state, the effect will not work. Chinese scientists F. F. Gong, H. M. Shen and Y. N Wang, from the National Laboratory of Solid State Microstructures in

Nanjing University confirmed this observation in their article "Fabrication and Characterisation of Sputtered Ni-rich NiTi Thin Films" on page 405:

"Amorphous NiTi thin films do not show the shape-memory effect."[31]

Further evidence also comes from the PowerPoint presentation prepared by Andrew J. Birnbaum PhD, Ui Jin Chung, a research scientist, and advisors Y. Lawrence Yao and James Im. Originally available online in February 2009, the presentation titled "Selective Laser Crystallization of Thin Film NiTi Shape Memory Alloys" stated:

"Shape memory responses stem from crystallographic shifts upon transformation. Thus, amorphous material systems do not exhibit shape-memory properties."

Setting up a crystalline lattice structure in nitinol is certainly one aspect in controlling the movement of the atoms, and hence the shape-memory effect. The other has to do with the flow of electrons. If there is absolutely no flow of electrons, the shape-memory response is totally eliminated.

Measurements on the electrical resistivity of amorphous nitinol by P. Lindqvist, A. Kempf and G. Fritsch working in Germany and France in 1991 revealed some interesting information in support of this "electron flowing" idea. In the article titled "Electronic Transport Properties of Amorphous NiTi Alloys", scientists noticed how an amorphous nitinol increases electrical resistivity, which is another way of saying it reduces electrical conductivity. Therefore, should the electrical conductivity go down, it will make it harder for an electrical current (i.e. a flow of electrons) to pass through the alloy and activate the shape-memory response.

In addition to this "electrons flow" idea, there is also a view that the number of freely available *valence electrons* in the outermost orbits play an important role as well.

Valence electrons are those that are able to be released with the least amount of energy (i.e., at the right temperature) by the metal atoms for moving around the crystalline lattice. They also play a crucial role in the formation of chemical bonds for holding atoms together in a crystalline structure, and also control things like the physical size of the

174

atoms (also known as the atomic radii of atoms). As Zarinejad & Liu stated:

> "It is known that in metallic bonding the valence electrons act like 'glue', bonding non-valence electrons and nuclei units together; whereas non-valence electrons contribute to the total atomic volume of the alloy."[32]

For example, in the equiatomic NiTi alloy, we know titanium has 22 protons and nickel has 28 protons in the nuclei. As protons are positively charged, this means that titanium needs 22 electrons and nickel needs 28 electrons to balance out the charge. However, the electrons are not permitted to fall into the nuclei to balance this positive charge out. The nuclei is also undergoing rapid spin and in the process emits a continuous and intense wave of electromagnetic energy to the surroundings that prevent the electrons from falling in or hanging around any position around the nuclei. Rather, the electrons are organised into specific orbits[33] (or shells) according to the following electronic configuration:

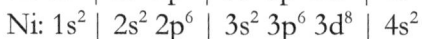

Ti: $1s^2$ | $2s^2 2p^6$ | $3s^2 3p^6 3d^2$ | $4s^2$
Ni: $1s^2$ | $2s^2 2p^6$ | $3s^2 3p^6 3d^8$ | $4s^2$

Thus we have four principal shells. Within each shell, there are subshells called s, p, d and f. Fortunately, there are no f subshells to worry about (this would appear in the fourth shell), so the most subshells we will find in Ni and Ti is in the third shell out from the nuclei where it contains three subshells. Also, each shell and subshell can hold only a certain number of electrons according to some rules in quantum mechanics. The number of electrons that can be held are shown below:

Energy Level	Subshells	Number of Electrons
1 (K-shell)	s	2
2 (L-shell)	s, p	8 (2 in s, 6 in p)
3 (M-shell)	s, p, d	18 (2 in s, 6 in p, 10 in d)
4 (N-shell)	s, p, d, f	32 (2 in s, 6 in p, 10 in d, 14 in f)

In addition, the further we move away from the nuclei, the easier it is to remove electrons. In the above electronic configuration for Ti and Ni, this means the two electrons (indicated by the superscript) in the s subshell of the fourth shell (or $4s^2$) is the outermost orbit. These require the least amount of energy to remove and hence are the ones that get released first.

However, there is another peculiarity of transition metals, in which titanium and nickel are a part of this group. It concerns how easily the electrons in the d subshell can be removed. More specifically, the amount of energy needed to remove the electrons from the d subshell has been found by direct observation to be not as great as originally thought. This strange property only occurs with transition metals and only when the d subshell is not properly filled up with the right numbers of electrons in order to attain the necessary stability after which it becomes much harder to remove the electrons. But while it is incomplete, electrons in the d subshell behave very much like valence electrons.

With this in mind, the outermost electrons that are most likely to get released and move about within the crystal lattice, or become shared with other atoms to form the essential chemical bonds, are those in the outermost 3d and 4s orbits. Therefore, the total number of valence electrons is:

Ti: 2 + 2 = 4 electrons
Ni: 2 + 8 = 10 electrons

Knowing these valence electrons are likely to play an important role in the shape-memory effect and other physical properties of SMAs (since these are the ones to flow through the crystalline lattice and form the chemical bonds between metal atoms), scientists like to know if there are similar patterns to be observed across different SMAs. The best tool for this is to use basic mathematics to calculate the average electrons per atom of the alloy. In the case of NiTi, this is (4 + 10)/2 = 7.

When this formula is applied to ternary (i.e., NiTi-X) and quaternary (i.e., NiTi-X-Y) SMAs, the ratio stays within a remarkably narrow range of between 5.86 and 7.60. This observation tells us that a large number of valence electrons are present in the elements making up the SMAs.

More importantly, they can be dislodged from the outermost orbits with relative ease to flow through the crystalline structure at the right temperature. With this in mind, it seems these electrons, when they are flowing through the crystalline structure in the high-temperature state, are performing some kind of integral activity in generating the shape-memory response. On the other hand, in the low-temperature state, these valence electrons quickly settle down and allow themselves to be shared to form the chemical bonds linking the metal atoms (and with it the familiar crystalline structure for creating a solid object) with the aim of ensuring the positive charge of the nuclei is properly neutralized as best they can.

Alloy (at.%)	Average electrons per atom
$Ni_{25}Ti_{50}Cu_{25}$	7.25
$Ni_{50}Ti_{45}Ta_5$	7.05
$Ni_{50}Ti_{47}Ta_3$	7.03
$Ni_{49.8}Ti_{40.7}Hf_{9.5}$	6.99
$Ni_{49.5}Ti_{40.5}Zr_{10}$	6.97
$Ni_{49.5}Ti_{35.5}Hf_{7.5}Zr_{7.5}$	6.97
$Ni_{49.8}Ti_{35.2}Hf_{15}$	6.99
$Ni_{49.5}Ti_{30.5}Zr_{20}$	6.99
$Ni_{49.8}Ti_{30.2}Hf_{20}$	6.97
$Ni_{110}Ti_{50}Pd_{40}$	7.00

Examples of NiTi-X and NiTi-X-Y SMA alloys with an average electrons per atom very close to or equal 7.00. (Source: Zarinejad & Liu 2008)

In the case of NiTi, there is a complete overlapping of the outermost electrons between the Ni and Ti atoms in the ground state (i.e., at low temperatures). This is ideal when neutralizing the positive nuclei of the metal atoms.

However, scientists have also noticed another interesting pattern through the ratio of average valence electrons per atom. It has to do with the way the number of valence electrons seems to affect the martensite (M_s) and austenite (A_s) start temperatures of those SMAs whose ratio is not precisely 7. In particular, should the average valence electrons per atom be less than 6.8, the M_s and A_s temperatures decrease with an increase in the ratio (but still below the 6.8 mark). Should the figure be above 7.2, there is also a tendency to slightly decrease the M_s and A_s temperatures with increasing ratio. If the figure hits the 7.0 mark or very close to it, a broad range of M_s and A_s temperatures is observed, covering from ambient temperatures right up to as high as 900°C for M_s and 950°C for A_s. It means that for SMAs with an almost precise ratio of 7, there are definitely other factors affecting the transformation temperatures and the shape-memory response.[34]

In determining what these other factors might be, scientists have uncovered yet another link between binary NiTi, ternary NiTi-X, and quaternary NiTi-X-Y SMAs having the average valence electrons per atom very near to or at 7.0 and the atomic radius. If the ratio is in the range of 6.97 to 7.0 and 20 at.% alloying elements are used, the thermal hysteresis increases with increasing atomic size. As Zarinejad & Liu stated:

> "[A] plot of transformation hysteresis as a function of the atomic size (radius) of the alloying elements in NiTi-based shape memory alloys...[shows that] the transformation hysteresis seems to be influenced by the atomic size of the alloying elements. By increasing the atomic size at almost constant e_v/a [the ratio for calculating average valence electrons per atom] the transformation hysteresis is increased."[35]

Could atomic size be another factor in controlling the shape-memory effect as well as the transformation temperature and hysteresis?

Yet another observation worth mentioning is how the nickel and titanium atoms are arranged in NiTi—it is not the same as in ordinary alloys. A typical alloy such as steel will have the elements of iron and carbon fairly randomly distributed on the crystal lattice. Not so with NiTi. With this and other SMAs, the atoms are highly ordered and in specific locations within the lattice. This ensures an alternating sequence of nickel and titanium atoms always appear throughout the crystal lattice.

And in case it isn't obvious by now, there must be two or more different elements to show this shape-memory effect. In other words, there is no metal in the periodic table that can perform the kind of dramatic shape recovery we see in SMA. It requires an *alloy* to be developed.

So far all this work is indicating just how important the flow of electrons, the presence of a crystalline structure, and having atoms of different elements alternating throughout the lattice really is when activating the shape-memory effect as well as how it controls transformation temperature and hysteresis. Despite all these interesting observations, the precise reason why the atoms of nitinol should actually move when electrons are flowing through a crystalline structure still remains a mystery. Clues are certainly emerging for how transformation temperature and hysteresis is controlled, but we can only speculate on the possible internal workings of the shape-memory response.

For example, it is possible the alloy could be experiencing, on the atomic level, a kind of strong electromagnetic force of attraction between the sea of flowing negatively charged electrons and the positively charged nickel and titanium atoms, which is helping to bring together the metal atoms into a tight-packed structure. Or maybe it is as simple as releasing additional weakly-held electrons in the outermost orbit around the metal atoms at the right temperature, and this is enough to reduce the atoms' overall atomic sizes and so make them more favorable in coming together into a more tight-packed crystalline structure.

Speaking of atomic sizes, a look at the geometry of the unit cell structure in NiTi could be just as important as how big the atoms are in NiTi.

To expand on this idea, think of the tetragonal unit cell for NiTi in the high-temperature state as containing 10 titanium atoms acting as a "cage" for holding in place the 4 nickel atoms in what should be the face-centred positions (please refer to the crystalline structure diagrams shown earlier in this chapter). Now if we don't apply any stress to the structure (such as bending it) but just let it cool, fewer electrons will flow through the crystalline structure. Then the outermost electrons (mainly in the d subshell while the other two electrons in the s subshell are relatively free to move around) simply get shared between the metal atoms to ensure they stay together and in a fairly rigid and solid way. However, if the shared electrons are sluggish in their movements, the chemical bonds holding the atoms together may not be very strong. It means the atoms may have a tendency to move away from each other slightly. Hence the titanium "cage" can naturally expand to form the martensite.

Then the unexpected happens at a critical temperature when one of these nickel atoms in the face-centred position (being slightly smaller than the titanium atoms) can suddenly slip out of the "cage", forcing all other nickel atoms in other "cages" (or unit cells) to do the same right down the line. Which nickel atom in the tetragonal unit cell does the slipping may not be immediately obvious, but it is likely there could already be a subtle slanting in the austenite unit cell structure that scientists cannot detect (i.e., there has been some stress applied to "train" enough of these unit cells to slant in the same direction). At any rate, should the slanting get exaggerated even more as the temperature is reduced, two opposite faces will naturally increase in area slightly and this may be enough to dictate which two nickel atoms will slip out first. At the same time, the direction of the slanting and choice of the two nickel atoms to slip out first and its ability to affect other unit cells "down the line"[36] may also determine the way in which the alloy will naturally bend in the cool state (and hence reveal its two-way shape-memory response).[37]

If this is true, it seems plausible that the shape-memory effect (at least the two-way shape-memory situation in the cool state) is an entirely physical process involving the different atomic sizes and geometry of the unit cell structure, and how the smaller nickel atoms are able to move about within the unit cell at low temperatures and so affect other unit cells down the line.

While this theory does not explain how the flow of electrons affects the movement of atoms at the high-temperature state, it is yet another piece in the puzzle for scientists to consider.

Another theory to make the rounds in the scientific literature can be observed in recent studies attempting to determine the crystalline structures of NiTi-X alloys where X is a third "compatible" element designed not to destroy the shape-memory effect. According to Prof. Mehrdad Zarinejad and Dr. Yong Liu:

> "There is a thought that in shape memory alloys, prior to martensitic transformation a softening of elastic constant occurs with lowering temperature. Limited experimental results, mainly in Cu-based and also in NiTi-based shape memory alloys [where a third or more elements are added to form NiTi-X], suggest that there is a critical value of elastic constant at which transformation takes place which is not sensitive to alloy compositions, and is only slightly dependent on temperature. Based on this, the elastic modulus of the alloy, which is dependent on both composition and temperature is thought to influence the transformation temperatures. However, it is not clear how the addition of different alloying elements to NiTi is linked with the elastic properties and, therefore, the transformation temperatures."[38]

In this alternative theory, certain non-titanium elements may be helping to soften the chemical bonds (or, for those with a propensity to learn new technical terms, a reduction in the *elastic modulus*) of the titanium crystalline "cages", and only gets softer as the temperature goes down. In this way, it becomes easier for the smaller atoms held within or at the faces of the "cages" to move. Any movement may be enough to see enough of these "cages" suddenly change shape and size and with it the ability to remember a certain shape for the alloy. This is an interesting theory that is currently receiving more attention at the present time.

In fact, studying alloys of the form NiTi-X has been a rather useful exercise for helping scientists to unravel the secret inner workings of the shape-memory effect. However, as Zarinejad & Liu have pointed out:

> "Despite numerous studies on modification of the NiTi transformation temperatures by adding ternary or quaternary elements, very little work has been done to understand the fundamental reason for these changes."[39]

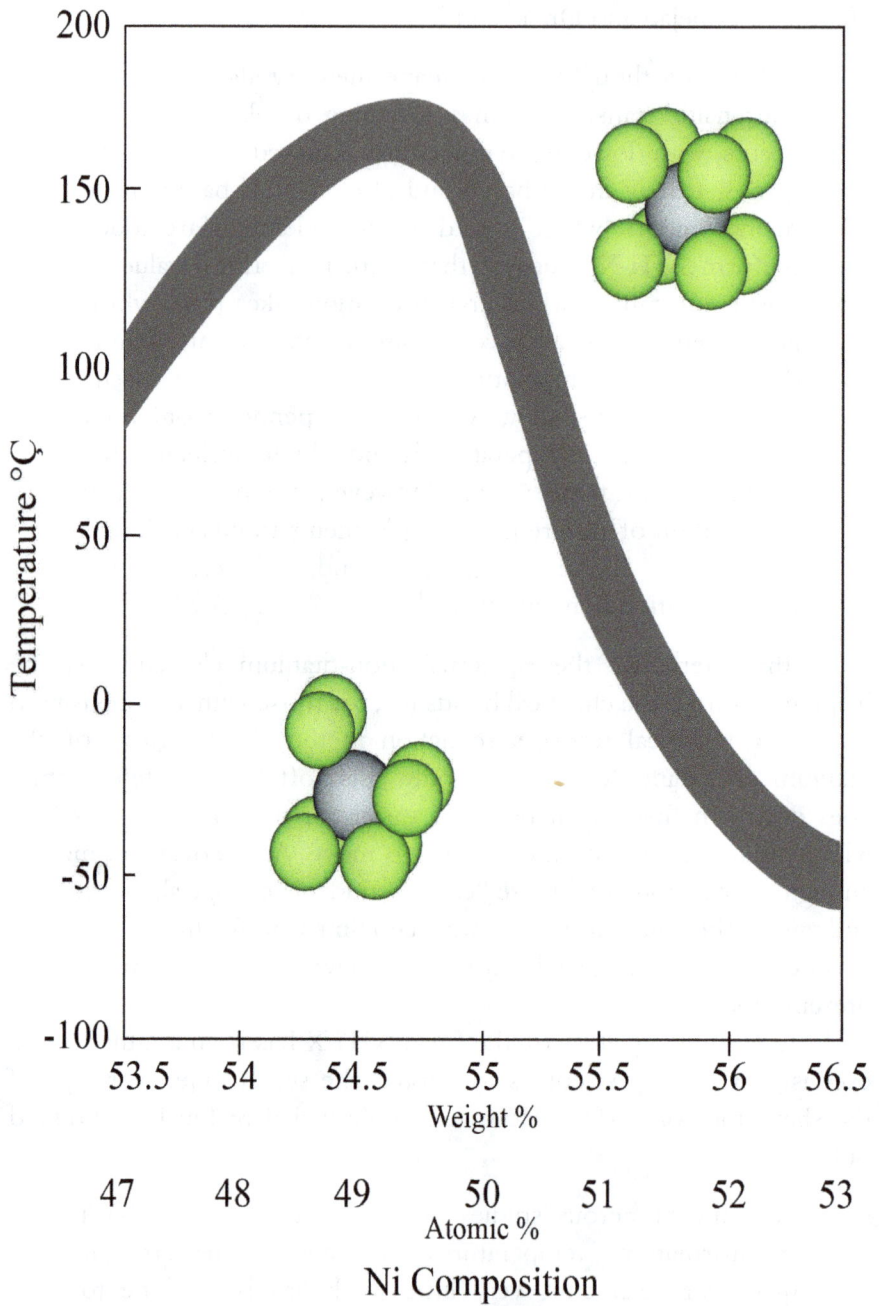

Graph of nickel composition of NiTi to temperature of shape-memory activation.

Since making this statement in 2008 (despite how many years NiTi has been around), scientists still do not know for sure how nitinol, or any SMA, works.[40]

Are there dual crystalline structures above the transformation temperature?

The controversy does not stop there. Debate continues within scientific circles over whether nitinol has a single or dual crystalline structures in the austenite phase (i.e., above the critical martensite transformation temperature). On page 2 of *Nitinol Characterization Study* (NASA CR-1433), William B. Cross et al. wrote:

> "There is much disagreement in the technical literature (ref. 3,4,5,6, and 7) in regard to the phases that exist in the Nitinol alloys from 1200–1300°F down to their transition temperatures (M_s). Some investigators believe that the alloy in this temperature range consists of an ordered single-phase NiTi matrix that has a complex structure with a 9Å repeat distance; others state that it is a single-phase (NiTi) ordered CsCl-type phase with a lattice parameter of about 3Å; while still others believe that two phases are present. One of the two phases is said to be NiTi with an ordered CsCl-type structure, and the other is either Ni_3Ti (for alloys that are slightly rich in nickel) or Ti_2Ni (in the case of alloys that are slightly rich in titanium)."

More work is continuing to this day to help elaborate on the exact crystalline structures of nitinol and how the shape-memory effect actually works.

Nitinol remembers two distinct shapes

Another peculiarity of nitinol we have already touched upon is its ability to remember two distinct shapes both in the low- and high-temperature states. That is, it develops a *two-way* memory response. If the process is repeated, say, about 20 to 30 times, the alloy will fully remember the shapes required in both the high- and low-temperature states. Keep repeating the process and the alloy gets uncannily effective in transforming between the two different shapes. The likely reason for this is a combination of the alloy's increasing hardness over time as the atoms get progressively closer together to form a compact crystalline structure, and how the electrons seem to flow more efficiently along a particular direction in the crystalline structure to ensure easy switching between the two shapes.

Controlling the temperature of nitinol's shape-memory response

In addition, the temperature for activating the shape-memory effect (known as the *transition* or *transformation temperature*) depends significantly on the percentage of titanium and nickel used to produce nitinol, and its purity. As confirmed by Holec et al.:

> "Temperatures of martensitic phase transitions (PTTs) are strongly dependent on the alloy composition and the impurity content."[41]

To be more precise, the transformation temperature, and how sensitive and quick the transition is from the martensite to the austenite phase when activating the shape-memory response, is controlled by three factors:

1. The level of impurities added to the crystalline lattice structure;
2. The amount of microdefects in the crystalline lattice structure; and
3. The composition of nickel and titanium.

In the case of increasing the purity and minimizing the crystalline microdefects of nitinol's lattice structure, this has the effect of reducing the difference in the start and end transformation temperatures (i.e., reducing the thickness of the line shown in the temperature graph, and hence its hysteresis). Generally, the fewer impurities and microdefects there are in the crystalline structure, the more responsive and dramatic is the shape-memory effect.

As for the actual transformation temperature, this can be controlled by the composition ratio of nickel and titanium. By adjusting the ratio, the shape-memory effect can be activated in the range from +160°C (approximately 54.8 weight % Ni or 49.5 atomic% Ni) to -50°C (56.5 weight % Ni or 53 atomic% Ni). While the equiatomic range does reveal some differences in the transformation temperature pattern, generally speaking, the more nickel there is in the alloy, the lower the transformation temperature.

As Fernandes et al. noted in the work by F. M. Sanchez and G. Pulos in 2008:

> "The composition of NiTi alloys induces the range of transition temperature and is responsible for variability in the number of electrons available for bonding. A very small excess nickel in structure can reduce TTR [transformation temperature range] and increase the permanent yield strength of the austenite phase by roughly threefold....Meanwhile, titanium-rich alloys contain a second phase Ti_2Ni in the matrix and have higher transformation temperatures than those of the nickel-rich or equiatomic NiTi alloys."[42]

So if the transformation temperature of a non-electrically charged nitinol is required to show a highly responsive "shape-memory" effect at room temperature of 25°C or below, a sample of highly pure nitinol with a composition of approximately 55.6 wt.% Ni (or 51.1 at.% Ni) or higher would be sufficient (as can be observed from the temperature graph).[43]

Phase 7: Superelasticity

Another curious property of NiTi is the existence of a *superelastic phase* observed above the A_f temperature with a superelastic window of roughly 0 to 40°C (occurs at $Ti\text{-}Ni_{50.8\ at.\%}$) in width, and is most dramatic when the alloy is machined into a thin wire or paper-thin sheet. When the alloy is in the superelastic phase (i.e., at the right composition and temperature), it can be easily deformed (i.e., more than 180°) and will spontaneously return to its original shape in a highly elastic manner as soon as the force to deform the alloy is removed. However, go slightly above this superelastic temperature range and the alloy loses this exaggerated elasticity and behaves more like stiff plastic or very hard armour in the austenite phase, making it extremely difficult to bend.

The term *superelastic* means a highly flexible phase where nearly all of the martensite has transformed into the austenite, but not completely due to the temperature being just low enough. It is at this point, just above the transformation temperature and below the temperature at which no martensite is present, where the alloy can bend easily and significantly through a stress-induced loading force, causing most of the austenite phase to transform into the martensite phase. As soon as the stress is removed, the alloy automatically transforms back to the near-complete austenite phase and with it the original shape of the alloy.

For another definition, try this one from Arkaprabha Sengupta and Panayiotis Papadopoulos:

> "[Superelasticity] is the capacity to sustain large deformations that can be completely reversed upon removal of the stress, and is due to a stress-induced solid-solid phase transformation at constant temperature."[44]

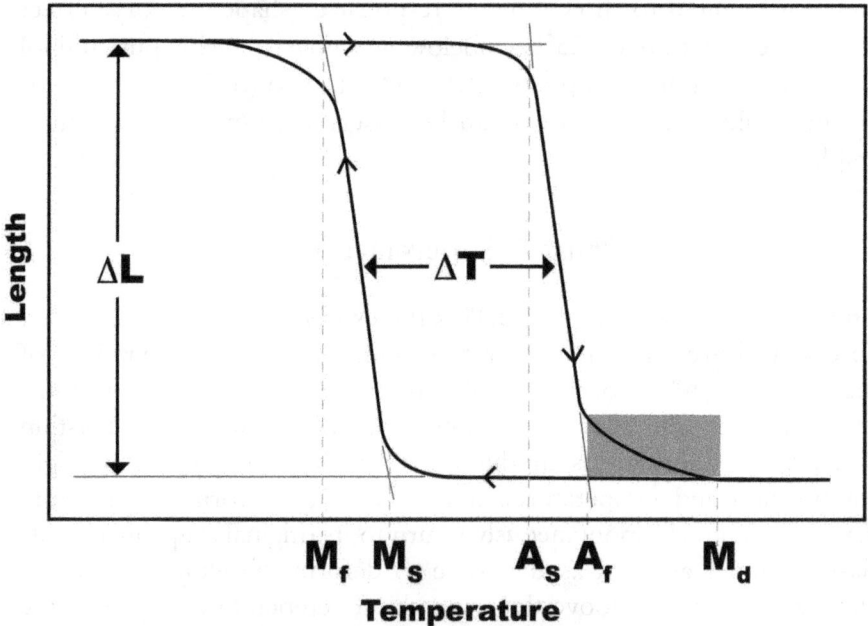

The region marked in grey is where superelasticity is most prominent.

To better understand where this narrow temperature range for superelasticity exists, if you refer back to the graph of uniaxial

186

expansion for NiTi, the transformation temperature is defined as M_s. The temperature range for superelasticity is, therefore, between A_f and M_d.

François Cardarelli gives support for this temperature range in his 2008 edition of *Materials Handbook: A Concise Desktop Reference*, in which he wrote on page 140:

> "Superelastic NiTiNOL can be strained several times more than ordinary metal alloys without being plastically deformed, which reflects its rubberlike behavior. This is, however, only observed over a specific temperature area. The highest temperature at which martensite can no longer be stress-induced is denoted by M_d. Above M_d, NiTiNOL is deformed like ordinary materials by slipping. Below A_s, the material is martensitic and does not recover. Thus, superelasticity appears in a temperature range from near A_f up to M_d. The largest ability to recover occurs close to A_f."

It should be noted that just because pure NiTi has a strong shape-memory effect, it does not necessarily imply it will display a superelastic effect. There is also a critical composition range for NiTi to reveal this superelastic effect. According to Michael Horzewski and Jeffrey Giba, they stated:

> "In general, binary compositions of Nickel (Ni) and Titanium (Ti), yield alloys with shape memory and superelastic properties....Their precise physical and other properties of interest are extremely sensitive to the precise Ni/Ti ratio used. Generally, alloys with 49.0 to 50.7 atomic % of Ti are commercially available, with superelastic alloys in the range of 49.0 to 49.4% , and shape memory alloys in the range of 49.7 to 50.7%."[45]

Thus the range has to be within:

$$Ni\text{-}Ti_{49.0\text{-}49.4\ at.\%}$$

or

$$Ti\text{-}Ni_{50.6\text{-}51.0\ at.\%}$$

Stay within this composition range and just above the transformation temperature and the elastic range of 10 to almost 40 times greater than that of a steel spring will not be uncommon for NiTi. The higher end of this elastic range is usually achievable when NiTi is at its highest purity and has virtually no imperfections in the crystalline structure. Move slightly outside this superelastic temperature or composition range and superelasticity is greatly diminished or non-existent.

In addition, the superelastic range is said to be between -20°C (at Ni-Ti$_{49.0 \text{ at.\%}}$) and 60°C (at Ni-Ti$_{49.4 \text{ at.\%}}$), with body temperature (i.e., 37°C) located at Ni-Ti$_{49.2 \text{ at.\%}}$. Very little additional preparation is required to be done to the alloy to reveal this superelastic phase other than to have the alloy "...cold worked 40% and annealed at 500°C (930°F) for 2 min"[46].

If there is any problem with making superelastic NiTi, it would have to be in the composition range. According to Duerig and Pelton, they stated:

> "Binary [NiTi] alloys with less than 49.4 at.% titanium are generally unstable. Ductility drops rapidly as nickel is increased."[47]

To counteract this situation, Duerig and Pelton recommended the following:

> "Aging of nickel-rich alloys increases austenitic strength to a typical peak strength of 800MPa (116 ksi). Surprisingly, ductility is also increased during the aging process."[48]

Otherwise, the only alternative option is to add a ternary element to NiTi to improve the ductility and keep it stable.

Duerig and Pelton further go on to say that while the superelastic composition range is in the region of 49.0 to 49.4 at.% titanium, they did add that shape memory alloys start at between 49.7 to 50.7 at.%. As they stated:

> "Titanium-nickel is extremely sensitive to the precise titanium/nickel ratio. Generally, alloys with 49.0 to 50.7 at. % titanium are commercially common, with superelastic

alloys in the range of 49.0 to 49.4 at.% and shape memory alloys in the range of 49.7 to 50.7 at.%."[49]

This leaves us with a narrow window of opportunity to create a more stable NiTi and still exhibit some kind of a superelastic response in the composition range of 49.41 to 49.69 at.% titanium, so long as there are ways to enhance and expand the superelastic temperature range.

In fact, it is not surprising for scientists to consider adding a ternary element to the binary NiTi alloy system to see if the superelastic phase can be enhanced and the temperature range extended. For example, Robert M. Abrams and Sepehr Fariabi of Abbott Cardiovascular Systems Inc have discovered an Ni-Ti-X alloy having the composition range:

$$Ti_{30 \text{ to } 52 \text{ at.\%}}\text{-}Ni_{38 \text{ to } 52 \text{ at.\%}}\text{-}X$$

where X is added in small quantities and consists of one or more of the following elements:

iron, cobalt, platinum, palladium, vanadium, copper, zirconium, hafnium and niobium

will not only enhance this superelastic phase, but also give it a wider superelastic temperature range. As a result of this discovery, a U.S. patent was approved in 2007.[50]

So far, the main uses for superelasticity in NiTi have been in the field of medicine. Here, thin superelastic nitinol wires can be packaged into a small size and delivered to specific parts of the body. Once there they are unpackaged to allow nitinol to do its job, such as unblocking blood vessels.

The power of NiTi-X alloys

The usefulness of NiTi-X alloys where X is another element added in small quantities without destroying the shape-memory effect can never be underestimated. Apart from learning more about how the shape-memory effect works in NiTi and other SMAs, it can help engineers to custom design new SMAs to do things that are not possible with the

189

standard binary NiTi alloy system. In particular, things like changing the transformation temperature to outside the standard -50°C to 160°C range as well as dramatically reducing or increasing the width of the hysteresis are all possible thanks to the physical properties other elements can bring to the table. Or what about enhancing the superelastic phase of NiTi?

The importance of adding a ternary element to the standard NiTi alloy system can be better appreciated through the following quote from Horzewski and Giba:

> "The addition of small amounts of third elements in the [NiTi] alloy can also have very significant effects on performance of the materials. Elements including, but not limited to, oxygen (O), nitrogen (N), iron (Fe), aluminum (Al), chromium (Cr), cobalt (Co) vanadium (V), zirconium (Zr) and copper (Cu), though having various effects on the Ni-Ti matrix, can have the tendency to increase strength, increase stiffness, control hysteresis and/or decrease or increase phase transition temperatures."

Reducing the width of the hysteresis

To give an example, scientists understand a standard binary NiTi alloy to have a hysteresis ranging from 20 to 50°C. The variability in these figures is more a reflection of how highly dependent the hysteresis is on composition, purity and the level of imperfections in the crystalline structure. For example, very high purity and lack of imperfections in the lattices of NiTi can see the width of the hysteresis reduce quite dramatically. However, if the idea of increasing the purity and removing all imperfections in the NiTi lattice seems a little too daunting with modern manufacturing techniques, Duerig and Pelton have recommended adding small amounts of either niobium (Nb) to increase hysteresis (over 100°C), or copper (Cu) to reduce hysteresis (around 5°C).[51] In the latter case, Liangliang et al. stated in their 2015 article, "Effect of Cu Content on Atomic Positions of $Ti_{50}Ni_{50-x}Cu_x$ Shape Memory Alloys Based on Density Functional Theory Calculations":

> "...the transformation temperature hysteresis becomes smaller with increasing Cu content in the TiNiCu alloy....We

also learned that when the Cu content reaches 20 at. %...the transformation hysteresis is reduced to as low as 5 K."[52]

Otsuka and Ren gave an explanation for why the hysteresis is small:

"The addition of Cu...[shows] the transformation type changes with increasing Cu content; Cu ≤ 7.5: B2→B19', 7.5 ≤ Cu ≤ 15: B2→B19→B19', Cu ≥ 15: B2→B19 only. The B2→B19 transformation is associated with a smaller temperature hysteresis than that for B2→B19', and the transformation hysteresis becomes smaller with increasing Cu content."[53]

In other words the energy needed to move the atoms from the tetragonal/cubic B2 to the orthorhombic B19 is considerably less compared to the monoclinic B19'. This is because there are additional frictional forces one has to overcome when moving enough atoms around to attain the more complex B19' structure. Generally, the less energy required to make the phase transformation, the smaller the hysteresis.

A similar reduction in the hysteresis has also been reported with additions of palladium (Pd). As K. V. Ramaiah et al. noticed in the high transformation temperature alloy $Ni_{49.7-x}Ti_{50.3}Pd_x$ (where x is 25 at.%) of 181°C, a narrow thermal hysteresis of around 8.5°C could be attained.[54]

Increasing the width of the hysteresis

Or to widen the hysteresis, we have already heard how the addition of small amounts of niobium to the NiTi alloy system can easily achieve this. As confirmed by Melton et al.:

"...the effect of Nb additions to Ni-Ti are reported, and it is shown that very wide hysteresis alloys can be produced."[55]

The alloy used in this study was Ti_{44}-Ni_{47}-Nb_9 at.%. In fact, the recommended composition ratio needed to maintain the shape-memory effect and the wide hysteresis should be of the form:

For another example, a small amount of hafnium performs a similar widening effect on the hysteresis of NiTi according to recent work done by NASA.

But perhaps what is not often realized in these discussions is how minor changes to the composition ratio of the standard NiTi binary alloy can have a rather significant impact on the width of the hysteresis. As Sehitoglu et al. noted in their 2004 article, "Hysteresis in NiTi Alloys":

> "In the low Ni alloy the thermal hysteresis expanded with increasing stress while in the high Ni alloy the thermal hysteresis contracted with increasing stress....The results are reported for 50.1%Ni (at.) and 51.5%Ni NiTi alloys in the aged conditions. We make the important observation that the thermal hysteresis grows with increasing stress in the 50.1%Ni case, while it contracts in the 51.5%Ni case. Although we report only on thermal induced transformations in this paper, an equivalent behavior has been observed for the stress-induced transformation under constant temperatures."[56]

The last sentence in this quote is thought-provoking, given that it implies that the superelastic range can be widened or narrowed with just minor variations in the composition of nickel in the NiTi system. Furthermore, this widening of the superelastic range seems to go hand-in-hand with a widening of the hysteresis. Not only that, but the amount of stress applied to NiTi can also affect the width of the hysteresis and with it the superelastic range as well. As Sehitoglu et al. stated:

> "With increasing applied stress levels the temperature hysteresis grows drastically for the 50.1%Ni case before reaching steady state levels. For the 51.5%Ni material the thermal hysteresis narrows over a wide stress range."[57]

Assuming there is a link between the width of the hysteresis and the superelastic range, then for the superelastic alloy:

it may be possible to widen the superelastic range by the mere application of stress to the alloy so long as the nickel composition is around the low range of 50.6 at.%. However, raise the nickel composition slightly (probably to around 51.0 at.%) and it is likely the application of any stress-induced behavior will reduce the superelastic range.

Or is it closer to the truth to say that anything above 51.0 at.% (which is outside the superelastic range) is when the hysteresis starts to contract and anything between 50.6 and 51.0 at.% (which is the superelastic range) is where the hysteresis naturally widens when applying the stress and so helps to enhance superelasticity? Clearly further work is needed here to determine the nickel composition that will extend the superelastic range for the standard NiTi binary alloy system, and by how much.

But why would a subtle composition change in NiTi make such a difference?

One theory is because when NiTi widens the superelastic range, it is, in fact, developing its R-phase. This is the transitional crystalline structure that makes it easier to move between the austenite and martensite crystalline structures. Is this true? According to Brachet et al., they stated:

> "It is well known that superelasticity properties are enhanced by the occurrence of premartensitic R-phase transition."[58]

Of course, *Wikipedia* will claim the R-phase on its own is not responsible for the superelastic effect:

> "The R-phase is a phase found in nitinol. It is a martensitic phase in nature, but is not the martensite that is responsible for the shape memory and superelastic effect."[59]

This is probably because the R-phase on its own does nothing to promote superelasticity. Rather, it is the transformation from martensite to the R-phase that is likely to be what defines superelasticity. This seems to be supported by Duerig & Bhattacharya when they stated in 2015:

"...it is the M $\Rightarrow$ R [Martensite $\Rightarrow$ R-Phase] peak that defines superelasticity."

Another thing we are told is important for developing the superelastic phase in NiTi is to cold work them and perform the subsequent annealing at 350 to 500°C. Well, this happens to be precisely what is required to develop the R-phase. Now it is looking like there is a link between the R-phase and the presence of superelasticity. Just so long as there is a martensite phase as well, superelasticity will always be revealed too.

However, Brachet et al. have also discovered much to their delight that it is possible to bring out this superelastic effect without cold work or performing any kind of annealing to NiTi. All one has to do is, not surprisingly, add a ternary element to NiTi to make this superelasticity happen. For example, after studying the ternary alloy $Ni_{48}Ti_{50}$-Fe_2, they stated:

> "These results show that it is possible to obtain superelastic properties at temperatures below 0°C on SMA of TiNiFe type, without cold-working and the subsequent annealing at 350–500°C as it is necessary for binary TiNi alloy."

It is clear we are at the early stages of this kind of work, but we can expect more interesting results on superelasticity to come out in the coming years.

A theory on how other elements affect the hysteresis

Why do certain elements affect the width of the thermal hysteresis? The precise reason is not fully known, but one of the factors that seem to affect this involves the size of the atoms. Basically, the larger one of the atoms in a SMA is, the larger the hysteresis. As Zarinejad & Liu stated:

> "A large atomic size of the additional alloying elements may result in more frictional force thus more energy dissipation leading to increased hysteresis."[60]

In the case of the narrow hysteresis shown by NiTi-Cu, we know copper and nickel atoms do have similar atomic sizes.[61] There seems to be a connection here. As for how the shape-memory effect works and why it would be highly responsive within a narrow hysteresis, this must come down to the specific atomic properties of copper compared to nickel.

Raising the transformation temperature

As mentioned before, slight variations in the composition of NiTi can dramatically raise or lower the transformation temperature. The best that scientists can achieve with the standard binary NiTi alloy system is between -50°C to 160°C. But what if you wanted the transformation temperature to go outside this range? Can this be done? Again, all this can be achieved with the help of adding a ternary element to NiTi. It is all a question of knowing which element to add to NiTi to make this happen.

To give an example of the effect of a ternary element on the transformation temperature of NiTi, Ronald Dean Noebe, Susan L. Draper, Michael V. Nathal, and Edwin A. Crombie from NASA in 2010 patented an NiTi-X alloy with a transformation temperature greater than 160°C. The elements recommended to be added to NiTi to achieve this are platinum (Pt) and one or more of the following elements:

gold (Au), palladium (Pd) and copper (Cu)

to the alloy NiTi-Pt-X according to the following composition formula:

$$(Ni+Pt_{10\text{-}30 \text{ at.}\%}+X_{0\text{-}10 \text{ at.}\%})_y Ti_{(100\text{-}y)}$$

where y is in the range of 49 to 55 at.%.[62]

NASA was not the first to make this discovery.

T. W. Duerig and A. R. Pelton from the Nitinol Development Corporation noticed a similar increasing transformation temperature trend with small additions of palladium (Pd) and platinum (Pt) on NiTi. Although a small amount of Pt and Pd in the order of 5 to 10% tend to decrease M_s, anything above 10% will increase M_s to as high as 350°C.[63] Further details can be found in the 1994 article, "Ti-Ni Shape Memory Alloys" of *Materials Properties Handbook Titanium Alloys*, pages 1035-1048.

NASA's interest in nitinol has gone back a long way, such as this 1972 report.

Here are more examples of high-temperature transformation SMAs[64]:

$Ni_{50-x}Ti_{50}Pd_x$ (where x is between 35 and 50 at.%)
$Ni_{50}Ti_{50-x}Zr_x$ (where x is between 22 and 30 at.%)
$Ni_{50}Ti_{50-x}Hf_x$ (where x is between 20 and 30 at.%)

196

Lowering the transformation temperature

On the other hand, if the aim is to reduce the transformation temperature of NiTi, then small additions of cobalt (Co) to the near stoichiometric TiNi (i.e., the 50:50 composition ratio) via the chemical formula $TiNi_xCo_{1-x}$ (where x is between 0 and 1) will drop the transformation temperature to -238°C and extend the shape-memory response range to 166°C.[65] This was further supported by Duerig and Pelton in 1994 after observing how small additions of iron (Fe), aluminium (Al), chromium (Cr), cobalt (Co), and vanadium (V) to substitute for nickel tend to sharply reduce M_s. As Duerig and Pelton wrote:

> "Fe, Al, Cr, Co, and V tend to substitute for nickel, but sharply depress M_s (Ref 14 to 16), with V and Co being the weakest suppressants and Cr the strongest. These elements are added to suppress M_s while maintaining stability and ductility. Their practical effect is to stiffen a superelastic alloy, to create a cryogenic shape-memory alloy, or to increase the separation of the R-phase from martensite."[66]

In addition, Nespoli et al. noted in "New Developments on Mini/Micro Shape Memory Actuators", page 35:

> "In particular, NiTiCu, with Cu substituting Ni in the 3-10at% range, or NiTiCo system are used respectively when the application requires narrow thermal hysteresis or high stiffness."[67]

Knowing that the shape-memory effect of NiTi-Co extends into the cryogenic temperatures (i.e., below 200°C) as well as having a narrow hysteresis, it can potentially make NiTiCo a particularly useful SMA for aerospace applications.

Adding a ternary element to NiTi to affect the transformation temperature is one thing. The choice of which element in NiTi to replace with the ternary alloy is just as important. For example, if tungsten (W) is used as the ternary element, scientists have made the following observation:

> "Apart from the selection of alloying elements, their replacement of either Ni or Ti, or both is also of

paramount importance. For instance, when W is added to replace Ni, the Ms temperature is above room temperature; whereas when it replaces Ti, the Ms temperature drops to below room temperature."[68]

The results in a nutshell

In summary of these results, Zarinejad and Liu have made a more extensive review of scientific literature to determine which elements can raise or lower the transformation temperature of NiTi without destroying its inherent shape-memory effect. Here is what they had to say:

> "A survey of the literature reveals that the addition of Fe, V, Mn, Co, Cu, Zr, Nb, Mo, Pd, Ag, Hf, Ta, W, Re, Pt, and Au affects the transformation temperatures of NiTi-based alloys. In particular, addition of Fe, V, Mn, Co, Cu and W for Ni, and Re and Cu for Ti, reduces the transformation temperatures; whereas addition of Hf, Zr, Ta and W to replace Ti, or Pd, Pt, and Au to replace Ni, increases the transformation temperatures considerably."[69]

Actually, the idea of adding a third element to the NiTi alloy system is not as silly as it sounds. As Kleinherenbrink and Beyer understood when they stated,

> "...the transformation temperatures are extremely sensitive to thermomechanical treatment and variations in concentrations of the constituting elements and of impurities..."[70]

there is usually a cost involved in "preparing" NiTi alloys to get them to exhibit shape-memory effects at the right transformation temperatures. To give an idea of the work involved in preparing NiTi to a high standard (and so produce reliable samples during scientific testing or for any precise work requirements), Kleinherenbrink and Beyer had this to say:

> "The alloys were melted from pure elements by arc melting under Argon. To ensure homogeneity, the buttons, of approximately 25 grams each, were turned and remelted

several times. Then they were vacuum annealed, 60 to 200 hours at 950°C, for homogenization. For further homogenization and in order to get workable dimensions they were hot worked in vacuum, at 860°C, in order to remove the casting structure. Then they were cold rolled to a diameter of 3.7 mm. After every rolling step the alloys were annealed at 800°C. Before every rolling step the alloys were cooled in liquid nitrogen to ensure the martensitic structure. When the above dimensions are reached, the alloys were annealed during 1 hour at 900°C to define the starting material."[71]

Sometimes, in such circumstances, it is probably better just to add another element to NiTi knowing the transformation temperatures can be attained and with reasonable precision without any further treatment to be done to NiTi-X alloys.

For details of the transformation temperatures achieved by NiTi-X alloys, see Appendix G.

At the present time, scientists are seeking new SMAs to further extend the transformation temperature to well above and below the standard NiTi range, as well as widening and reducing the width of the hysteresis.

Further NiTi observations

Other interesting observations worthy of mentioning include the following:

1. Adding an extra one percent nickel will increase the yield strength of the austenite phase as well as reduce the phase transformation temperature.
2. Continuously cold-working the martensite phase of the alloy (when it is in the pliable martensite state) will harden it and give the austenite phase greater tensile strength and hardness when heated. It will also give the alloy a pronounced two-way shape-memory effect.
3. When switching between two different shapes through the two-way shape-memory effect, the strain limit on the alloy is reduced

and gets harder to bend with each cycle (i.e., martensite ro austenite and back again). This makes for an excellent armour-plating material, but can break if the applied strain exceeds the limit. For example, maximum transformation strain for a freshly made NiTi straight from the oven is 8%, for 100 cycles it is 6%, and for 100,000 cycles it is 4%.

4. To get the alloy to cycle between two shapes in a superelastic way (where the strain limit is much larger) and to reduce the hardening effect, adding a small amount of copper and a dash of cobalt will keep the alloy flexible and able to withstand a much higher number of cycles (exceeding 10 million times) without suffering metal fatigue. This is known as an *ultralow-fatigue SMA*.[72]

5. Oxygen and carbon are contaminants to nitinol. They will dampen the shape-memory effect and reduce mechanical properties (such as strength). These contaminants should be avoided.

6. To reset the parent shape, or original shape, you bend nitinol into the shape you desire, keep it in this shape, and heat the alloy to around 500°C for about 10 minutes. This rearranges the nitinol atoms into a new rigid, compact cubic crystalline structure. Then the alloy is allowed to cool in air or water.

7. Nitinol can be used in medical applications or devices because the body does not reject the material.

8. Nitinol does not corrode and can be fully recycled (meaning it can last forever).

Other SMAs exist

Since the discovery of nitinol, more SMAs have come to light. For example, Marek Novotny and Juha Kilpi stated in their article titled "Shape Memory Alloys (SMA)":

> "Shape Memory Alloys, for example Ag-Cd, Au-Cd, Cu-Al-Ni, Cu-Sn, Cu-Zn-(X), In-Ti, Ni-Al, Ni-Ti, Fe-Pt, Mn-Cu and Fe-Mn-Si alloys, are a group of metallic materials having ability to return to a previously defined shape when subjected to appropriate thermal procedure."

Of all the SMAs discovered so far, only five are two-way SMAs. These are Cu-Al, Cu-Al-Zn, In-Ti, Ti-Ni and Ti-Ni-Cu-Al-Mn, of which just *three* contain titanium. But when it comes to the most powerful shape-memory effect and toughest "aerospace-useful" materials (especially in the superelastic phase), only the titanium alloys containing Ti-Ni in significant quantities possess this feature. Not even the other titanium alloy In-Ti is good enough because of its lower melting point (under 1,000°C) compared to Ti-Ni. If you have to choose the best two-way SMA and the toughest, Ti-Ni (or Ti-Ni-X where X is one or more other elements added in small amounts) is the one to go for.

This means there are three main groups covering the complete range of two-way SMAs:

Cu-X where X=Al, Al-Zn
Ti-X where X=In
Ti-Ni-X where X= no elements, or Cu-Al-Mn

Cu-Al is the lightest of all the two-way SMAs (and the cheapest to make), and Ni-Ti is the lightest titanium-based SMA (and the most expensive to make—unless, of course, you want to include those SMAs with gold as a major constituent). Out of these two lightweight alloys, only Ni-Ti is suitable for aerospace applications as it can resist high temperatures, and possesses other important engineering qualities such as hardness and strength in the austenite phase as needed to build tough flying objects traveling at high speeds through the atmosphere and in space.

Even if we were to include one-way SMAs, Ni-Al is the only other alloy lighter than Ti-Ni, but it isn't a *two-way* SMA. Neither is Ni-Al suitable for aerospace applications due to its low melting point. Therefore, from the available list of known SMAs, only those containing titanium in their crystalline structures are considered useful for aerospace applications. And Ni-Ti is the most powerful titanium-based shape-memory alloy.

As for titanium-based two-way, and standard one-way, SMAs, the main subgroups identified by metallurgists at the present time are:

Ti-In (titanium-indium)
Ti-Zr (titanium-zirconium)
Ti-Pd (titanium-palladium)
Ni-Ti-X (nickel-titanium-various)
Ti-Nb-X (titanium-niobium-various)
Ti-Mo-X (titanium-molybdenum-various)

where X contains a few other elements (or none at all) of which Ti-In, Ti-Ni and Ti-Ni-Cu-Al-Mn are the only titanium-based two-way SMAs we know of today.

When it comes to the lightest titanium-based SMA[73] and the most powerful, only Ni-Ti meets this criterion. Alternatively, the addition of a small amount of another element (probably under 1.5 wt.%) to Ni-Ti can enhance its existing physical properties and/or add additional new properties for just a marginal increase in the weight of this ternary Ni-Ti-X alloy. Otherwise, there is absolutely nothing in the entire SMA family that comes close to showing the kind of striking and pronounced shape-memory effect that nitinol is famous for. And it is also one of the toughest SMA alloys around. Nitinol has been, and still is, the most powerful[74] and lightest titanium-based SMA known to science.

For a more complete (but not exhaustive) range of alloys identified as having this curious shape-memory ability, see Appendix G.[75]

Summary

Here is a summary of the facts as we know them for nitinol and SMA in general:

1. The reason for the high melting point of nitinol is due primarily to the significant amount of titanium. With a melting point for titanium of 1,650°C, adding nickel reduces the melting point to between 1,240°C and 1,310°C in the equiatomic range. The flame of a propane blowtorch burning with oxygen available in air can reach 1,995°C. However, all objects subjected to a blowtorch dissipate heat, which in turn lowers the flames' temperature. Thus a propane blowtorch can just about melt copper at 1084.62°C. But it would not be enough to melt nitinol. The temperature of a

202

blowtorch can be improved in pure oxygen with flame temperatures reaching as high as 2,820°C. But the way nitinol has the eerie ability to dissipate heat so efficiently suggests that even this type of blowtorch would have trouble melting a very pure sample of nitinol.

2.   Nitinol can attain a remarkable hardness through regular cold-working (e.g., bending) of the alloy. The hardness can reach a point where it can resist cutting by scissors and cannot be ripped by hand, as well as resist dents from the heavy blows of a sledgehammer.

3.   Nitinol is the world's lightest titanium-based SMA.

4.   Nitinol is the world's most powerful titanium-based SMA, and the most powerful superelastic alloy.

5.   Nitinol possesses a distinct dark-grey color. No other shape-memory alloy that doesn't have the nickel and titanium combination in its chemical composition in reasonable quantities has this color. It would require some form of surface anodization or adding an oxide layer to create a color similar to nitinol.

6.   Nitinol can duplicate the shape-memory effect and superelasticity observed in the Roswell foil if the composition range is Ti-Ni$_{50.6}$ to 51at.%. The lowest martensite transformation temperature in this range is between 20°C and 25°C. However, any small additions of a third "compatible" element to the NiTi system can easily lower the transformation temperature to well below 0°C (simply adding a bit of chromium or cobalt can bring this down to cryogenic temperatures) and affect the width of the hysteresis in such a way as to expand the temperature range of the superelastic phase.

7.   Only a couple of SMAs contain aluminium—Ni-Al, and Cu-Ni-Al. However, the alloys are not tough enough compared to Ni-Ti to resist the high temperatures of a blowtorch or the cutting by scissors in the "newspaper thin" form. Furthermore, Ni-Al and Cu-Ni-Al are not dark-grey like Ni-Ti.

When considering the Roswell foil, it can be seen to have much in common with nitinol. In terms of color, hardness, high temperature resistance, temperature of shape-memory activation, and extremely lightweight, it is hard to go past nitinol. In fact, nitinol (or possibly

NiTiCo or NiTiCr) remains the most likely candidate as the metal the Roswell witnesses saw in the mysterious foil in early July 1947.

But is it the alloy observed by the witnesses? And did the USAF actually make nitinol or something close to it for whatever secret military experiment they had allegedly conducted?

To answer this question, we will need to look at the history of nitinol and see at what point the USAF may have begun their investigations into the alloy.

CHAPTER 7

The Official Scientific History of Nitinol

I S NITINOL the world's first shape-memory alloy (SMA)?

The following abstract, taken from *An Encyclopedia of Metallurgy and Materials,* does gives this impression[1]:

> **Shape memory alloys.** (Also termed "Marmem", from Martensitic Memory). These alloys, if deformed at a temperature below a particular transformation temperature, will revert to the original shape when heated above that transformation temperature. A martensitic type of tansformation is involved, the crystals having a higher symmetry above the transformation temperature than below it. The deformation is accompanied by internal twinning, which returns to the single orientation higher symmetry structure on heating above the transformation temperature.
>
> Originally discovered in the near equi-atomic nickel titanium alloys (nitinols), the effect has been found also in Au/Cd, Fe/Pt, Cu/Zn, Cu/Al/Ni and In/T alloys. Variations in the composition of Nitinol alloys have produced alloys with transformation temperatures in the range − 190 °C to + 100 °C.

What the abstract doesn't tell you, however, is that nitinol is the world's first SMA to be studied by scientists for its shape-recovery effect, but it isn't the world's first SMA. As we shall see in this chapter, there are other alloys that have been observed to have this curious shape-memory effect.

This is important because it might open the door of opportunity for others like the USAF to claim they had already known about SMAs and had been studying them before July 1947 and, therefore, could have developed and used a SMA in a secret military experiment.

Let us take a closer look at the official history of SMAs from a scientific perspective based on available scientific literature to see how likely this is.

The world's first SMA

After consulting the metal journals from the Research School of Physical Sciences at the Australian National University (ANU), it is clear that the world's first SMA is the gold-cadmium (AuCd) system[2]. This alloy was first produced in 1932 by Swedish physicist Arne Ölander.

In *Chemical Abstracts*, Volume 26 (1932), Number 2, page 5467, we find information concerning the Swedish physicist and his work on AuCd. A copy of the abstract is shown below:

> **The crystal structure of AuCd.** ARNE ÖLANDER. *Z. Krist.* **83**, 145-8(1932).—In the system Au-Cd there is an α-phase around 50 at. %. Electrode potential measurements indicate a transition point for this phase at 207°. Alloys contg. 47.5 at.% Cd annealed in H at 220° and 430°, resp., for 40 hrs. and quenched gave the same x-ray pattern, corresponding to a unit cell with a = 3.144, b = 4.851 and c = 4.745 A. U., contg. 4 atoms. The space group is V^5_h. Cd atoms are at 000 and 0 ⅝ ½, Au at ½ ½ 0, ½ ⅛ ½. This is a slightly deformed face-centered structure and also corresponds to a distorted CsCl lattice. A photograph at 400° gave the CsCl structure with a = 3.34 A. U.   L. S. RAMSDELL

On closer inspection, there are no obvious indications of a shape-memory effect. However, other sources obtained online suggest that Ölander did notice what is today described as a *pseudoelastic effect* (or in its most broadest terms, a *shape-memory effect*) A document titled "Biocompatibility Evaluation of Nickel-Titanium Shape Memory Metal Alloy", published on May 7, 1999 by Orma Ryhänen at the University of Oulu, stated:

"Ölander discovered the pseudoelastic behavior of the Au-Cd alloy in 1932."

Does this mean pseudoelastic and shape-memory behaviors are the same? According to Wikipedia (not always the most credible source, but the information provided is reasonable on this occasion):

"Pseudoelasticity, sometimes called superelasticity, is an elastic (reversible) response to an applied stress, caused by a phase transformation between the austenitic and martensitic phases of a crystal. It is exhibited in shape-memory alloys."[3]

This strongly suggests the two terms are the same. In fact, one could even bundle in *superelasticity* into the mix and assume all the terms are the same according to the above quote, except it isn't clear what is the difference between "pseudo" and "super" when dealing with an elastic response. Why are the words used so interchangeably in scientific discussions about the shape-memory effect and SMAs in general? If there is meant to be different strengths in the electricity of alloys, why don't we call them weak, moderate, and strong shape-memory effects?

There is also further variation on this definition. For example, ChemEurope.com defines the two terms as follows:

"A shape memory alloy (SMA, also known as a smart alloy, memory metal, or muscle wire) is an alloy that "remembers" its shape. After a sample of SMA has been deformed from its original crystallographic configuration, it regains its original geometry by itself during heating (one-way effect) or, at higher ambient temperatures, simply during unloading (pseudo-elasticity or superelasticity). These extraordinary properties are due to a temperature-dependent martensitic phase transformation from a low-symmetry to a highly symmetric crystallographic structure. Those crystal structures are known as martensite (at lower temperatures) and austenite (at higher temperatures)."[4]

Here we see three interesting facts:

1. Pseudoelasticity and superelasticity are seen as one and the same thing.
2. Pseudoelasticity (or superelasticity) requires a minimum temperature to initiate a martensite-to-austenite transformation leading to the familiar elastic response we see in SMAs.
3. The term pseudoelasticity (or superelasticity) is made to look different from the term shape-memory effect despite both performing the same behavior of returning to its original shape. The difference is that one requires no heating (or there is no change in temperature), and the other requires heating.

As further support, Prof. Marcelo A. Savi in Mechanical Engineering from the Federal University of Rio de Janeiro, and Ph.D. student Luciano G. Machado from the Military Institute of Engineering in Rio de Janeiro, Brazil, gave their definition for *pseudoelasticity* in their 2003 article, "Medical Applications of Shape Memory Alloys" as follows:

> "Pseudoelasticity occurs whenever an SMA [Shape Memory Alloy] sample is at a temperature above A_F (the temperature above which only the austenitic phase is stable for a stress-free specimen)."[5]

Another variation on this definition can be seen from Dr. François Cardarelli, an electrochemist from the Materials and Electrochemical Research (MER) Corp. in Arizona, USA, and the author of the 2008 edition of *Material Handbook: A Concise Desktop Reference*. He wrote on page 140 of that book:

> "Superelasticity, or pseudoelasticity, refers to the ability of NitiNOL [or any SMA] to return to its original shape upon unloading after a substantial deformation."

Notice how Cardarelli is giving the impression that pseudoelasticity and superelasticity are the same terms. Furthermore, he does not mention temperature in his definition. A person taking this quote at face value (as the public would) could be forgiven for thinking temperature is irrelevant when defining *pseudoelasticity*, and that it is equivalent to *superelasticity*, despite how confusing it is to see the words "pseudo" and "super" in the same definition for both terms. Because of this, it is possible to find further confusion among other online sources when defining the term *pseudoelasticity*. For example, the website ExplainThatStuff.com gave its interpretation for the above definition of pseudoelasticity as recently as August 2016 as follows:

"Pseudo-elasticity is similar, but no temperature change is needed to make the object return to shape after you deform it. If you bend a pair of shape-memory eyeglasses, the stress you apply makes the titanium alloy from which they're made flip into an entirely different crystalline structure; let go and the crystalline structure reverts back again, so the glasses spring back to their original shape."[6]

Technically it is true that no temperature change is required. However, a pseudoelastic response still requires a sufficiently high-temperature to get the crystalline structure of SMAs into the austenite phase. This temperature-dependence of AuCd is proven when Venkata Suresh Guthikonda and Ryan S. Elliott wrote in their article, "Toward an Effective Interaction Potential Model for the Shape Meory Alloy AuCd":

"The SMA properties of Au-47.5at%Cd were identified by Chang and Read (1951) using an X-ray analysis of the orientation relationships, electrical resistivity measurements, and motion picture studies of the movement of boundaries between the two phases during phase transformation. From the observations of this experiment, it was concluded that Au-47.5at%Cd undergoes a diffusionless transformation from a high symmetry B2 cubic structure to a low symmetry B19 orthorhombic structure when it is cooled to about 60°C. The reverse transformation occurs from the B19 structure to the B2 structure at 80°C as the alloy is heated (Chang and Read, 1951)."

So what has happened here is that Cardarelli assumed a high enough temperature had already been reached, except he did not state it implicitly. To be strictly correct, the alloy must always be heated above what scientists call the critical temperature leading to the formation of the austenite crystalline structure such that the removal of a load that had initially caused the alloy to bend in the first place will result in the alloy returning to its original shape. Heat always play a crucial role when it comes to activating the pseudoelastic or superelastic properties of alloys of this type.

And this includes all SMAs as this definition from the SmartLabs website published by Texas A&M University, College Station, USA, shows:

> "Shape Memory Alloys (SMAs) are a unique class of metal alloys that can recover apparent permanent strains when they are heated above a certain temperature."[7]

Thus, the only observable difference between SMAs and pseudoelastic alloys is that a SMA must be heated to activate its elastic response, whereas a pseudoelastic alloy has already been heated. But whether it is actually pseudoelastic or superelastic depends on the amount of bending that is achieved and how well the alloy regains its original shape once the deformation force is completely removed. That's the only reason one can fathom from scientists wanting to use the words "pseudo" and "super". One must be a strong elastic response, and the other is weak.

Similarly, when we find in the online Oxford dictionaries the following definition for *thermoelasticity*:

> "Relating to elasticity in connection with heat."[8]

again this is the same as shape-memory, pseudoelasticity, and superelasticity. All these terms specify that heat is required to activate the alloy's natural elastic response.

So why don't we lump them all under a broader definition of *SMAs*? For example, Clive Barnes wrote in *Shape Memory and Superelastic Alloys*:

> "The shape-memory effect was first observed in AuCd in 1951 and since then it has been observed in numerous other alloy systems."[9]

Clearly Barnes wants to see AuCd as a shape-memory effect. Certainly makes life so much simpler for most non-technically-minded individuals.

Or why not characterise all such presumably reversible elastic responses as *pseudoelastic*? After all, that was the original term used by Ölander in 1932 to describe this type of alloy. And if we really wanted to be pedantic, add *superelastic* to the definition mix to categorise those alloys showing a higher degree of elasticity.

Better still, we can just stick to the term *elastic*. Seriously, material experts do not use terms like pseudohardness and superhardness to describe the degree of hardness a material possesses. It is just called hardness. Ditto for *strength* and other terms describing a material's physical properties. To know the difference in hardness or some other physical property in a variety of materials, you only need to know a number. Then we can say in plain English that this material is harder than another material, and the numbers tell us by how much. Easy isn't it?

So why complicate the term *elasticity* when discussing SMAs? An online source defines *elastic* in the following manner:

> "A material is said to be elastic when it exhibits a stress-strain curve which is fully reversible."[10]

Or to put it another way,

> "The removal of stress eliminates the strain. The energy stored in the material when under stress is fully recovered."[11]

There is considerable overlap in the terms used to describe these types of flexible alloys in the scientific literature. We see the one fundamental term that keeps cropping up, which we can all understand because of the words "fully recovered" and "fully reversible", is clearly *elastic*. So why deviate from this term? Just call it for what it is: an *elastic alloy*.

Indeed, all materials possess a certain amount of elasticity, some more than others. For example, a marshmallow has a certain amount of elasticity. Gently squeeze a marshmallow between two fingers and you can see how its original shape is deformed. In crystallographic terms, any variation in the shape is effectively a distortion of its crystalline structure by the applied strain. When the strain (i.e., your fingers) is instantly removed, the marshmallow tries to return to its original shape, as if the chemical bonds between the atoms act as microscopic springs to restore them to their original positions. However, on carefully examining this movement of the marshmallow, you will notice how the shape it tries to return to is not exactly perfect. In these circumstances, it is best not to call it an elastic material. However, if the amount of deformation is minimal and within what scientists call "certain strain limits", it is possible for the marshmallow to return to its original shape. When this happens, we can call it an elastic material. Indeed, why not call it a shape-memory material? Of course, it is not as elastic as NiTi heated above a critical threshold temperature to activate its elastic response, or even a piece of rubber (now they are really elastic). Nevertheless, you can still describe a marshmallow as elastic within certain prescribed limits.

Likewise, the shape of a piezoelectric crystal may show no perceptible change to its crystalline structure when a reasonably high degree of force to deform it is applied. However, scientists know that the shape must be deformed (no matter how slight this may seem to the naked eye). The deformation is enough to cause a change in its crystalline structure because electricity can be measured on the surface. If the applied strain was not there, then the crystal would not produce electricity. This means there has to be enough flex in the chemical bonds and movement in the atoms to permit scientists to measure this electrical energy. So, while this flexing behavior exists, even inside a seemingly hard material such as a crystal, and it can be repeated and is fully reversible without signs of damaging the crystalline structure to create this electricity, scientists call this an *elastic* response. And since the crystal can return to its original shape upon removal of the deforming strain or stress without damage, it can be described, incredibly enough, as a shape-memory, and hence an elastic, material.

Similarly, we know how easily a piece of highly flexible rubber can return to its original shape. So long as the rubber is within prescribed strain and temperature limits, the material can be bent significantly and will return to its original shape as soon as the applied strain is removed. Rubber is definitely a shape-memory material.

With this in mind, both the piezoelectric crystal and the piece of rubber are *elastic* materials having a *shape-memory* property so long as they are within certain strain limits set by the materials' own crystalline structures. In the case of rubber, scientists are happy to further classify this as a *superelastic material* because it can be repeatedly bent by at least 180 degrees and still return to its original shape, but this is not so in the case of a piezoelectric crystal. In the latter case, we should call it a *pseudoelastic material.*

Then there are metals and alloys that can tolerate varying degrees of applied strain. For example, position a straight, one-meter long, stainless steel ruler over the edge of a table, hold one end in place with one hand and push down with the other hand on the other end. The ruler, of course, bends. As soon as you suddenly remove the strain created with your hand, the ruler quickly returns to its original position (and, hence, technically speaking, its shape, which is straight and flat). So, in a sense, the ruler can be described as elastic, even shape-memory, dare we say it.

In the case of pseudoelastic and superelastic materials in the world of SMAs, not only is there a certain amount of bending which these alloys can withstand just like with the stainless steel ruler, but if the strain exceeds a certain physical limit, stretching of the chemical bonds holding the atoms together in a crystalline structure will take place. For the steel ruler, such bond stretching would only leave behind a permanent bend, and no amount of heat or electrical energy would cause it to return to its original shape unless it is melted and molded back into the correct shape. Not so for SMAs. The application of heat or electrical current to these alloys can move the atoms back towards their original positions even after the chemical bonds have been stretched beyond the physical strain limits of most standard metals and alloys. It is during this moment of the atoms moving into the original positions is what gives these alloys a remarkable sense of elasticity.

However, not all SMAs are the same. You will find varying degrees of elasticity depending on the elements used and their relative amounts. For example, we know Ni-Ti is more elastic than the sample of Au-Cd produced by Ölander in 1932. Noting these differences, scientists try to extend the usefulness of certain scientific terms, not just to give them a sense of historical importance (since the terms were mentioned before by other scientists), but also to indicate the relative elasticity of various alloys.

For example, if a SMA is not perfectly elastic (i.e., it cannot return to its original shape after the application of a deforming force), then the alloy has been strained beyond its normal limits. If it shows no signs of elasticity (technically incorrect since all materials have a certain amount of elasticity within certain strain limits), then we do not have a SMA. It is just another alloy or metal. However, an alloy that does attempt to return to its original shape in the austenite phase, even if not entirely successfully, is a SMA, although it is more precisely termed a *pseudoelastic alloy*. Similarly, if a SMA is bent less than 180 degrees and the shape-recoverability seems perfect in every way, scientists may call it a *pseudoelastic alloy* so long as the temperature is above a critical point for the austenite phase to exist, thereby allowing the elasticity of the alloy to be observed. On the other hand, an alloy that can essentially regain its original shape is not only a *SMA*, but if it can be bent significantly, say, by at least 180 degrees, and still achieve the remarkable feat of returning to its original shape no matter how many times this is performed, scientists will call this a *superelastic alloy*.

In summary, all materials are elastic to some extent and, therefore, will show a shape-memory response so long as they are within certain physical strain limits and temperatures as determined by their crystalline structures.

Regarding alloys, we just need to take great care in our use of the various terms and definitions available to us from the scientific literature. We must also remember that scientists like to be very precise in their definitions. While trying to recognise and often retain some semblance of historical importance of certain terms, the definitions are also there to help scientists to communicate with each other in a more precise manner.

For the public, perhaps a simpler way to explain these terms is to say:

1. All materials are elastic within certain strain limits.
2. Any alloy heated (whether by a person with a blow torch or the surrounding air) above a critical temperature to reveal an elastic response should be classified in the broadest sense as a shape-memory material.
3. How well the shape-memory effect works and whether the original shape is attained is covered by the terms pseudoelastic and superelastic alloys.
4. A pseudoelastic alloy is just a weaker form of the superelastic variety.

Beyond that, the mechanics behind the elastic response remain the same no matter what terms are used to describe these types of SMAs. A minimum temperature is always required to enable an elastic response. Furthermore, any recovery of its shape within certain prescribed strain limits is always described as *reversible*. As ExplainThatStuff.com explains it:

> "What's happening with shape-memory and pseudo-elasticity is that the internal structure of a solid material is changing back and forth between two very different crystalline forms: in other words, its molecules are rearranging themselves in a completely reversible way."[12]

Thus the earlier quote from Barnes suggesting Au-Cd is a SMA is technically correct. In fact, all materials show a shape-memory response within certain strain limits. For SMAs this implies a wider elastic range compared to standard metals and alloys. However, in the case of Au-Cd, we must ask whether this range would be closer to that of a pseudoelastic alloy or a superelastic alloy. One thing is certain, the shape-memory effect allegedly seen in AuCd was not enough for Ölander to start the SMA movement in 1932. The scientific community's lack of excitement over the supposedly rubber-like elastic response of the alloy would have to imply he saw nothing more than a pseudoelastic response, a kind of weaker form of the shape-memory effect.

So when you see Darel E. Hodgson of Shape Memory Applications, Inc., Ming H. Wu of Memory Technologies, and Robert J. Biermann of Harrison Alloys, Inc., state the following in the history section of *Shape Memory Alloys*:

> "The first recorded observation of the shape-memory transformation was by Chang and Read in 1932 (Ref. 1). They noted the reversibility of the transformation in AuCd by metallographic observations and resistivity changes, and in 1951 the shape-memory effect (SME) was observed in a bent bar of AuCd."[13]

it does not mean the study of SMAs began in 1932. It just means someone noticed the elastic (or more accurately a pseudoelastic) effect and wrote about it in a scientific article together with a few other observations when the alloy was heated. The actual work of understanding what was going on at the crystalline level as part of a coherent theory to explain it was yet to be uncovered. And with no one else excited by the observations (scientists at the time put the weak elastic effect of $Au_{47.5at.\%}Cd$ down as an anomaly.), it would take another alloy with a more pronounced shape-memory effect to grab the attention of the scientists.

In that case, when did interest in SMAs begin? As we will see in the next section, official study of SMAs did not commence until after 1958. The term *shape-memory* was not even invented until after that year. In 1951, Chang & Read were the first to use the term "shape recovery", but not shape-memory. Prior to 1958, the only thing scientists knew was that Ölander did nothing more than report the peculiar rubber-like behavior of AuCd at a meeting of the Swedish Metallurgical Society on May 27, 1932 and that's about it. He had essentially ignored afterwards what was, in his mind, a rather weak pseudoelastic response[14]. Instead, he investigated the crystalline structure of the alloy through a combination of x-ray diffraction and electrochemical techniques, and published his results. From this rather modest work, Ölander was able to identify with reasonable confidence the respective presence of the B2 cubic (austenite) phase and the B19 orthorhombic (martensite) phase in AuCd when he raised and lowered the temperatures.

As for L. C. Chang and T. A. Read, the significance of this shape-changing behavior and how to optimise it was not the focus of their work on $Au_{47.5at.\%}Cd$ in 1932. Rather, in the comfort of their laboratory at Columbia University, they recorded important metallographic observations and carefully measured the resistivity[15] at different temperatures. From this work, the scientists discovered one important fact: the martensite phase transformation in AuCd was reversible. In other words, the crystalline structure observed at one temperature can repeatedly change into another at a higher temperature, and back again, with no obvious signs of a defect creeping into the crystalline structures. Of course, this does not necessarily mean they had a SMA. Without bending it and heating the alloy to see what happens at the large scale and making the connection between its elasticity and changes at the crystalline level, there is no way of telling it was a SMA. In fact, this is what happened. Until someone else would make this link (namely a couple of Russian scientists as we shall see later), just observing the change in the crystalline structure at different temperatures would not prove conclusively that they had a SMA. However, by 1951, they had read the latest information from the Russians stating there is a link between this elastic behavior and the phase transformation at the crystalline level. This time they did have a look at this behavior. And from this, Chang and Read were able to confirm that Au-Cd can show an elastic response within certain physical strain limits and would coincide with the martensite phase transformation. Unfortunately, those limits were not great as Ölander discovered. As Chang and Read already knew, significantly deform the Au-Cd wire (by bending it to 180°) would easily exceed its normal strain limits to the point where the alloy did not return to its original shape perfectly. However, even within the right strain limits, the shape-memory effect was not exactly riveting or spectacular in their minds that they wanted to do an all-out study into SMAs in 1951. Far from it. What was more important to Chang and Read at the time was to deduce the crystalline structures of AuCd in 1932, and to confirm the results from the Russians in 1951.

If what has been said so far is untrue, and the study of SMAs began much earlier and with gusto among the scientific community, we would know about this very quickly. Because in modern times (i.e., after 1958), scientists have adjusted the composition ratio of the gold and cadmium and realized the elastic effect is actually more significant than previously thought, to the point that it can be called a superelastic SMA. Chang and Read did not try to improve the elastic response by changing its composition. The most that can be determined from this work is that the scientists focused on looking at the surface of the alloy with a microscope (which is what metallographic techniques actually involve), and when they saw details suggesting the effect was reversible (i.e., the surface changed in a repeatable way), they confirmed it using the more accurate resistivity data, which showed no signs that the crystalline structure was developing any kind of defect with each martensite-to-austenite phase transformation as the temperature of the alloy increased. The reversibility of the crystalline structures was confirmed beyond any reasonable doubt. Beyond that, an official explanation for *why* this phase transformation occurs at all was not given in 1932, and only a tentative explanation along the lines of atoms moving in a solid substance was suggested in 1951, after confirming the results of the Russians.

Not even German scientist Carl Axel Fredrik Benedicks, with his 1940 article titled "On the Elasticity of Solid Solutions, in particular those of AuCd" (published by Almqvist & Wiksell in Stockholm and

Friedlander in Berlin) would be sufficiently excited to look for other similar alloys and make a comprehensive study of SMAs in his part of the world. He seemed content to see the alloy he was studying as an anomaly.

Similarly, Anders Bystrom and Karl Erik Almin from the Institute of General and Inorganic Chemistry at the University of Stockholm wrote the article titled "X-ray Investigation of Gold-Cadmium Alloys Rich in Gold" in 1947, and still neither of them considered conducting a study of all known SMAs. It is almost as if scientists needed another example of a more powerful elastic alloy to persuade them to take a more serious notice of what they were looking at.

Yet remarkably, AuCd was not the only pseudoelastic alloy to be studied in 1932.

Not known to Swedish and American scientists at the time, another German scientist named Erich Scheil had worked independently on a second pseudoelastic alloy known as FeNi. Assisted by his colleague Friedrich Forster, Scheil conducted experiments to measure the small voltage and resistance variations caused by the martensitic transformations in FeNi. According to a Google-translated summary of the study:

> "The transformation of austenite to martensite cannot be explained by the thermal exchange space of atoms. The conversion is considered to be a special deformation mechanism. The influence of the voltages on the conversion has been studied in irreversible nickel steels."[16]

In plain English, the formation of a particular crystalline structure called a martensite is not controlled by temperature, but by applied stresses (such as by bending). Actually, to be strictly correct, it probably is controlled by temperature. It is just that the alloy was already in the austenite phase when Scheil attempted to bend it. At any rate, what Scheil noticed is how bending can reverse the transformation back into the martensite phase. As soon as he removed the strain, a return to the austenite phase was noted and with it a certain elastic response (apparently weak in his view). In trying to determine what causes this, he noticed how the use of voltages would affect the transformation process (mainly to force FeNi to the austenite phase in a stronger way). Take away the voltages and the strain and somehow the transformation to the austenite phase would persist. So he may have thought that the voltages were still present at the crystalline level to force the atoms to move.

In other words, FeNi could be acting like a piezoelectric crystal when it is bent. The martensite phase, where the atoms are naturally spaced apart and the unit cell structure is not symmetrical, could be generating a tiny voltage between the negatively-charged electrons forming the chemical bonds of the crystalline structure and the more positively-charged atoms. As soon as the stress is removed, the voltages get nullified (or electrical energy by way of the electrons in the chemical bonds get back into a stable state with those orbiting the atoms and so reducing the positive charge of the nucleus) through a movement of the atoms into what is effectively a compact and highly symmetrical B2 crystalline structure (i.e., the austenite phase), and with it the original shape.

Whether this is how it works, Scheil's results were unassumingly published in the *Journal of Inorganic Chemistry* under the title, "Über die Umwandlung des Austenits in Martensit in Eisen-Nickellegierungen unter Belastung" ("On the Transformation of Austenite to Martensite in Iron-Nickel Alloys under Load").

In another study conducted in 1929, Scheil also looked at the martensite phase transformation of hardened steel, noting, among other things, the "clearly audible noise"[17] he could pick up during the conversion from the austenite to the martensite phase. Further details of Schiel's work can be found in "Über die Umwandlung des Austenits in Martensit in gehärtetem Stahl" ("On the Transformation of Austenite to Martensite in Hardened Steel") published in the same journal.

However, in studies of both steel and FeNi, Schiel did not focus on the reversibility of the shape-recovery effect (in fact, there is no such effect in steel when the chemical bonds are stretched beyond certain strain limits, and it appears to be irreversible in FeNi showing how weak the pseudoelastic effect was for this alloy at the composition ratio chosen). His interest, as with the Swedish and American scientists who worked on AuCd, was to look at the martensite phase transformation in terms of the crystalline structures being formed, the reversibility of those structures at different temperatures, and to find a reasonable scientific explanation.

Several more years would pass without anything special to kick-start work on the "SMA" front.

Not even in 1938 when scientists Alden B. Greninger and Victor G. Mooradian from Harvard University "observed the formation and disappearance of a martensitic phase"[18] by decreasing and increasing the temperature of $Cu_{60at.\%}Zn$ and $Cu_{74at.\%}Sn$, a phenomenon known today as *thermoelasticity*, would a SMA be revealed. It appears the scientists were not looking for this shape-memory effect with a view to refining the composition ratio to see how it could improve the effect. True, the martensite (or, strictly speaking, the austenite-to-martensite) phase transformation is considered an important first step towards identifying a potential shape-memory effect in these alloys, and scientists later confirmed its existence. However, there is again no evidence that scientists at the time had tested the alloys for shape recoverability, including making essential adjustments to their composition ratio to see if this would maximise the effect.

Today, scientists can confidently state that CuZn is a SMA after making the attempt to cool the alloy to below the martensite transformation temperature (under $-10°C$), bend the alloy, and then heat it to above the transformation temperature needed to activate the shape-memory response and so allow the alloy's movement to be observed.

For example, in 1974, A. J. Perkins wrote an article titled, "Residual Stresses and the Origin of Reversible {Two-Way} Shape Memory Effect" in *Scripta Metals*. During his investigation, he discovered the shape recovery effect of a wire (approximately 1 x 1 x 90mm) made of $CuZn_{38.5-39.8wt\%}$ after being bent to a U-shape (or 180°) below the martensite transformation temperature (which for this alloy is below room temperature). He later heated it to room temperature and found the process to be completely reversible (i.e., it reverted to a straight configuration). However, should the amount of zinc lie outside the 38.5 to 39.8 atomic percent range, the reversibility was diminished or non-existent.[19]

Since Greninger and Mooradian had used a sample of CuZn containing 40 percent zinc. Given the sensitivity of the alloy's shape-memory effect to tiny changes to its chemical composition ratio, this might explain why the study of SMAs did not commence in 1938. The

Microradiographic study of the distribution of alloying components in a nickel solid solution. K. A. Osipov and S. G. Fedotov. *Doklady Akad. Nauk S.S.S.R.* **78,** 51–3 (1951).—An exptl. study was made of binary alloys of Ni contg. 5 at. % W, Mo, Nb, Ti, and Ta. Ingots (100 g.) were made by melting electrolytic Ni and alloying element of tech. purity in a corundum crucible in a high-frequency furnace. Coarse grains were produced in the alloys both by slow cooling during crystn. and by prolonged annealing at 950 to 1100°. Plates of the undeformed alloy 0.05 to 0.025 mm. thick were radiographed with Fe radiation at 24 kv. This thickness was nearly that of a grain. The Ni-W alloy was inhomogeneous after annealing for 96 hrs. at 950° and somewhat less so after an addnl. 50 hrs. at 1100°. The dendrite arms were rich in W. About 50% of the grain-boundary area absorbed x-rays weakly, possibly because of shrinkage porosity, resulting in broken lines in the microradiograph. The Ni-Mo alloy after annealing for 96 hrs. at 950° was rich in Mo at the grain boundaries and also showed Mo-rich flakes and needles. The needles, but not the flakes, were also seen under the microscope. The dendritic segregation in the cast Ni-Nb alloy was removed by annealing for 50 hrs. at 1100°, but the Nb was still segregated near the grain boundaries and a region of low x-ray absorption existed at the grain boundaries. The Ni-Ti alloy contained 1.8 at. % Ti and was made by using a 4 wt. % Ti master alloy. Segregation of Ti was present both in the cast alloy and after annealing for 50 hrs. at 1100°. Areas of low absorbing power were removed by annealing. Dendritic segregation in cast Ni-Ta alloy was greatly reduced by annealing. In all these alloys microscopic study showed the segregation in the cast state but not after annealing. A. G. Guy

Earliest interest in NiTi by the Russians confirmed by this abstract published in 1951.

diminished nature of the shape-memory effect in their sample of CuZn was enough to view AuCd, CuZn and probably CuSn as anomalies.

Nevertheless, the establishment of a link between "the formation and disappearance of a martensitic phase" with changes in temperature, and the observed pseudoelastic movement of AuCd, eventually came with the publication of an article in 1949 by Russian scientists G. V. Kurdjumov and L. G. Khandros. However, once again, the specific explanation for how the pseudoelastic (or thermoelastic) effect works remained a mystery.

In 1951, Chang and Read noticed the work from the Russians and decided to look again at AuCd in "Plastic Deformation and Diffusionless Phase Changes in Metals—The Gold-Cadmium Beta Phase"[20]. This time, the scientists gave their first tentative explanation of how AuCd is able to reveal a pseudoelastic effect during the martensite phase transformation. According to Yong Liu and Zeliang Xie in the introduction to their 2007 book, *Progress in Smart Materials and Structures*:

> "...Swedish physicist Arne Ölander [1932a;b] discovered the rubberlike behavior in an Au-47.5Cd alloy which was later explained by Chang and Read in 1951 as due to 'reorientation' of martensite twinned lattices."[21]

This is a sophisticated way to say there is a movement in the metal atoms leading to the formation of a different crystalline lattice structure in both the austenite and martensite phases. Whatever the explanation, confirmation for the work into AuCd and CuZn since 1938 is given by Ryhänen:

> "Greninger & Mooradian (1938) observed the formation and disappearance of a martensitic phase by decreasing and increasing the temperature of a Cu-Zn alloy. The basic phenomenon of the memory effect governed by the thermoelastic behavior of the martensite phase was widely reported a decade later by Kurdjumov & Khandros (1949) and also by Chang & Read (1951)."

The results of Chang and Read on CuZn were also confirmed in *Shape Memory Alloys,* edited by K. Otsuka and C.M. Wayman:

"This [pseudoelastic, and later the shape-memory] effect [through a change in temperature] was first found in a Au-47.5 at% Cd alloy by Chang and Read [in 1951]."[22]

In 1953, another alloy would join the exclusive elastic (and hence shape-memory) alloy club when T. A. Read teamed up with M. W. Burkart to study the indium-thallium (InTl) alloy system, which at first exhibited a similar pseudoelastic effect. Other scientists later (kind of like after 1958) discovered the reversible nature of the elastic response in this alloy[23]. However, again the response was not significant enough to blow the minds of the scientists who saw it. InTl, just like the few other similar alloys in the past, was relegated into the realm of scientific oddities.

Further details about this low melting temperature InTl alloy can be found in the article titled "Diffusionless phase changes in the indium-thallium system"[24].

Another study into the martensite phase transformation of CuZn would take place in August 1954 by E. J. Suoninen for his Massachusetts Institute of Technology (MIT) thesis, *Investigation of the Martensitic Transformation in Metastable Beta Brass*, along with Erhard Hornbogen and Glinter Wassermann[25] in 1956 in the article "Phasenumwandlungen in β-CuZn".

Finally, C. W. Chen studied the martensite transformation in CuAlNi in October 1957. The article for this work is titled "Some Characteristics of the Martensite Transformation of Cu-Al-Ni Alloys"[26].

Despite these early works, nothing would get scientists quivering with excitement over these alloys. It was almost as if more examples were needed, or these pseudoelastic alloys had to have a more pronounced elastic effect before commanding the attention of scientists and inspiring them to establish a new branch of science to study the shape-memory behavior. Certainly neither AuCd, FeNi, CuZn, CuSn, InTl nor CuAlNi would motivate scientists to study the shape-memory effect. So clearly having more than one alloy to show this elastic effect was not going to make a difference. It would require an alloy demonstrating a much more significant shape-memory effect to come along, something scientists would later describe as more superelastic and, therefore, impossible to ignore. Until then, none of the alloys displaying this curious pseudoelastic effect would grab the interest of any scientist (let alone the USAF, at least on an official level) to predict the existence of a more powerful titanium-based SMA like NiTi. Without actually seeing a reversible shape-memory response from a significantly deformed alloy, noticing how quickly it returns to its original shape, and being able to provide a detailed theory of how the shape-memory effect works, scientists could merely note a few past observations, perhaps made a link in those observations, speculate on a basic theory in 1951, and essentially ignore the alloys as scientific oddities. Beyond that, someone would have to see a reason to study SMAs more seriously. For the scientific community that reason would eventually come not in the 1930s, or the 1940s, but in the early 1960s and only after someone noticed something interesting in one specific alloy in the late 1950s and decided to mention it outside his military laboratory. And the alloy to launch the interest in SMAs in a big way was NiTi. Let us explain further.

When was nitinol discovered?

In 1938, the world's first known sample of an NiTi alloy was produced and studied by scientists as confirmed by this abstract:

792   ALPHABETICAL INDEX OF WORK ON METALS AND ALLOYS

Scherrer camera. The lattice spacings were calculated by a method described by them using the Taylor and Sinclair (Nelson and Riley) extrapolation function.

NITI has a b.c. cubic, A2 type of structure, a = 2.980 kX, M = 1, *Im3m* (DUWEZ and TAYLOR, 1950); a = 3.005 kX (TAYLOR and FLOYD, 1952). No evidence of ordering was found, although LAVES and WALLBAUM (1939*) reported that the phase had an ordered B2 type of structure and POOLE and HUME-ROTHERY (1955) also referred to the structure as ordered, giving a = 3.0070 ± 5 kX at 22°C for the length of the unit-cell edge. DUWEZ and TAYLOR reported that the phase decomposed at a low

At the time, the sample produced by Laves and Wallbaum was not called nitinol. The name nitinol would come much later. Instead, F. Laves and H. J. Wallbaum from the Mineralogical Institute at the University of Göttingen modestly called the alloy NiTi, according to the results published in the German metallurgical journal *Die Naturwissenschaften* in October 1939 under the title, "Zur Kristallchemie Von Titan-Legierungen". According to this work, the scientists noted that NiTi seemed to have a B2 type of structure (the same found in CsCl, or cesium chloride) at around the equiatomic composition of the two metals. As they stated:

"In the intermediate titanium range, we found the following compounds were of the CsCl type: FeTi, CoTi, NiTi, and ZnTi."[27]

This was precisely the same CsCl structure observed earlier in AuCd at 400°C. Apart from this, other crystalline structures of NiTi were observed (mainly in the nickel-rich end of the NiTi composition spectrum). As Yurko et al. stated:

"Laves & Wallbaum (1939) stated that the Ti_2Ni phase had a complex face-centered cubic structure with 96 atoms per unit cell. No value of cell constant was given."[28]

There was a recommencement into understanding the full range of crystalline structures of NiTi from 1949 to 1955. The handful of scientists that initially started work in this area focused mainly on the Ni_2Ti phase in an attempt to confirm Laves and Wallbaum's results as well as look more closely at the titanium-rich end of the alloy at the request of the Battelle Memorial Institute where the data was limited. Only three other scientists, one of whom emerged from Wright-Patterson AFB, would focus on the equiatomic range for NiTi, and this time using much higher purity of the alloy. Further details about this to come later.

In the meantime, the question to ask is, did Laves and Wallbaum discover the shape-memory behavior of NiTi when the B2 crystalline structure was observed at this time? The abstract and an inspection of the full article when it became available in 2016 showed no evidence of it whatsoever.

Then again, this could be a bit like the work of the Swedish physicist Ölander discussed earlier. Maybe Laves and Wallbaum already saw a fairly weak pseudoelastic effect in NiTi, but didn't mention it in this article? However, the reality is, after conducting further research into the work of these scientists, it has become clear that neither scientist was looking for the effect nor were they aware NiTi was a SMA. Indeed, the lack of any verbal mention of a possible pseudoelastic effect in NiTi to prestigious scientific delegates at a meeting (which is what Ölander did when he observed the effect in AuCd) would strongly support this claim. As far as can be ascertained at the present time, the interest was primarily in determining the crystalline structure of NiTi (located near the equiatomic range of the

composition) and Ni₂Ti (located at the nickel-rich end of the composition) and in producing the world's first partial phase diagram for the alloy.

composition) and Ni_2Ti (located at the nickel-rich end of the composition) and in producing the world's first partial phase diagram for the alloy.

Even if the scientists had a secret reason to believe NiTi could possess a shape-memory effect, they had no hope of noticing it in the sample of NiTi because their sample was not pure. The scientists were working with the best available sample at the time (around 99.5 percent pure), but a purity of at least 99.995 percent was required to reveal the property. Unfortunately, as we shall discuss later regarding the history of titanium, the technology wasn't refined enough in 1939 to produce this level of purity in titanium, NiTi, or any other titanium-based alloy. The scientists had virtually no idea what they had in their hands, or how special NiTi was going to be. NiTi was seen as just another ordinary alloy.

Purity is the name of the game

suggest that while turbine power is still slightly cheaper, the difference is narrowing.

Nitinol is still in an early, even primitive stage of development. If the material is strained beyond certain broad limits, it can be permanently deformed or develop fatigue. It is also expensive, about $200 a half-kilo, and tricky to make. To achieve a desired transition temperature, the proportions of nickel and titanium must be accurate to one part per thousand; production requires a vacuum furnace and sophisticated support equipment to ensure purity. Some researchers say that all the material produced so far is crude and that

Continued Page 119

MIRACLE METAL
From Page 13

properly refined Nitinol could be vastly more powerful, able perhaps to respond to temperature differentials as low as three or four degrees Celsius.

Recently metallurgical advances are making possible new levels of precision, purity and consistency. Near Utica, New

clos
R
ma
prol
E
beer
wid

Nitinol must be extremely pure (at least 99.995 percent) for scientists to notice its behavior. There is no other good scientific reason why anyone would want to make extremely pure NiTi. The importance of making pure NiTi is officially confirmed from the above paragraph taken from the original 1982 Australian edition of *Omega Science Digest* article about the alloy.

Thus, in the days when Laves and Wallbaum were studying the alloy, it was so impure that it would have been impossible for them to have noticed anything unusual.

But why the difficulty in manufacturing nitinol in pure form? Surely, melting a bit of titanium and nickel, mixing them up, and letting the sample cool down couldn't be that technically challenging for the scientists. Well, in fact, it was. The reason for the difficulty lay in the manufacturing of a particular metal making up nitinol to a pure form which had stopped scientists from fully appreciating nitinol's fascinating property. The metal is known as *titanium*.

You see, titanium is extremely difficult to manufacture in a pure form because of its high susceptibility to absorbing impurities from the air at temperatures above 700°C (especially oxygen), which is well below the melting point of the metal. The temperature would need to be more than doubled (i.e. around 1,675°C) in order for titanium to melt and combine with other metals. Unfortunately, this would result in titanium absorbing large amounts of impurities from the air, thereby ruining whatever special properties a typical titanium-based alloy like nitinol may have to show to the scientists.

So, when did the level of purity needed to reveal nitinol's shape-memory property become a reality? To answer this, we need to look more closely at the history of titanium, how the metal was refined, and at what purity was the metal attained, and when. Because if titanium could reach a minimum of 99.995 percent purity using the right technology, then, technically speaking, anyone could make a titanium-based SMA like nitinol if they knew which elements to combine and somehow discovered its miraculous shape-memory property.

The question is what year would this occur?

In other words, if the correct purity was attained by, say, late June, or early July 1947 at the very latest, it might be possible to explain the Roswell metallic foil as a secret military test by the USAF using nitinol (i.e. not Project Mogul), leaving us with the question of who the victims were. But at least we can safely assume nothing extraterrestrial in the nature of the Roswell object. But if the required purity can't be reached by the USAF or any other country on Earth at the right time, then this *artificially* manufactured Roswell foil can't be a man-made titanium-based SMA. And, if it turns out to be the latter, the Roswell object that crashed in July 1947 must clearly be scientifically significant and we have ourselves something remarkable, and worthy of scientific study.

The history of titanium

The history of titanium began in 1791 when William Gregor (1761-1817), a clergyman and amateur British chemist, observed an unusual and unidentified reddish brown calx (later called *ilmenite*) together with some gun powder-like black sand in a local river in Menachan Valley, Cornwall, England.

In 1795, a well-respected and highly renowned German chemist, Martin Heinrich Klaproth (1743-1817), identified a white oxide material extracted from a sample as rutile from Hungary. On closer inspection, he managed to identify a new element. He called it *titanium*.

Why the name *titanium*? As Klaproth explained:

> "Whenever no name can be found for a new fossil which indicates its peculiar and characteristic properties (in which situation I find myself at present) I think it best to choose such a denomination as means nothing of itself, and thus can give no rise to any erroneous ideas. In consequence of this [...] I shall borrow the name for this metallic substance from mythology, and in particular from the *Titans*, the first sons of the earth. I therefore call this new metallic genus Titanium."[29]

Over the next 150 years, the history of titanium would involve considerable effort on the part of scientists to isolate the element from its compounded material.

For example, Jöns Jacob Berzelius (1779-1848) attempted to isolate titanium in 1825 through melting but succeeded only to a very impure state. A more successful method came in 1887 when Lars Fredrik Nilson (1840-1899) and Sven Otto Pettersson (1848-1941) managed to reach a purity of 95 percent.

In yet another classic attempt, Prof. Henri Moissan (1852-1907) of the University of Paris learned of a different method, which resulted in the purity going up to 98 percent. However, this method still relied on heating the metal.

In 1910, the New Zealand-born U.S. metallurgist Matthew A. Hunter (1878-1961), from the Rensselaer Polytechnic Institute in Troy, N.Y., in cooperation with General Electric Company[30], managed to successfully produce a small sample of 99.5 percent pure titanium *without* heating the metal above 700°C. He did this by isolating the titanium atoms from its naturally-occurring oxide form using chlorine and later applied the reducing agent sodium (Na) to chemically reduce $TiCl_4$ in an airtight steel cylinder to create the metal titanium. The reaction is:

$$TiCl_4 + 4Na \rightarrow Ti + 4NaCl$$

Confirmation for the purity level reached by Hunter was revealed in the 1983 edition of *Encyclopedia of Chemical Technology*:

> "In 1910, 99.5% pure titanium metal was produced at General Electric from titanium tetrachloride and sodium in an evacuated steel container...."[31]

Despite the almost pure ductile titanium metal being produced, it was not enough to start the titanium industry. This is because the reducing agent sodium (Na) is rare and expensive to produce in pure form. A better alternative, and one that promised lower costs, was to use another more common metal, calcium (Ca), as the reducing agent. This was realized by metallurgist and inventor Dr. Wilhelm Justin Kroll (1889-1973) while working in his private laboratory in Luxembourg in 1932.

By 1938, Kroll isolated a world-breaking 50 pounds (or roughly 23 kilograms) of 99.9 percent pure titanium with the help of calcium as his preferred reducing agent. Despite his incredible effort, the quantity was still not enough to construct an aircraft of any size (even for one pilot). Indeed, if Kroll had his heart set on developing a titanium-based SMA like nitinol, which it wasn't, the metal had to exceed the 99.995 percent pure form. As this was not his aim, we can see why Kroll's sole intention was entirely to increase the quantity of titanium at a lower cost. Because if he could reduce the costs sufficiently, a new business in manufacturing titanium could be established where he would make himself very wealthy indeed (the natural aspiration of most men at the time).

Thus, the assumption for Kroll was that titanium was sufficiently pure at 99.9 percent, so why make it any purer? Unless one had a good scientific reason, Kroll considered the 99.9 percent pure titanium to be sufficient for the time.

With quantity being the initial pressing scientific issue for Kroll in the late 1930s, it soon became clear to him that in order to make much greater quantities of titanium using the titanium tetrachloride approach, an alternative reducing agent had to be sought.

As Kroll searched for a suitable reducing agent, world events would conspire to put a slight dampener on his work by slowing him down somewhat as World War II began. As a result, Kroll fled Europe to find support for his titanium extraction and purification work in the United States.

In early 1940, Kroll was invited to present a paper about his work to members of the Electrochemical Society. Enough people attended the presentation to the point where several of them were impressed by the idea. Realizing his work's potential and the opportunity to make money, Kroll decided it was time to protect his intellectual property. His perseverance was soon rewarded with a patent[32] on June 25, 1940. As soon as he became a permanent resident of the U.S., the doors of opportunity began to open for him.

Kroll's first offer was a consultancy position at the research laboratory of Union Carbide and Carbon Corporation at Niagara Falls, New York.

Kroll moved to the U.S. Bureau of Mines at Albany, Oregon, in 1944 where his idea was put to fervent use in the extraction and purification of a new metal, zirconium, for use in atomic reactors. For Kroll, it seemed natural to apply his chemical method of extraction and purification to zirconium as it was in the same group in the periodic table as titanium.

Kroll's love affair with titanium would return after moving to the new Bureau of Mines plant in Boulder City, Nevada. It was here where he made the decision in 1946 to use magnesium (Mg) as the reducing agent for $TiCl_4$ because with magnesium, he could at last produce relatively pure titanium at reasonable cost.

Here is the chemical reaction he found that finally made him famous in the world of metallurgy:

$$TiCl_4 + 2Mg \rightarrow Ti + 2MgCl_2$$

With further refinement of the process, Kroll succeeded in recovering nearly all of the magnesium and chlorine for reuse in another batch of titanium processing through electrolysis via the chemical reaction:

$$MgCl_2 \rightarrow Mg + Cl_2$$

As a result of his outstanding efforts in this field, Kroll would later receive a second U.S. patent[33] on September 19, 1950 as confirmed by this abstract[34]:

Titanium alloys. Wm. J. Kroll. U.S. 2,522,679. Sept. 19, 1950. $TiCl_4$ is a good solvent for other metal chlorides. Thus anhyd. metal chlorides can be prepd. by dissolving the oxide or water-contg. metal chloride in $TiCl_4$; the $TiCl_4$ hydrolyzes to TiO_2 with liberation of HCl. The technique also leads to prepn. of a new type of getter material in electronic or gas-discharge tubes. Highly oxidizable metals, such as Zr, Hf, Th, Y, Nb, Ta, Cr, and U are dissolved as chlorides in $TiCl_4$, and after filtering, the soln. of the chlorides is reduced in an inert gas atm. with molten Mg, followed by compacting and sintering steps. $TiCl_4$ will dissolve 14–15% $TaCl_5$ at 25° and 40% $TaCl_5$ at its b.p., 136.4°. $NbCl_5$ dissolves to 0.24% at 25° and 2% at the b.p. $ZrCl_4$ dissolves to 30% in $TiCl_4$ at 25°. The getter alloys formed are substantially free of embrittling oxides, nitrides, or hydrides.   Frederick C. Nachod

Despite being famous, it was still not enough to make Kroll a rich man from his work. Apart from the U.S. Bureau of Mines developing the equipment to commercialize the Kroll method, the only thing missing was finding a willing buyer to purchase titanium in vast quantities, which luckily for Kroll didn't take long. The buyer who was showing more than a fleeting interest in the metal in 1946 turned out to be the USAF in a study called Project Air Force RAND; this would be the first *official* interest in the metal by the U.S. military. Why the interest? Because it seems the Bureau of Mines responsible for developing the titanium purification technology based on the Kroll process was promoting the metal as the new successor to aluminium and stainless steel.

Just as the 1983 edition of the *Encyclopedia of Chemical Technology* stated:

"Titanium, termed the wonder metal, was billed as the successor to aluminium and stainless steels."[35]

231

Before we explain the world's first involvement in titanium by the USAF, let us present a step-by-step guide to the process of purifying titanium. By presenting this information, we can better appreciate what was, and still is, the best available technology for creating reasonably pure titanium.

Step 1: Find a titanium-rich source

The first step involves finding a titanium-rich deposit for mining. The best sources for titanium are in minerals called *rutile*, which are approximately 95 percent titanium oxide (TiO_2), and *ilmenite* ($FeTiO_3$), which contains between 50 and 65 percent TiO_2.

Although titanium is the ninth most abundant element in the Earth's crust (it makes up an estimated 0.62 percent of the earth's crust) and is the fourth most abundant structural metal after aluminium, iron and magnesium, which suggests that titanium oxide and ilmenite minerals should be readily available throughout the world, the highest and most concentrated sources of titanium can be found in Australia, the United States, Canada, South Africa, Sierra Leone, Ukraine, Russia, China, Norway, India, Malaysia, Brazil and a few other countries.

Step 2: Turn TiO_2 into $TiCl_4$ (titanium tetrachloride)

The second step is to extract the titanium in the naturally-occurring oxide form known as TiO_2. While TiO_2 in pure form using a sulfate purification process is useful as a pigment for creating a brilliant white paint, as stated in the 1983 edition of *Encyclopedia of Chemical Technology*:

> "In the early 1900s, a sulfate purification process was developed to commercially obtain high purity TiO_2 for the pigment industry..."[36]

the extraction of titanium to form a pure metal requires a different method. This is done through a chlorination process designed to replace the oxygen atoms with the more reactive chlorine atoms through an exothermic reaction to yield a stable chemical compound known as titanium tetrachloride ($TiCl_4$), as revealed in the following chemical reaction:

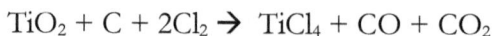

$$TiO_2 + C + 2Cl_2 \rightarrow TiCl_4 + CO + CO_2$$

The carbon (C) added to this reaction is supplied by a fluidized bed of petroleum coke. The temperature is increased to between 850°C and 1,000°C to help speed up the extraction of oxygen by the coke to form carbon monoxide (CO) and carbon dioxide (CO_2), leaving behind free titanium atoms to chemically bind with chlorine to form $TiCl_4$ in a gaseous form[37]. Next, the raw $TiCl_4$ is passed through a filter to remove fine particles of coke and titanium ore. This allows $TiCl_4$ to be liquefied, followed by a final cleaning up of volatile impurities of both high and low boiling points which is performed through a distillation column. What remains is a colorless $TiCl_4$ liquid of a purity of at least 99.9 percent.

Step 3: Extract the metal

The next step is to chemically reduce $TiCl_4$ using a reducing agent in an airtight steel cylinder. For a choice of a suitable reducing agent, one can use sodium (Na), in which case the reaction is:

$$TiCl_4 + 4Na \rightarrow Ti + 4NaCl$$

This is the process discovered by Hunter early in the 20th century, and would allow him 99.5 percent pure titanium metal to be consistently produced. However, the most successful method of creating large amounts of pure titanium at low cost came when William Kroll used magnesium as the preferred reducing agent. As noted previously, the chemical reaction for this is:

$$TiCl_4 + 2Mg \rightarrow Ti + 2MgCl_2$$

First involvement in titanium by the USAF

As alluded to earlier, the USAF would express the first interest in titanium with the commencement of Project Air Force RAND in 1946.

Of course, showing interest and actually *using* the metal are two distinct issues. The study would have to reveal a good reason to use titanium. Should a reason be found, it had to occur before July 1947 given the way the Roswell foil behaved and its high toughness factor, such as temperature resistance to a blowtorch, and with talk by one

233

military general that the foil may have contained titanium. Yet, the official history of titanium for the U.S. military did not indicate this at all. As we shall learn, the official military titanium industry for producing high quantities of titanium needed to build at least one USAF aircraft would not come until after the commencement of the official civilian titanium industry, which was after 1947. It was not until 1948 when the U.S. Bureau of Mines finally finished its commercializing of the titanium production plant to allow other civilian-based U.S. companies to manufacture the new metal. Even by 1950, nothing was officially done by the U.S. military to build an aircraft made of titanium. In fact, it was in 1952-53 that the USAF built the world's first official titanium military aircraft known as the X-3. In other words, 1950 would be the earliest year for the official commencement of the military titanium industry. Or, more technically correct, 1952-53 was really the moment when the military titanium industry began to permit the USAF to build a titanium aircraft. Yet, incredibly, the USAF would never use a titanium-based SMA for its structural or outer skin component of the X-3, which is odd considering a super tough nitinol-like (almost certainly titanium-based) SMA was observed in early July 1947.

Why?

Furthermore, it was only in 1948 when titanium had reached the required 99.995 percent purity and in limited quantities for experimental purposes, according to a titanium article in the April 1949 issue of *Scientific American*. It resulted in U.S. scientists getting uniformly enthusiastic about the metal, especially in the alloy form, leading to a sudden status change for the metal and a realization of its new engineering benefits for the aerospace industry; but never has such enthusiasm been seen before 1948, or more specifically, prior to July 1947 as far as we can officially tell. Even if the USAF had been working in this area in secret, they weren't officially jumping for joy at the prospect of using highly pure forms of the metal for their secret alloy work during 1947, and neither were any of the scientists which we must presume had been helping the military in this field. It was only after 1947 that saw the USAF requesting scientific assistance from the Battelle Memorial Institute to study titanium and certain alloys of interest to the USAF. Then suddenly, the enthusiasm among scientists over titanium could not be contained in 1948, even if the work had been conducted in secrecy on behalf of the USAF. Someone was not doing a good job at keeping secrets.

We see this in two recently declassified USAF/Battelle Progress Reports into the development of titanium and its alloys for the 1949 period (with all work commencing on May 18, 1948). From it, we discover the sudden renewed interest by the USAF after 1947 in titanium and its alloys, followed by considerable effort on the part of Battelle scientists to continue finding ways to refine the metal to a higher level of purity even by 1949 (shouldn't the refinement have been achieved in 1947?). Not only that, but scientists had to study up to four notable titanium-based SMAs after 1947, including NiTi, on behalf of the USAF at Wright–Patterson AFB. As for revealing a shape-memory effect in certain titanium-based alloys, Battelle scientists were not up to the task of explaining, let alone predicting, this class of alloys. Battelle scientists required the help of world-renowned chemist Dr. Linus Pauling to give a lecture speech on February 7, 1951 on the latest theory on metals and alloys, especially in the understanding of the chemical bonds holding together the atoms inside these substances. If Battelle had any prior knowledge of a shape-memory effect in certain titanium-based alloys and a theory to explain how it works, it would certainly not come in 1947, or even before 1951. This naturally begs the question: why was this work not carried out in 1947, or more specifically before July 1947, to help explain the shape-memory metallic foil observed near Roswell (and so support a man-made explanation of the event)?

Before we discuss this further in the next chapter, let us look at the evidence in support of the official scientific commencement of the titanium industry.

When did the modern titanium industry begin?

A review of titanium articles has uncovered two important years for the official commencement of the modern titanium industry.

The first is 1948. This was the year when the civilian titanium industry first took off. The second is 1950, when the U.S. DoD provided incentives to start the titanium military industry; and by 1952, there was enough titanium to build a substantial flying object for the USAF.

We see evidence supporting the 1948 commencement year and the U.S. military's interest in the metal in the 1969, 1980 and 1988 editions of the *Encyclopedia Britannica* and in the 1987 edition of *Encyclopedia Americana*.

For example, the following quote taken from the 1987 edition of the *Encyclopedia Americana* showed that the U.S. military, specifically through the USAF, was the principal instigators for the serious interest in titanium:

> **Uses.** The high cost of titanium metal often limits its use to military purposes. Because of its lightness and strength, titanium is used as a structural material in high-speed aircraft, rockets, guided missiles, and recoil mechanisms for artillery. Titanium is often used in the chemical processing industry because of its resistance to corrosion. This resistance is probably due to a thin coating of titanium dioxide, which protects the metal from further corrosion. The metal has unusually good resistance to corrosion by salt water, and so it is used in propeller shafts and other parts exposed to the sea.

In other words, by claiming the expensive nature of producing titanium was restricted to the U.S. military, it would imply the U.S. military, with their oversized annual budgets (especially so close to the end of World War II), had to be the first to ask, and so benefit, from titanium.

As for 1948 being the year of commencement, this was written in some earlier encyclopedias as a sudden change in the status of titanium after 1947. As quoted from the 1969 edition of *Encyclopedia Britannica*:

> "TITANIUM: a metallic element, changed after 1947 from a rare metal to an important structural metal. Because of its lightweight and high strength, particularly in alloy form, it is in demand for use in structural parts in high-speed airplanes....No other structural metal has been studied so extensively nor has advanced in technical stature so rapidly."

The 1980 edition of *Encyclopedia Britannica* reaffirms this fact:

"After 1947 titanium changed from a laboratory curiosity to an important structural metal....Known since the late 18th century, titanium was not produced in commercial quantities until the 1950s. Its strength at high temperatures and its light weight are valued in the aerospace industry."

Occurrence, properties, and uses. Titanium is widely distributed and comprises 0.44 percent of the Earth's crust. The metal is found combined in practically all rocks, sand, clay, and other soils. It is also present in plants and animals, natural waters and deep-sea dredgings, and meteorites and stars. The two prime commercial minerals are ilmenite and rutile. The metal was isolated in pure form (1910) by the New Zealand-born U.S. metallurgist Matthew A. Hunter by reducing titanium tetrachloride (TiCl4) with sodium in an airtight steel cylinder. After 1947 titanium changed from a laboratory curiosity to an important structural metal commercially produced by the Kroll process (magnesium reduction of the tetrachloride).

Here is another abstract, taken this time from the 1988 edition[38] of the *Encyclopedia Britannica*, under the general heading of "Titanium", showing the year of sudden interest.

Not mentioned in this extract is who showed the interest. Of course, it had to be the U.S. military showing the first and greatest interest going by the *Enbcyclopedia Americana* quote mentioned previously. Nevertheless, who specifically in the U.S. military had began the interest has already been determined from Air Force RAND. But after 1947, we again see who re-commenced the interest into titanium and titanium-based SMAs as we shall see in the next chapter.

Although this last abstract shows the year when titanium was "discovered", it is worth noting that more recent editions of this now U.S.-based encyclopedia[39] no longer mention the year, suggesting nothing dramatic happened in the U.S. after 1947. It is like titanium is now seen as just another metal making its usual appearance in the U.S. or some other part of the world scene and never really garnered the interest of the U.S. military or anyone else for that matter. Why remove this important information?

Further evidence for the change in the status of titanium after 1947 can be seen in the number of abstracts published on titanium and titanium alloys in scientific literature. An analysis has found a significant jump from around 12 and 11 for 1946 and 1947 respectively to more than double, or 31 abstracts to be more exact, in 1948 (as mentioned in the *Industrial Arts Index* for 1946-48)[40].

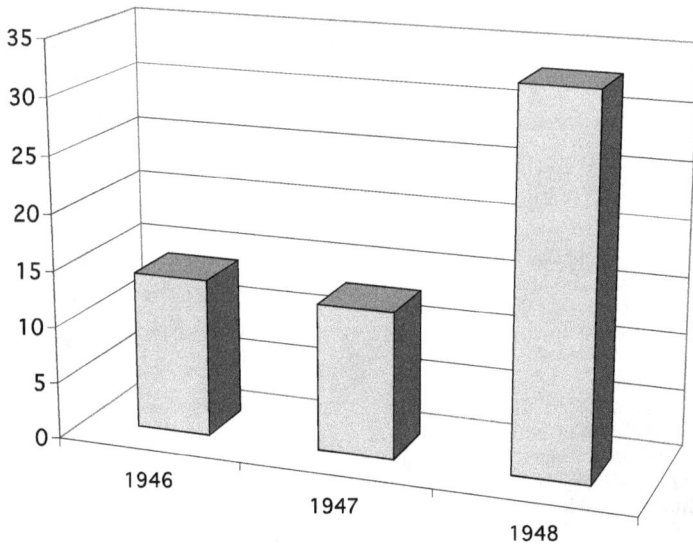

Why 1948? It is because the U.S. Bureau of Mines had completed its work on commercializing the Kroll method, made samples available to the industrial community, and was licensing the technology to interested U.S. companies willing to manufacture the metal in 1948. Because as soon as it was ready, E. I. du Pont de Nemours & Co., Inc., quickly seized on the technology where it soon announced the availability of commercial titanium and so commencing the civilian titanium industry in 1948. In fact, the Du Pont company was doing so well, it succeeded in producing (by late 1948) around 100 pounds of titanium per day at a cost of US$5 per pound[41]. Obviously not cheap in those days, but certainly affordable to the U.S. military compared to 1946 or 1947. As the 1983 edition of the *Encyclopedia of Chemical Technology* stated:

> "In [1948], E.I. du Pont de Nemours & Co., Inc., announced commercial availability of titanium and the modern titanium metals industry began."[42]

As for the U.S. Bureau of Mines, it was able to produce batch sizes of 104 kg of titanium by the end of 1948.[43]

But this wasn't the only change to take place in 1948. There was also the question of purity in titanium, as well as the interest expressed by the USAF to get U.S. scientists to look at titanium and certain titanium-based alloys.

The USAF request for greater purity in titanium

According to Henry K. Adenstedt and acting USAF project engineer, First Lieutenant William R. Freeman, in their Wright Air Development Technical Report 53-109 dated April 1953 titled "The Tentative Titanium-Silver Binary System", it is claimed the New Jersey Zinc Company produced 99.9+ percent pure iodide titanium[44] considered the purest available. Before the iodide method was used, the company operated facilities that converted the traditional titanium tetrachloride gas into liquid for various industrial uses at a time when the Du Pont company was manufacturing Process A titanium using the Kroll method.

Still, the purity of iodide titanium was not enough to satisfy the stringent requirements of the USAF. For some reason the USAF had a need for much purer forms of titanium. It seems someone at Wright-Patterson AFB had an insatiable quest for developing highly pure titanium-based binary alloys.

So the Battelle Memorial Institute in Columbus, Ohio, was issued a new contract (the previous being Air Force Project RAND) by the USAF commencing on May 18, 1948 to gather data on selected titanium-based alloys, and to look at ways of attaining higher levels of purity for this new study.

The work was not officially known to other scientists at the time. It was carried out in secret, and several classified and unpublished reports were written by Battelle for the military.

For example, we now know Battelle scientists were asked to focus on the chemical structure called *titanium tetraiodide* (simply known as iodide titanium) to determine just how well the structure could hold up to different temperatures over time. There was a concern that oxygen could still be potentially reacting with the titanium in the supposedly stable structure. All the while, the USAF was not supplying any of its own (it would have to be secret) knowledge or technology to Battelle to help speed up the process and explain how to make more pure titanium and its alloys. Quite remarkable considering we are told the Roswell flying object is supposed to be a secret military experiment (either a weather balloon, or something else the public has yet to be told about). The existence of what appears to be a titanium-based shape-memory foil for the skin of a presumed "man-made" flying object should be sufficient grounds to suggest the USAF had already figured out the issues of making pure titanium and its alloys. Yet, this does not appear to be the case. It seemed the USAF was not happy with the purity of titanium alloys by early 1948. And the military could neither continue working in secret on their own in this area, nor did they supply its own knowledge or technology to Battelle scientists to make sure titanium-based alloys for the new study were produced in sufficiently pure form. Something odd happened here.

We will discuss this further in the next chapter.

At any rate, as soon as this request from the USAF came through, and Battelle made headway into higher purity titanium, U.S. scientists finally began to appreciate the new engineering benefits of titanium, especially for the aerospace industry such as rocket and aircraft components, when the purity exceeded the unprecedented 99.99 percent. This is why the 1983 edition of *Encyclopedia of Chemical Technology* stated:

> "Titanium metal has become known as a space-age metal because of its high-strength-to-weight ratio and inertness to many corrosive environments"[45]

And on page 126:

> "Titanium metal was first established as a material for aerospace, 'Metal-to-air' applications. In the late 1970s, it was developed as 'metal-for-sea' uses."

In other words, the useful engineering properties of pure titanium can be expanded or improved significantly simply by adding the right amounts of other elements.

Evidence for this higher level of purity attained in titanium (resulting in excitement among U.S. scientists and a sudden status change in the metal) can be seen in the article "Titanium" in the April 1949 issue of *Scientific American*:

> "...[It is] a rather common element that has been known for more than 150 years....Yet as a useful metal, titanium has

been discovered only within the past decade....It now seems to be on the threshold of a brilliant career....Metallurgists and engineers believe that titanium will soon become one of the world's most important metals.

Titanium is a new metal because it has only recently been refined sufficiently to reveal its properties. It is still produced on an experimental basis...but the metal is under investigation in some 100 laboratories, and the investigators are almost uniformly enthusiastic about its possibilities. Their studies of course include the alloying of titanium with other metals."

One should bear in mind scientific articles of this nature are normally published six months after the authors have researched and written them to maintain a certain sense of relevance to readers. Therefore, this article published in April 1949 must be referring to events in 1948, not 1949 or 1947.

Additionally, as the article suggests, U.S. scientists "in some 100 laboratories" were "almost uniformly enthusiastic" over the metal's potential, because they were able to sufficiently refine the metal to an unprecedented level of purity. We are not talking about the standard 99.95 percent as the U.S. Bureau of Mines had achieved in 1946, but at least 99.99 (and probably exceeding 99.995) percent by no earlier than 1948 with the work being particularly useful in "the alloying of titanium with other metals".

So whatever the USAF was doing in 1947, we definitely know on an official level that the purity level needed to produce nitinol and its shape-memory effect was achievable by 1948, or more specifically sometime after May 18, 1948. The only thing is, could this purity level have been reached by July 1947? The answer will depend on what level of technology and knowledge was available to the USAF and Battelle to make this happen. The fact that the USAF had to ask Battelle for help again after 1947 to look at this problem would strongly suggest the technology and knowledge was not available to the USAF or Battelle in 1947.

And more concerning in this regard has to be the quantity. Even if the USAF could have found a way to make highly pure titanium alloys in 1947, the quantities observed in the Roswell foil while it was strewn about on the desert floor would reveal another technological hurdle for the military. As we will discover in the next chapter, the world's first titanium aircraft was built by the USAF between 1952-53. Why wait for the quantities to be available and do it all again if the quantities had already been achieved in 1947? With no one officially making highly-pure titanium alloys in commercial quantities to build a flying machine carrying a number of "victims" in 1947, it raises another interesting dilemma for the USAF.

Were there other methods for making pure titanium?

Maybe the USAF knew of another secret way of producing pure titanium, and they didn't tell Battelle about it. Perhaps. But as the U.S. Court of Claims Decision No.112-58 dated January 24, 1964 stated on page 7:

> "Dr. John P. Nielsen, plaintiffs' expert, head of the Department of Metallurgic Engineering at New York University, specializing in titanium, defined the Kroll process as the reduction of titanium tetrachloride by the use of magnesium at an elevated temperature under controlled atmosphere or vacuum. He estimated that from 1948 to 1961, 98 percent or more of the titanium produced in the United States was produced by the Kroll process."[46]

This leaves us with roughly two percent of the titanium being produced by another method. However, the only other method for making higher purity titanium is by using titanium iodide. And as we shall see in the next chapter, the USAF contract with Battelle to refine the metal after 1947 was not on developing a brand new technology to make super pure titanium in vast quantities. The focus for the scientists was on the titanium tetraiodide structure to determine how stable it was at various temperatures over time. No radically new method of purifying titanium was discovered. Not even Prof. Nielsen knew of another way. The Kroll method remains the only way to make reasonably pure titanium. And with slight modifications, the presence of iodide allowed higher purity titanium to be produced to the point where a shape-memory effect in a titanium-based alloy can eventually be detected.

242

If there were other ways of extracting and purifying titanium, the USAF wouldn't or couldn't say. Instead Battelle was allowed to focus its attention exclusively on the titanium tetraiodide structure. It suggests the Kroll method and titanium iodide were the best available scientific method, certainly at the time, for producing the purest titanium and its alloys.

Titanium quantities

As mentioned earlier, the American conglomerate E.I. du Pont de Nemours & Co., Inc., (commonly referred to as DuPont) was known to be producing titanium at roughly 100 pounds per day in late 1948. With this fact in mind, it seems natural to ask how much titanium had been produced for the entire year of 1948? And what about before and after 1948?

Two online sources have confirmed the production of titanium in 1948 was 3 tons[47]. By 1953, more than 900 tons was available. And, in the 1980s, it reached 20,000 tons[48]. Nothing is mentioned for the 1947 period or earlier, suggesting either no titanium was manufactured, or the amounts were too limited to produce anything substantial such as a large flying machine.

Considering some titanium had to be supplied for analysis in Air Force Project RAND, perhaps it would be safer to assume there was some titanium produced, but not enough to build anything substantial. What little was produced was restricted to small samples for experimental purposes—namely to analyze the physical properties of some alloys. But, even if the quantities were high enough on an unofficial level to build a secret titanium flying machine, was the titanium produced in 1947 of sufficient purity to make a titanium-based SMA of the likes of NiTi?

It appears not on an official scientific level. Firstly, scientists only became aware and expressed enthusiasm about titanium and its alloys after reaching a critical purity level exceeding 99.99 percent in 1948. And secondly, the quantity of highly pure titanium were not sufficient to build a flying object of the type to carry several small pilots even by late September 1949. As mentioned by inventors Schuyler A. Herres and Thomas K. Redden in their patent titled "Method of Aging Titanium base Alloys":

"Pure titanium metal which, itself, is not commercially obtainable, is very soft and ductile."[49]

This statement was made at the time when the inventors submitted their application on September 29, 1949.

But what about unofficially?

To answer this, we need to analyze two Battelle/USAF Progress Reports on the development of titanium and its alloys, declassified since 2008; and to look at one scientific article written by the head of the Battelle titanium research grouped contracted by the USAF, Dr. Charles Craighead, to determine the state of technology for making pure titanium in the first half of 1947. Further details will come in the next chapter.

What was happening in the world of SMAs?

Leaving aside these interesting titanium developments and link with the USAF, on the SMAs front we find lots of interesting snippets of information emerging from scientific literature, with inexorable links to the U.S. military after 1947.

In 1948, we begin to see evidence of a potential interest in SMAs through an emphasis in martensite transformation studies. Scientific literature for 1948 would reveal for the first time a request for scientists A. B. Greninger and A. R. Toriano to obtain a complete review of all pertinent literature on *martensite transformation*. Greninger is the same scientist who happened to have worked with Mooradian on the SMA CuZn in 1938.

It is unclear who made the request, but we see the work had to be done *after* 1947. The first *official* work into understanding the mechanics of shape-memory effects by focusing attention on martensite transformation in various alloys by Greninger and Toriano is titled "The Martensite Transformation", and was published in *ASM Metals Handbook*, 1948, on page 263. However, no mention of NiTi was given in this article.

The Russians were a bit slow off the mark in this field, with evidence of interest in one known SMA in 1949 as previously mentioned by Ryhänen:

"The basic phenomenon of the memory effect governed by the thermoelastic behavior of the martensite phase was widely reported a decade later by Kurdjumov & Khandros (1949) and also by Chang & Read (1951)."

As we recall, thermoelastic behavior is another term for te shape-memory effect. Yet the term "shape-memory alloys" was not yet in the scientific vocabulary at the time. Thermoelastic behavior (like pseudoelastic, or "shape-recovery") was still seen as an anomaly and not even the Russian scientists had figured out the mechanics of how it worked. All they knew was that the behavior was related to the martensite phase.

It certainly reveals the infancy of SMA research at the time in the sense that no scientists were discussing whether to classify CuZn and the original world's first AuCd pseudoelastic alloy under a new branch of science called *SMAs*. The alloy chosen for the Russian study would remain a scientific oddity compared to the range of other presumably more useful engineering metals and alloys for commercial and military use at the time, including those with titanium.

Nevertheless, one thing is certain. The Russian scientists G. V. Kurdjumov and L. G. Khandros (named in the article) made sure the link between martensite phase and thermoelastic behavior was firmly established by 1949.

With all this talk of martensite transformation studies in 1948 in the U.S. followed by the Russians joining in with an article linking the martensite phase with pseudoelastic alloys in 1949, the next pressing question that naturally comes to mind is, why not earlier? Why not, say, 1947, or mid-1947 to be more relevant to this discussion? Maybe we should be asking the USAF (and why not Battelle?) for the answer if it is willing to explain.

Before we get to that point, let's discuss the alloy the Russian scientists were studying.

On closer investigation of the Russian work published by Kurdjumov and Khandros in *Doklady Akad. Nauk (Russian Academy of Sciences) SSSR*, Volume 66 (1949), pp.211-214, we noticed the Russian scientists were actually studying the peculiar thermoelastic properties of the martensite phase of the AuCd alloy. There was definitely no study of NiTi. Does this mean no one in the world was aware of a possible shape-memory effect in certain alloys within the titanium-base family?

Leaving aside the USAF for the moment (until the next chapter), there is no evidence in the official scientific literature of any U.S. scientist outside of a USAF contract studying a titanium-based SMA—not even NiTi.

The Russians, on the other hand, would eventually make a mention of NiTi. This would come in 1952 as this abstract[50] shows:

> The Ni-Ti alloy contained 1.8 at. % Ti and was made by using a 4 wt. % Ti master alloy. Segregation of Ti was present both in the cast alloy and after annealing for 50 hrs. at 1100°. Areas of low absorbing power were removed by annealing.

This is the earliest known involvement in nitinol (NiTi) by the Russians, according to *Chemical Abstracts*, but not for nitinol's shape-memory property. Indeed, the lack of a technical term such as martensite transformation, pseudoelastic effects, or something along those lines for NiTi is unusual, considering the Russians had already made the connection with AuCd in 1949. Therefore, one must conclude that the Russians had no knowledge of SMAs in the titanium-based family by 1952.

Unexpectedly, an important footnote for NiTi was discovered in C. J. Smithells' (editor) *Metals Reference Book* (London: Buttersworth Scientific Publications) of the 1950s. This footnote would have important implications to the Roswell research. More specifically, it would prove an undeniable link between NiTi and secret research into this alloy by Wright-Patterson AFB by 1949. This is well before anyone else in the world knew the importance of NiTi. Apart from making a crude sample of NiTi by 1939, this is the next moment we hear about NiTi and the one closest to the 1947 event. And it is remarkable to see the USAF is somehow involved in this alloy. If we recall, the dark-grey Roswell foil with its elastic or shape-memory behavior was sent to Wright-Patterson AFB in July 1947 for scientific analysis. With enough military witnesses agreeing it was a metal or alloy of some sort, we are definitely dealing with an SMA. And with one General who had gone on the record as claiming the Roswell foil probably contained titanium to explain its toughness and high temperature resistance, it just makes the foil look all the more interesting. Because when we look at scientific literature for a distinctly dark-grey SMA, the only one we can find happens to be NiTi. Or it could be an NiTi-X alloy with small additions

246

of a ternary element. So long as the quantities of Ni and Ti are substantial, it is the only known SMA having the naturally dark-grey appearance. And it just so happens to be the world's most powerful of its type (certainly at the time the USAF starting looking at it by 1949). One would be hard pressed to find another dark-grey SMA or to overlook this rather obvious connection.

Is this a coincidence?

The discovery of this important footnote began in 1994 when an Australian researcher approached Russian metallurgist Dr. Andrzej (Andrew) Calka. While working at the ANU between 1987 and 1997[51], Dr. Calka had been specializing in amorphous alloys and later methods of combining metals to form alloys under high mechanical pressure without heat at the time. The researcher asked for Dr. Calka's insights into NiTi in terms of how it worked, how to manufacture the alloy in pure form, the latest and most detailed phase diagram of the alloy and crystalline structure details, and any information he may know about the history of the alloy. Dr. Calka mentioned to the researcher that he did combine Ni and Ti under pressure to create NiTi. The researcher was shown the machine he used in another room. He was also aware of its shape-memory effects but thought it was old knowledge.

The researcher asked if he had articles on NiTi. Dr. Calka went into another office and brought back with him several metallurgy books including the one from Smithells.

On inspecting the footnotes of the NiTi article in the Smithells book, the researcher discovered footnote number 10—it showed clear and undeniable official involvement by the USAF in nitinol research by 1949, and potentially as early as 1948 through an unknown first progress report. Shown below is the footnote:

1053   Ni-Tl

9. J. R. Long, E. T. Hayes, D. C. Root, and C. E. Armantrout, *U.S. Bur. Mines Rept. Invest.* 4463, 1949; J. R. Long, *Metal Progr.*, **55**, 1949, 364–365.

10. C. M. Craighead, F. Fawn, and L. W. Eastwood, Battelle Memorial Institute, Second Progress Report on Contract AF 33 (038)-3736 to Wright Patterson Air Force Base, 1949.

11. H. Margolin, E. Ence, and J. P. Nielsen, *Trans. AIME*, **197**, 1953, 243–247.

12. A. D. McQuillan, *J. Inst. Metals*, **80**, 1951-1952, 363–368.

13. A. D. McQuillan, *J. Inst. Metals*, **82**, 1953, 47–48.

The specific contribution provided by footnote 10 to the article is as follows:

"According to unpublished work by [10] on Mg-reduced titanium-base alloys, the eutectoid point would be located (by extrapolation) at 7wt. (5.8 at.) % Ni and 765°C."

We now have official scientific confirmation for the involvement of Battelle scientists under contract with the USAF at Wright–Patterson AFB to study NiTi. Moreover, it shows the USAF were indeed revealing more than just a fleeting interest in a notable titanium-based SMA. We also find at the same time purity was a major issue for the military for some reason (unfortunately this interest would come after 1947) as indicated by the use of a magnesium-reduced form of titanium to help extract titanium to a pure form (which is probably just another term for applying the Kroll method in extracting and purifying titanium).

Just how pure are we talking about?

According to the date of the Second Progress Report as shown in this footnote, this was 1949. That's about a year after U.S. scientists could officially produce 99.995 percent titanium for experimental purposes.

Realizing the importance of this discovery, the researcher asked Dr. Calka to read and give his comments on the witnesses' quotes in Berlitz and Moore's book, *The Roswell Incident* (1980 edition), concerning the color and behavior of the metallic foil. Calka stated that nitinol could behave and look in this way, but was not certain based on the quotes alone. He was also not aware of the full history of nitinol other than saying the alloy was discovered in the 1960s. With nothing much further he could add, the researcher left with photocopies of the NiTi articles to assist with this research.

There is absolutely no doubt that the USAF was officially involved in NiTi research in 1949 and almost certainly before this time with the existence of an *unpublished* First Progress Report, suggesting more information about NiTi may exist. So where is this First Progress Report, and when was it produced?

An even bigger coincidence was the involvement of Wright–Patterson AFB in the study of NiTi. As we recall, Wright–Patterson AFB officially received the Roswell wreckage in July 1947. We are talking about wreckage consisting mostly of a high-temperature resistant, hard and strong dark-grey metallic foil that looked and behaved like nitinol.

In fact, the use of a blow torch to test the Roswell foil and the kind of high temperatures this piece of equipment can reach would quickly put the foil into the category of a metal (or alloy). There is no known polymer (or plastic) even by today's standard that can withstand such high temperatures. And apart from multilayered graphene (we will discuss more about this material later in this book), no polymer in the newspaper-thin dimensions would have the required hardness to resist cutting and piercing with a sharp instrument. Depending on the way SMAs work at the crystalline level, we know NiTi gets very hard in the austenite phase when heated. This is due to the way the atoms get into the most compact crystalline structure possible. Well, this is the sort of thing you would need to make a material to resist cutting and piercing. Therefore, SMAs have the potential to explain all the observations of the Roswell foil.

However even if the Roswell foil is a metal (or more accurately an alloy because of its shape-memory property, as no metal made of one element shows this property) we must again ask, is this Roswell foil-NiTi link a coincidence?

As we continue the work into the history of SMAs, further research has uncovered another gem; this time from reputed top metallurgist Pol E. Duwez, who published a metals article titled "The Martensite Transformation Temperature in Titanium Binary Alloys". It can be found in *ASM Transactions*, 1953, Volume 45, p.934. This article indicates titanium-based binary alloys (of which NiTi is an example) were getting the special treatment (in terms of understanding the martensite transformation temperature) for some unknown reason. Who was pushing for this kind of research, and why titanium binary alloys?

Then we see another U.S. military link to this research through the work of Y. C. Liu.

The young graduate, Liu, of the Research Engineering Division at NYU, wrote a Technical Report No. DA-30-069-ORD-823 of interest to the U.S. DoD in November 1953 titled "Mechanism of Martensitic Transformation of Titanium Alloys". It was later revised on June 1, 1954 (currently available from DTIC, an arm of the U.S. DoD).

Liu also authored the article "Martensitic Transformation in Binary Titanium Alloys" in *Transactions of the Metallurgical Society of AIME (Journal of Metals)* in August 1956 (pp.1036-1040). The reference section for this article shows Liu had access to Report No.22 to Wright Air Development Center at Wright–Patterson AFB by authors R. P. Elliott and Prof. William Rostoker on contract No. AF-33-038-8708. He also authored an Interim Technical Report No.1 on titanium to Watertown Arsenal on contract No. DA-30-069-ORD-823.[52]

This is another link to the USAF in terms of titanium-based alloys having this familiar martensite transformation mechanism. This would imply an interest not only in certain titanium-based alloys, but also those that may exhibit the shape-memory effect under this type of transformation by 1953.

Does this mean the U.S. DoD, via the USAF, had been indirectly pushing for this kind of research without explaining the real reason for their interest to the scientists—that is, in the shape-memory effect of certain titanium-based alloys they have discovered? If so, how did it all begin? And why? In other words, did the USAF combine certain elements on their own in pure form (how it was achieved in 1947 is anyone's guess) and observed something interesting in certain titanium-based binary alloys and wanted to know how it worked (and hence the scientific talk on martensite transformations we see subsequently)? Or did a scientist at Battelle discover it and told the USAF about it?

Or would it be a little too close to the truth to say that the USAF discovered something interesting from the Roswell foil when it was picked up and analyzed at Wright–Patterson AFB? The only problem is, if the foil was "discovered" by the USAF, then who exactly made it? How? And where?

Apart from mentioning NiTi in an unpublished and classified USAF/Battelle Second Progress Report (as revealed in a scientific footnote), none of the articles uncovered from scientific literature provide a definitive picture as to whether the USAF had known about SMAs. What is clear from scientific literature is that there was an interest after 1947 into martensitic transformations (where we know there is the potential to show a shape-memory effect) leading up to identical work into titanium-based binary alloys. It is here where we find the USAF showing the greatest interest in this work with titanium alloys, including the world's most powerful SMA, NiTi, in 1949. Also named in scientific literature in association with NiTi research is Wright–Patterson AFB—the very same people who received the dark-grey Roswell foil with its alleged shape-memory effect, as observed by the witnesses for analysis. If we didn't know witnesses saw a nitinol-like metallic foil in action in July 1947, all this would be telling us is that we cannot categorically deny the possibility that the USAF had no interest in SMAs, including those containing titanium as the base ingredient by 1949 and as early as July 1947. In fact, it may be possible to argue that the USAF saw a metallic shape-memory polymer or some non-nitinol SMA, so long as the material can be identified and tested to see if it would satisfy the observations of the witnesses. But since witnesses have mentioned it, and there are plenty of military witnesses willing to state it was a metal or alloy of some sort, and the foil was sent to this specific military base in Ohio for scientific analysis only to discover NiTi was an alloy of interest, makes it highly unlikely this was a coincidence, nor is it likely we are dealing with a non-nitinol material (not even a non-metal). Actually, we would have to say the USAF had known about SMAs, whether it was an alloy it picked up off the desert floor in New Mexico, or something it had created as part of a secret military experiment. In fact, the USAF should have known about this shape-memory effect of alloys by July 1947 because enough military witnesses were confident the foil was a metal or alloy of some sort (well, to be more accurate, it had to be an alloy considering no metal in the periodic table has been observed to display a shape-memory effect on its own and certainly not in the dramatic way shown by the Roswell foil), and the witnesses who observed the foil were certain it displayed a shape-memory effect. So once the foil reached Wright-Patterson AFB for analysis, there is virtually no way the USAF could not have seen this effect as well.

The only question is, was it a titanium-based shape-memory alloy?

Well, we have seen the quote from Brigadier-General Arthur E. Exon where he claimed the foil may have contained titanium. That is about the closest one can get to figuring out the likely composition of the Roswell foil. So far the USAF is not yet willing to confirm if the foil does contain titanium. And no, we cannot expect the USAF to serve to us on a platter the remaining element(s) making up this Roswell SMA to the public. That is something other people will have to find out.

But what about more specifically NiTi?

And is there evidence of other SMAs that the USAF had an interest in and had been studying at the time to help make this situation seem more plausible?

Certainly the dark grey color shown by NiTi and the Roswell foil is rather telling. And certainly both the Roswell foil and NiTi are tough materials. If one had to choose a known SMA today that could match the toughness and color mentioned by the witnesses including the blow torch test, NiTi would be way up there as a prime candidate. But can we be sure about this? Or is it possible for another SMA to be responsible for the observations in 1947?

Let us see what more we can learn.

The work of Prof. John P. Nielsen

Further scrounging around in scientific literature has revealed three more U.S. scientists involved in NiTi research at around the right time, all working at NYU, with one having a direct link with the USAF. They were Prof. John P. Nielsen (1911-1989), Dr. Harold Margolin and his associate Mr. E. Ence, together with assistance from an MIT student earning a PhD in metallurgy at NYU, Mr. Edward Roy Stover.

Nielsen[53] is a person of interest. According to his obituary[54] in *The New York Times* on August 15, 1989, Nielsen worked at Phillips Laboratory[55] prior to the summer of 1947 while earning a PhD in metallurgy from Yale University. A check online reveals that Phillips Laboratory was part of Wright–Patterson AFB, located at the time at Kirkland AFB, New Mexico. During his tenure at USAF, Nielsen became involved in space-related research where he applied his considerable scientific skills in x-ray diffraction and other metallurgical knowledge while completing his PhD.

Nielsen eventually completed his PhD sometime in the first half of 1947. Not long after, he came out of the laboratory and entered NYU to become a full professor in the Department of Metallurgy. According to former NYU archival assistant Steven D'Avria, this occurred in the "summer of 1947".

On rechecking his claim, the new NYU archivist assistant, Stephanie Schmeling, was unable to directly confirm this summer period. To make it more difficult, the NYU did not retain in archive copies of a list of candidates at the time that may have applied for this prestigious position. Instead, she supplied a copy of the NYU Staff Directory for spring (March 1947) and autumn (October 1947), which showed that Nielsen's name only appeared in the latter directory. Since Nielsen was still working at Phillips Laboratory and finishing his PhD in 1947, it seems reasonable to suggest he received his PhD sometime in the spring period.

It is also reasonable to suggest Nielsen would not have arrived at the university before the summer period as students were still finishing their studies and would probably require access to academic staff, including the previous departmental professor (and say their goodbyes); certain administrative papers would also have had to be prepared and signed before Nielsen's arrival; and, of course, the previous professor would need to vacate his office before Nielsen could set up his new office. All these would take time. If this had been an ordinary take-up of his position at NYU and nothing else, it would have been better to wait for the summer period, during the northern hemisphere public holidays, to make his grand entrance into university life seem seamless and unobtrusive.[56]

However, did this occur before or after July 1947? The information gathered so far is not yet clear. All we can say is that Nielsen had definitely taken up his new position sometime between March and September, with the most probable period being sometime during the summer if one can rely on the advice of the previous NYU archival assistant. We must assume nothing unusual here.

As for the reason to go to NYU, we learn from further research that it is because the university had just acquired the latest state-of-the-art vacuum furnace for producing particularly *pure alloys*[57]. But in order for him to have unimpeded access to the equipment, he was allowed to take up the position as professor in the metallurgical and engineering department. By rapidly reaching the position of professor and taking up his post at NYU, Nielsen could freely conduct his own research into whatever alloys were of interest to him at the time.

So, which alloys did Nielsen had an indelible need to focus his sights on at the university? A check of the scientific literature[58] reveals a rather strong focus on titanium alloys, of which NiTi was definitely among them. In fact, the WADC technical report by Adenstedt and Freeman confirms this when it stated on page 1:

"The major portion of the melting and alloy preparation was conducted by New York University, Contract AF 33(616)-22, under the sponsorship of the Materials Laboratory. A description of melting techniques may be found in a paper entitled 'Titanium-Nickel Phase Diagram', by Nielsen and Margolin."[59]

Interesting to see NiTi crop up as an alloy of interest to Nielsen. What else did Nielsen study?

Based on the articles he published during his time at NYU and reports released by DTIC, Nielsen and his early assistants (primarily Dr. Harold B. Margolin, and later Mr. E. Ence) initially studied the phase diagrams of Ti-O (titanium-oxygen), Ti-C (titanium-carbon), and Ti-N (titanium-nitrogen) alloys[60] by August 25, 1949 under USAF contract number W 300 69 ORD 4477 before moving onto the NiTi alloy containing some carbon[61] in 1950. In their article, "Report on the Effects of Carbon, Oxygen, and Nitrogen on the Mechanical Properties of Titanium and Titanium Alloys", Ogden and Jaffee noted the main reason for looking at Ti-O, Ti-N, and Ti-C:

"The elements carbon, oxygen, and nitrogen have long been recognized as the most important and most troublesome impurities in titanium. Along with hydrogen, which much more recently has been found to be an undesirable impurity, they are the only elements which have significant interstitial solubility in titanium."[62]

The results obtained after carrying out this work are best explained by Ogden and Jaffee:

"These interstitial elements [carbon, oxygen, and nitrogen] strengthen titanium, although this effect is lost at elevated temperatures. At room temperatures the interstitials have little effect on tensile ductility, but at subzero temperatures they promote embrittlement. They also have a deleterious effect on toughness, notch sensitivity, weld ductility, machinability, and bend ductility."[63]

Also understood at the time was how the transformation temperature for moving between the two different crystalline structures

Prof. John P. Nielsen. (Courtesy NYU Archives)

at around 882°C in pure titanium can be changed in a titanium alloy. As Margolin and Farrar stated when they summarised the work done in the early 1950s in their 1969 article, "The Physical Metallurgy of Titanium Alloys":

"Pure titanium exists in two allotropic forms, α-Ti [the alpha phase], which is hexagonal close-packed and exists below 882°C and β-Ti [the beta phase], body-centred cubic above this temperature. The temperature at which alpha is stable is known as an alpha stabilizer and has higher solubility in alpha. An element which lowers the temperature at which beta is stable is a beta stabilizer and had higher solubility in beta....

Three important alpha stabilizers, which occur for the most part as contaminants in titanium, are oxygen, nitrogen and carbon. They are important primarily because they lower ductility of titanium."[64]

There is also another physical property of titanium alloys that Nielsen was looking at. It can be seen in the following statement from the Ogden and Jaffee report:

"The hardening effect [on titanium] of each of the interstitials follows the same trend as noted for their strengthening effects. That is, about 0.3 percent carbon, 0.2 percent oxygen, or 0.1 percent nitrogen provide about the same hardening effect."[65]

A similar situation exists for iron. By adding carbon, we know it hardens iron. Adding a little bit of carbon (or nitrogen or oxygen) can do the same with titanium.

In the case of the Roswell foil, the ability to resist cutting and piercing requires considerable hardness. Yet highly pure titanium is soft and ductile. Even a fresh sample of highly pure NiTi is soft and ductile. Indeed, too soft to resist cutting with a sturdy pair of steel-toughened scissors if the alloy is manufactured at the newspaper-thin range. The Roswell foil, on the other hand, showed an extremely high level of purity in order to reveal its shape-memory response, yet somehow the material displayed great hardness. The question is, how was this possible?

It must have taken a while, but eventually Nielsen realized that the best way to tackle the problem was to go back to basic principles: impurities from oxygen, nitrogen, and carbon naturally increase the hardness of titanium and virtually any kind of titanium-based alloy. So what happens if he added just one of these impurity elements to high-purity titanium alloys? How would each of the elements making up the impurities of the alloy affect titanium in terms of its hardness, strength, and other physical properties? How much of the impurities were needed to maximise hardness? And if the amount can be calculated, what would happen if, say, an alloy like NiTi was given a little bit of one of these impurities from the right element? Would it retain the shape-memory effect while increasing the hardness? If so, it might be possible to develop a theory of how titanium alloys at a general level can be hardened.

The work into impurities of titanium was carried out in 1949. By August 25, 1949, Nielsen and another colleague named H. K. Work, under contract to the USAF, produced a report looking at the phase diagrams of Ti-O and Ti-N. Later, Nielsen would try to look specifically at Ti-C on his own.[66]

The idea of adding carbon extended to NiTi. Nielsen wanted to see what the effect might be on hardness among other mechanical properties. However, he decided to let a PhD student do much of this work.[67]

Nielsen got his PhD student Edward Stover to study the NiTi-C system and write his thesis in the same year. Part of it was completed in 1950. A more complete version arrived in May 1952 when Stover submitted his thesis, titled *The Binder Phase in Titanium Carbide-Nickel Cermets*. Before he did, his name also appeared with several others from MIT in the March 1952 formerly classified WADC Technical Report 52-103 for the Wright Air Development Center under the title *Titanium Carbide-Nickel Cermets: Processing and Joining*[68] as if this work on NiTi-C had interested the USAF.

During the time Stover was doing his research, he received financial support from Wright–Patterson AFB according to his enlightening thesis.

Part of the reason for letting a student do this work can be seen in 1950 when Duwez and Taylor thought they would re-confirm the results from Laves and Wallbaum, refine the phase diagram for NiTi near the equiatomic and titanium-rich end of the composition, and publish the latest results. As Yurko et al. explained for some of the work done by these scientists:

> "Duwez & Taylor (1950) investigated several Ti_2X phases and found Ti_2Ni to be face-centered cubic with a density of 5.77 $g.cm^{-3}$ and lattice constant of 11.310 kX units— giving 96 atoms per unit cell."[69]

Otsuka and Ren gave more details, stating:

> "Duwez and Taylor first reported the decomposition of TiNi into Ti_2Ni and $TiNi_3$ at 800°C (and at 650°C)."[70]

In other words, the scientists were coming up with intriguing results around the equiatomic range for NiTi, especially at higher temperatures, which could be important for anyone interested in this area.

Nielsen was one person who could not ignore these results. He felt it was necessary to conduct a more thorough study into the crystalline structures in the equiatomic range for NiTi at the high-temperature range. He also knew he had the advantage of being able to produce higher purity NiTi at NYU than any other scientist in the world.[71] Here was an unprecedented opportunity for him to settle any debate in the scientific literature about crystalline structures associated with NiTi as well as to ensure his own phase diagram on NiTi was the best available.[72] By late 1950, Nielsen had focused his attention on the crystalline structures of NiTi, with assistance from Margolin. Of course, this work did not escape the interest of the USAF, specifically those at Wright Air Development Center. On realizing what he was going to do, Nielsen was quickly put under a USAF contract to perform the work.

The study was conducted in two parts. Strangely, during the study, a decision was made to not involve the USAF for the second part of the study. This was the time when a new metallurgist named Mr. E. Ence joined the team to help with the remaining aspects of the work. So while the USAF appeared as one of the research organizations involved in Part 1 of the study (the other being NYU), with Nielsen and Margolin being the only ones to have their names included, Part 2 only showed NYU. The names of Nielsen, Margolin, and Ence would appear in this final part of the report.

When the Technical Report was completed, Part 1 was given the publication date of December 1, 1950, and Part 2 in October 1951. The Report numbers for Parts 1 and 2 were AF-TR-6596 (Pt.1) and AF-TR-6597 (Pt.2).

As for the situation with Duwez and Taylor, Otsuka and Ren were able to determine the following:

> "...Margolin et al, who used higher purity alloys, did not find any evidence of such decomposition."[73]

Still, the problem of determining the crystalline structures above 500°C remained unsolved when in 1955 Poole and Hume-Rothery re-affirmed the conclusion of Duwez and Taylor by stating:

> "...(in X-ray diffraction of filings), four diffuse lines could be attributed to Ti_2Ni and five extremely faint lines were identified with those of $TiNi_3$."

In 1961, Purdy and Parr attempted to resolve the differences "by employing high temperature X-ray diffraction and metallographic techniques".[74] The phase diagram proposed by these scientists was probably the best compromise between the work of Nielsen et al. and Poole and Hume-Rothery. However, the scientists did add one important detail. As Otsuka and Ren explained it:

> "Furthermore, they found that the "TiNi" phase transforms to a phase called "π phase", tentatively indexed hexagonal at 36°C, and that the transformation is reversible. They noticed that the transformation occurs at a temperature low enough to prohibit diffusion-controlled process. Although they did not use the term 'martensitic', this is the first observation of martensitic transformation in the Ti–Ni alloy."[75]

This was probably the closest the scientific community ever got to uncovering a potential SMA in NiTi. About a year later, the U.S. Navy revealed NiTi's shape-memory effect, and almost immediately the study of SMAs took off in a big way for the scientific community. Many more years would be spent by the scientists in trying to refine the crystalline structures around the equiatomic range, this time below the 160°C mark where the transformation temperature range for transitioning between two crystalline structures would be seen, in an attempt by scientists to explain the effect.

Leaving aside all this phase diagram work and returning to the work of Nielsen, we learn that towards the end of 1951, the U.S. Army expressed interest in Dr. Nielsen's expertise by requesting that he look at a range of titanium alloy systems with a view to creating new types of armour. The U.S. Army's Watertown Arsenal group published his preliminary results under the title *Equilibrium Diagram of Titanium Alloy Systems* in March 1952.[76]

As Nielsen responded to the U.S. Army's request for suitable high-hardness, titanium-based alloys, he continued his work on binary titanium alloys with either carbon, nitrogen, or oxygen added to them, as well as the NiC binary alloy system (again we see carbon added to help with some particularly persistent problem he had at the time). He then extended his work by returning to the familiar NiTi with carbon added to it (once again he saw a need to return to this ternary alloy). Again, the professor received generous financial support from Wright–Patterson AFB for this work, which is believed to have been conducted in May 1952.

More work took place on NiTi in February 1953 with Margolin and Ence taking the reins to look at the phase diagram and crystalline structure of NiTi once again, as if trying to establish more accurate data. There was no indication of support from the USAF for this work on this occasion.[77] The results were published in *Transactions of the Metallurgical Society of AIME (Journal of Metals)* in February 1953 under the title "Titanium-Nickel Phase Diagram".

At the same time as this work was being published, the U.S. Army received the final report on titanium and titanium alloys from Nielsen under the title "Final Report to Watertown Arsenal on Contract No. DA-30-069-ORD-208". He later published yet another scientific article on the Ti-C alloy in the *Journal of Metals* in February 1953, again raising more than a casual interest from the USAF through another contract.

Nielsen started to get a little more ambitious with ternary alloys of titanium containing carbon and oxygen, carbon and nitrogen, and nitrogen and oxygen later in 1953, and he eventually added boron (B) to titanium in his mix for a more complete study in 1954. The latter alloy system again raised the auspicious interest of the USAF at Wright–Patterson AFB for the same reasons mentioned in the Ti-C alloy system.

Stover made a comeback by either completing or updating his thesis on the NiTi-C system. The thesis was submitted in February 1956, with Dr. Nielsen's signature appearing on the front page confirming the work was completed. The title for this thesis was *The Nickel-Titanium-Carbon System*.

Margolin performed further independent work on behalf of the USAF to look at the rapid decomposition of the beta to alpha phase of active eutectoid titanium alloys, as it was considered important for the military to understand the rapid crystalline changes that take place in, say, a SMA as well as ways to stabilise the two phases (if the aim is to find other SMAs). Perhaps this was the time when the USAF was developing a theory about SMAs. For example, any "decomposition" of phases is just another way of saying that there is a movement of atoms causing one crystalline structure to change into another. The is not quite a complete theory, of course. A far better explanation as to why atoms should move in a heated solid and an understanding of how to increase the hardness of NiTi would be required, but unfortunately these remained elusive for the USAF. All this interesting work was published in October 1958 for the USAF under contract number AF 33(616)-3942, but not in any scientific journal.

Not long after this, Nielsen focussed his attention on grain growth in metals and published an article, "The encounter mechanism in grain growth" in the *Journal of Metals* (Volume 18) and a book, *The Grain Coalescence Theory, in Recovery, Recrystallization and Textures* (edited by Margolin), in 1966. He decided at some point that he would move on from NYU and made an illustrious career in dentistry while living in New York until his death.

As for Margolin, he continued working at NYU and eventually became a professor in the Department of Mechanical Engineering, of Polytechnic University, located in Brooklyn, New York. From 1958 to 1967, Margolin assisted the USAF and later the U.S. Army with further titanium alloy work. In the case of the U.S. Army, he teamed up with Paul A. Farrar to conduct work under contract to find tougher, and more high-impact strength alloys. The alloy that was developed in the late 1950s for the Army combined titanium, aluminium, and vanadium to create Ti-6Al-4V. During the 1960s, this alloy was modified by Margolin and Farrar to attain higher yield strengths and an ability to cope with a wider range of environmental conditions, such as avoiding embrittlement in seawater.

In the 1970s, Margolin was free to conduct his own personal titanium research, such as grain growth and the interface phase in alpha-beta titanium alloys..[78] Although retired, Margolin became professor emeritus in the Department of Engineering Science of Yale University in July 2004, and he has continued working in the area he loves and specializes in. He still contributes to articles on titanium and dual alpha-beta phases of other alloys, such as the 2005 article, "Mechanical Properties of Alloys Consisting of Two Ductile Phases". He now lives in relative solitude in his home in New Rochelle, New York, speaking to no one about the work he did with Nielsen and NiTi (not even to his family members—Deborah, Elaine and Amy Margolin) until, one day, when he was contacted by telephone by an American researcher in the late 1990s.

As the final chapter will mention, Margolin gave a rather brief and terse remark to an American researcher who mentioned a connection between the Roswell case and NiTi. It is one of the most interesting things to come from a scientist about what should have been nothing more than a look at new titanium-based alloys for the military. Far from trying to sound perplexed and asking a question about what the connection is between NiTi and the Roswell case, he wanted the researcher to "not go there" without giving a reason. It is almost as if he was admitting a much deeper insight into the purpose of studying NiTi and its relationship to the Roswell foil and why he was working with Nielsen at the time.

Does this mean the strong emphasis on NiTi with links to the USAF at around the right time really did involve the Roswell object of July 1947? If so, did Nielsen learn the purpose of studying NiTi during his time with the USAF? And was this before or immediately after early July 1947? Or was he asked to study NiTi without any knowledge of the Roswell case after early July 1947 because the military understood the importance of this alloy to the composition of the Roswell foil and already knew him as someone the USAF could trust? Certainly his expertise was in the right area—metallurgy. He was available at the right time—working for the USAF. And he was young with a brilliant mind.

The perfect combination. The USAF must have thought he was the right man for the job.

But why NiTi specifically? What importance did this alloy hold for the USAF?

For an alloy to be given so much focus by Nielsen and Margolin and with interest from the USAF, it seems as if NiTi was chosen because there was something about it that the USAF could not figure out. Was this to do with the shape-memory effect? Or was it to do with how to increase the hardness of NiTi? Or is the interest just part of a normal request from the USAF to study any kind of titanium-based alloy it just happened to come across? The only problem is, the shape-memory effect and the importance of hardness in NiTi is precisely what was seen in the Roswell foil. If that is not enough, both materials are considered very tough, and the color of the Roswell foil seems to match remarkably well with NiTi. No need to anodize or paint the surface a dark-grey color. NiTi is already the right color. Otherwise, it is an amazing coincidence. Somehow this is unlikely to be the case. Therefore, it is looking like the military did have a secret interest in SMAs in the late 1940s.

Does this mean the interest by the USAF in SMAs was through its own hard work and ingenuity, resulting in the discovery of the knowledge behind SMAs and how they work and later tested a new SMA by early July 1947?

The biggest problem for the USAF had to be the lack of scientific knowledge of how SMAs worked before July 1947. Obtaining an accurate phase diagram would have been essential when working out what might be happening at the crystalline level. Apparently that was the job of Nielsen when he arrived at NYU, with the results arriving after 1947 if the published scientific articles on NiTi are anything to go by. Since it should be before July 1947, this is not a good start for the USAF. Then there was another problem of how to harden NiTi, which again seems to be the focus of Nielsen's work right up to the mid-1950s. Again, not good if the USAF wanted to explain its interest in this field before July 1947. In fact, we see all this work on a notable SMA known as NiTi being conducted after July 1947. Why? If the USAF had made NiTi and tested it in July 1947, should the work have not been conducted prior to July 1947?

Of course, all this assumes that the reason for the continued interest into NiTi is because the USAF knew it was a SMA. What other reason could there be? Everything mentioned by the witnesses about the Roswell foil, including the shape-memory effect, the observation that it was a metal or alloy of some sort, and its location, as well as Wright-Patterson AFB's sudden interest in NiTi after July 1947 and its need for Nielsen and later Battelle to look at the alloy are far too much of a coincidence. It the witnesses saw a shape-memory effect in the Roswell foil, we have to see this claim as likely to be true. Therefore, there is virtually no way the USAF could have missed it or claim it is not a SMA. It might suggest a metallised shape-memory polymer to get out of the problem, but there are problems with this argument as we shall see later in this book. And if the military did make this alloy, it had to know something about SMAs. Furthermore, was it NiTi or some other SMA? And how did it manage to make this material using the technology available by July 1947?

Actually, it is unlikely the Roswell foil could be another SMA with a drastically different combination of elements. Otherwise, what happened subsequently would not make sense. As history tells us, the USAF decided against using this SMA. Clearly it was not a bad SMA—far from it. It worked brilliantly. The only problem was the way it attracted lightning, but this can happen with any metal or alloy. It would have made more sense for the USAF to solve the lightning strike problem and move on from there. Then we would have expected to see this dark-grey SMA put to use in vast quantities to build new fighter jets. Unfortunately, this is not what happened. The SMA that was developed and allegedly tested in early July 1947 was never used again. Why? And why continue the secrecy over this SMA in the Roswell case if it was not going to be re-used ever again? Is it really that unique given what we have seen in the scientific literature on AuCd and CuZn, other than how strong the effect was?

But if the SMA was so secret, why decide to look at SMAs again after 1947 and then choose another SMA to look at outside of the USAF using NYU equipment? Very risky indeed for national security. And what was wrong with the original SMA? The USAF could mass-produce it using whatever technology was available by July 1947. And the physical properties were virtually the same as what could be achieved with NiTi. Choosing NiTi over this other SMA would only make it more expensive and extremely difficult to manufacture given the high degree of purity required. It does not make sense to use NiTi when the original is easier to make and has the same physical properties —unless, of course, the Roswell foil was not another SMA, but NiTi itself, and it was the only SMA the USAF knew at the time.

Or perhaps NiTi-X was the real alloy. However, it would be one where the composition was principally NiTi to give it its distinctly dark grey color, and this was enough for the USAF to ask Nielsen to focus his attention on understanding the binary NiTi alloy system (it was a little simpler and still had the highly pronounced shape-memory effect to keep the USAF busy trying to understand effect once it was manufactured at NYU—probably Nielsen was not told about it other than to produce the purest form of this alloy and give it to the USAF). Maybe the USAF had hoped Nielsen would discover the shape-memory effect in NiTi (and not NiTi-X) and later explain how it works. Then the USAF, with its name all over the contracts with Nielson and NYU, could have easily moved onto ternary or quaternary NiTi-based alloys to give the final explanation for the Roswell foil and then claim it did the work in secret and already knew how SMAs worked. Then the Roswell object could be seen as man-made, so it would not matter where the foil came from or who made it. But again, as history shows, Nielsen did not discover the shape-memory effect. And the more time passed, the harder it became for the USAF to argue it had invented the alloy. Without the theory of how SMAs work or the technology to mass-produce highly pure titanium alloys in 1947, the USAF cannot claim it did the work on its own. It needed outside scientific assistance and equipment, and that was only to start the process of making small alloy samples. It was nothing like trying to build a large flying object made with a titanium alloy.

Furthermore, the use of another SMA would not explain the decision not to use it and later return to studying a titanium-based SMA in NiTi after 1947. It had to be NiTi or some variant that was used in the Roswell object. However, without the technology to make this alloy in adequate quantities before July 1947, the USAF would have trouble explaining how it did the work. More of a reason to keep NiTi and any other work on SMAs a secret (until an equally powerful and easier to make SMA could be found).

As for predicting a titanium-based SMA, one must have a theory of some sort to explain how SMAs work. But as we have seen, Nielsen (and later Battelle) was asked to look at NiTi after 1947. Yet the Roswell foil had already been made, which means that someone already knew about SMAs by no later than early July 1947. So clearly the theory to predict a titanium-based SMA must have been known before Nielsen got involved. However, if the USAF did know about SMAs, why was Nielsen kept in the dark about the shape-memory effect? If he was told to check the prediction, he would have discovered the shape-memory effect within days of making the alloy at NYU. The scientific literature would then have acknowledged Nielsen as the person who discovered the world's most powerful SMA in NiTi (since we see Nielsen had published some articles on NiTi in the late 1940s and early 1950s) or another titanium-based SMA, and the study into SMAs by the scientific community would have commenced by the end of 1947. Unfortunately, this is not what happened.

And there really was no point in asking Nielsen to look at NiTi if the USAF had the technology and the knowledge to make an SMA in the titanium family, unless the USAF wanted to find out something about the alloy with Nielsen's help, but didn't have the equipment or know how it works. In which case, there was no theory to predict the shape-memory effect in a titanium-based alloy or how to make it by July 1947.

This leaves us with a sobering conclusion that the Roswell foil was made by someone else, not the USAF, Battelle or NYU. What was found near Roswell had been an unexpected discovery for the USAF. Any decision to seek help outside of the USAF to study a titanium-based SMA followed by a request to Battelle and Nielsen to look at NiTi means that the Roswell foil was not only made substantially of NiTi, but the foil was made by someone else That is the only explanation for focusing on NiTi after 1947 and asking for outside help.

It is starting to look like as if this Roswell foil was not invented or manufactured by the USAF.

Is this sounding a bit too disturbing? You are not alone.

Whatever the truth, it seems there is a connection between the work of Nielsen on NiTi and the mysterious Roswell foil at Wright-Patterson AFB.

Did Battelle know about the shape-memory effect in the titanium-based class of alloys?

As mentioned before, Battelle could not have made the Roswell foil because the equipment to mass-produce it was not available by July 1947. But that is not all. Not even Battelle had the expertise to predict the existence of shape-memory titanium-based alloys in 1947 (this was also true for the USAF, NYU, and anyone else). Such knowledge would require a little more than x-ray crystallographic data and microscopic analysis of the surface of alloys. Someone had to devise a new theory for chemical bonds to explain how the atoms of metals and alloys stay together and form the various crystalline structures and how these structures might change with temperature. The only time scientists at Battelle could have gathered this knowledge was after September 1950, or perhaps as early as April 1949 if some of the scientists could read French.

On January 11, 1951, Battelle director Mr. Clyde E. Williams invited the world-renowned chemical expert, Dr. Linus C. Pauling (1901-1994)[79], to visit the Institute for the Phi Lambda Epsilon lecture series on February 7, 1951. Williams stated that his technical staff would be interested to hear Pauling speak on a subject of his choice. Hinting at a suggested topic, Williams remarked that the staff already had made numerous complimentary remarks on the speech he delivered to the Meeting of the American Chemical Society on September 6, 1950 titled, "Principles Determining the Structure of Intermetallic Compounds."[80]. Clearly this would connect rather nicely with whatever alloy research, secret or otherwise, Battelle was conducting at the time. The visit would also be seen as an opportunity to meet with old friends and view some of the research being conducted at the institute. Pauling replied to Williams on January 16, 1951 confirming his acceptance of the invitation.

During his visit, Pauling chose a topic to give a lecture on to Battelle staff. It turned out to be, not surprisingly, his famous work on metals and alloys.

After the lecture, Williams wrote back to Pauling on February 15, 1951 to thank him for visiting the institute and giving his speech to his enthusiastic staff. Williams also expressed his agreement with Pauling's suggestion that "a chemical approach to the structure of metals and intermetallic compounds should be a part of modern metallurgical training".

If all this is true, then it would suggest that Dr. Pauling's theory of covalent bonds (and the existence of valence electrons) that he developed and later extended to include metals and intermetallic materials (i.e., alloys) was not available to Battelle (or the USAF) in 1947. This came after 1947. In fact, the principle of electroneutrality considered essential to understanding chemical bonds between elements was only formulated by Pauling himself in 1948. And by September 1950, Pauling extended the theory of covalent bonds to include metals and alloys. It was only after September 1950 that enough Battelle scientists had the knowledge to attempt a task such as devising a theory on how SMAs work. Dr. Pauling probably never knew anything about SMAs at the time. The visit by Dr. Pauling to the institute was mainly to acknowledge his genius and re-affirm his theory on chemical bonds in metals and alloys (as well as give a good chemistry lesson to the Battelle scientists), for which he received a Nobel Prize about three years later.

Until Pauling came along to indirectly help Battelle scientists with their own alloy research, a proper and full understanding of SMAs leading to a way to predict a potentially more powerful SMA in the titanium family would have been virtually impossible for Battelle at the time the Roswell event occurred. So how likely is it that the USAF could have figured it out on its own? Sometimes it would be better to believe in Father Christmas having come down a chimney and dropped numerous pieces of a dark-grey SMA inside a very large Christmas stocking of the military than to believe the USAF had the scientific knowhow to create a titanium-based SMA on its own. It just seems highly unlikely.

When combined with the lack of a proper high-purity furnace within the USAF to randomly combine different elements (if it had to use titanium) unless it used the NYU equipment (or why not at Battelle?), the lack of suitable equipment to mass-produce highly pure titanium-based SMA in a short period of time and another machine to pump out vast amounts of a newspaper-thin sheet[81] of the titanium-based alloy to build the skin (and presumably the structural supports[82]) of a large flying object, it is starting to look like as if whatever was recovered in New Mexico was not made by the USAF—or anyone else for that matter.

A rather disturbing thought indeed.

Study of SMAs officially begins for the scientific community

Nitinol's amazing properties were discovered in 1958 at the US Naval Ordnance Laboratory (NOL). Hence the name: Ni (nickel), Ti (titanium)—plus NOL. The discovery was an accident. When the first Nitinol ingots came out of the furnace, William Buehler, then chief metallurgist for the American navy, routinely tapped the first two finger-size bars against each other, producing a flat, leaden sound. No special surprise. But only minutes later, he found that the next two bars from the same melt rang like a bell when tapped. The only difference was the temperature: the second pair of bars was still warm from the furnace.

Not long after, at a meeting of Navy scientists, Buehler demonstrated another peculiar property of Nitinol: it can be bent repeatedly without showing signs of metal fatigue. And although Nitinol gets warm at the bend point when bent, like any other metallic alloy, it becomes cool when bent back to its original shape.

A puzzled Navy scientist who had just lit his pipe put his lighter to a piece of Nitinol wire that had buckled into a concertina shape. The metal sprang straight. "That," recalls Buehler, "was the turning point."

While the USAF should have known about SMAs after July 1947, scientists outside the USAF were still ignorant of NiTi's interesting property.

In fact, years passed without anything spectacular to note for the scientific community on NiTi when unexpectedly, in 1958, scientists at the Naval Ordnance Laboratory (NOL)[83], White Oak, Maryland, USA, made an extraordinary discovery. As this article[84] in the 1984 Australian edition of *Omega Science Digest* reveals, a U.S. Navy scientist and chief metallurgist had discovered a truly amazing thing about NiTi never seen before on an official scientific level.

It began in January 1958 when chief metallurgist Dr. William J. Buehler (born in Detroit, Michigan on October 25, 1923) of the Magnetism and Metallurgy Division at NOL (and until recently the Naval Surface Weapons Center) in Silver Spring, Maryland, was looking for another project after completing research on a series of iron-aluminium alloys.

As Dr. Buehler said:

"I found within the U.S. Naval Ordnance Laboratory an ongoing materials project which had the goal of developing metallic materials for the nose cone of the U.S. Navy Polaris reentry vehicle. The in-house project was under the direction of Mr. Jerry Persh, an aerodynamicist. I was able to attach myself to this project, and my initial task was to

269

provide physical and mechanical property data on existing metals and alloys for computer-assisted boundary layer calculations. These calculations were to simulate the heating, etc. of a reentry body through the earth's atmosphere."

Naval Ordnance Laboratory, White Oak, Maryland, USA in 1953

Dr. Buehler sought to apply his own creativity to his work by investigating newly developed alloys possessing a wide temperature-range tolerance (especially those with a high melting point) and other properties, as he stated:

> "My informational role in this project very quickly became somewhat boring, and I almost immediately began to think in terms of possibly tailoring newly developed alloys that might better satisfy the drastic thermal requirements of the reentry body."

Sometime in the summer of 1958, Dr. Buehler decided the time was ripe to combine some nickel and titanium in the equiatomic range in pure form and under controlled conditions using the latest arc-melting vacuum furnace available at NOL.

Why NiTi?

The nickel-titanium system was one of 12 intermetallic alloy compounds selected by Dr. Buehler from a list of 60 he compiled after spending a week at the U.S. Library of Congress reading a lengthy book titled *Constitution of Binary Alloys*, edited by Dr. Max Hansen. This is believed to be the 1958 Second Edition read by Dr. Buehler and is part of the McGraw-Hill Metallurgy and Metallurgical Engineering Series from New York.

William J. Buehler

In honing in on the right alloy, Dr. Buehler made finger-sized alloy ingots of the twelve, measuring 0.5 inch by 2.5 inches, which were large enough for him to perform tests. Among the ingots he made were two pieces of NiTi. Dr. Buehler grabbed these two pieces out of the furnace. The first test he did was to purposely drop the ingots on the concrete laboratory floor to determine the damping capacity of the alloy. He heard a "flat, leaden sound", as he recalled, suggesting good damping properties.

The next test determined how brittle the alloys were by striking the ingots with a hammer. This simple test showed NiTi had the highest impact resistance. It was enough for Dr. Buehler to focus his attention exclusively on the alloy.

On further investigating the alloy, Dr. Buehler discovered something unusual. Previously, when the ingots were cool, they produced an ordinary sound. However, when the next two nickel-titanium ingots came out a few minutes after the first ones, they were still warm. This time the ingots "rang like a bell when tapped together".

Dr. Buehler said:

> "Following this, I literally ran with one of the warmer bars (that rang) to the closest source of cold water—the drinking fountain—and chilled the warm bar. After thorough cooling the bar was again dropped on the floor. To my continued amazement it now exhibited the leaden-like acoustic response. To confirm this unique change, the cooled bars were heated through in boiling water—they now rang brilliantly when dropped upon the concrete floor."[85]

Dr. Buehler asked his assistant if he did anything to make the ingots different in some way. As he stated:

> "Subsequent discussions with my melter-assistant revealed that he had in no way mixed or altered the alloy compositions during repeated melting. This immediately alerted me to the fact that the marked acoustic damping change was related to a major atomic structural change, related only to minor temperature variation."

To determine what was going on, Dr. Buehler decided he would make a smaller sample of the alloy—a long strip of about 0.01-inch thickness—to play with. It was enough for him to check on some of its properties in greater detail.

A short while later, Dr. Buehler discovered yet another peculiarity of the alloy: he could bend the metal strip in different directions as many times as he liked, and it would not suffer metal fatigue. Buehler had never seen anything like it.

In 1960, Dr. Buehler was joined by Raymond C. Wiley. Wiley would generate the data for understanding the properties of the NiTi system.

Over the next few months, the data that was being gathered revealed a picture of a useful alloy in the NiTi equiatomic range, but nothing told Dr. Buehler this was a SMA and a particularly powerful one at that.

Unable to attend, Dr. Buehler sent his assistant Wiley to a laboratory management meeting in 1961 to review ongoing NOL projects. Wiley showed Dr. Buehler's NiTi strip to his colleagues and mentioned the things Dr. Buehler had observed about it. Everyone began playing with it and all agreed it was interesting. Then, quite by accident, one of the associate technical directors, Dr. David S. Muzzey, a pipe smoker, took out his cigarette lighter to heat the strip to see what happened. As the strip was heated, the strip rapidly returned to its original shape.

Once everyone saw how pronounced and significant this shape-memory effect was (and was reproducible), U.S. scientists would never turn back.

After Dr. Buehler learned of what happened at the meeting and within days of the discovery, he nicknamed the alloy nitinol in accordance with the name of the laboratory he worked for and the names of the metals he used to make the alloy (i.e. Nickel, Titanium, and the Naval Ordinance Laboratory, or NiTiNOL). Afterwards, Dr. Buehler began filling his mind with a range of nitinol applications for the U.S. Navy. Soon, other scientists outside of the U.S. Navy quickly learned about it, and almost immediately the study into a new field of alloy research swung into gear.[86]

At the time the study into SMAs began, it wasn't actually called "shape-memory alloys". Dr. Buehler had his own term: mechanical memory alloys. But this didn't really catch on. By the time NiTi emerged from NOL into the civilian world for other scientists to observe and be amazed, the term shape-memory alloys became the preferred term and stuck in everyone's mind more easily. No matter what term was used, it was clear that nothing would turn back the tide into this scientific field once nitinol was discovered. But how did it work?

Dr. Buehler was receiving questions along these lines and he needed to find an explanation.

The answer came from Dr. Buehler's colleague, Dr. Frederick E. Wang (born 1932), an expert in crystal physics who joined NOL in 1962. Dr. Wang was the first to open the door to further possibilities when he explained to Dr. Buehler how the alloy probably worked and the sorts of wide-ranging peacetime commercial applications the alloy could have outside the influence of the U.S. military.

Dr. Wang's explanation of how nitinol worked was based on a solid-to-solid phase transformation where the atoms of the alloy moved from one crystalline structure to another (a kind of decomposition of one phase structure into another and vice versa) at a critical transformation temperature, not unlike the way the crystalline structure of graphite made of pure carbon can be converted into a different structure as we see in a diamond under the right high level of heat and pressure.

Fair enough. But what causes the atoms to move when a SMA is heated? No one knew. Not even Dr. Wang had an answer for this. The exact mechanics of how atoms move in a SMA remained a mystery.

In 1958, a modest demonstration of the power of SMAs was shown by Chang and Read at the Brussels World's Fair[87]. As people observed, they were intrigued by a sample of AuCd cyclically lifted a load mass at different temperatures. The imagination of the public began to fire up. It wasn't long before people learned about a more powerful SMA called nitinol being studied by U.S. metallurgists. Sometime between 1961-63 the alloy eventually came out of the laboratory and into the public mainstream.

In the online *Wiki Encyclopedia* (2008), the emergence of nitinol into the public is considered the beginning of the nitinol age:

> "The nickel-titanium alloys were first developed in 1962-63
> by the Naval Ordnance Laboratory."

Does this mean someone at NOL had leaked information about Dr. Buehler's work to other scientists, and eventually members of the public? It is unclear, but one thing was certain, Dr. Buehler had to publicize the alloy in *The Journal of Applied Physic*[88] on May 1, 1963 in collaboration with his colleagues J. V. Gilfrich and R. C. Wiley under the title "Effects of Low-Temperature Phase Changes on the Mechanical Properties of Alloys near Composition TiNi". Otsuka and Wayman confirmed this when they stated:

> "This [shape-memory] effect...was publicized with the discovery in Ti-Ni alloys by Buehler et al." [89]

Over the next year or two, Buehler found it hard to concentrate on his laboratory work. The interest he generated by his work saw him answering literally thousands of letters and hundreds of telephone calls from the enthusiastic public and other interested individuals and groups, including the more curious members of Congress and certain high-ranking military officials[90], about nitinol and its interesting shape-memory effect.

As early as 1969, Drs Buehler and Wang worked with John D. Harrison of the Raychem Corporation in Mento Park, California. Special Cryofit "shrink-to-fit" nitinol pipe couplers were invented and sold to the USAF as a successful solution to totally sealing hydraulic lines in the F-14 fighter jets.

In 1974, Dr. Buehler left NOL to work at Virginia Polytechnic Institute. Today, Dr. Buehler has retired to Florida saying he has never ceased to be amazed at the increasing numbers of new products relying on nitinol.

As for Dr. Wang, he left NOL in 1980 to form a company of his own called Innovative Technology International, Inc. (ITI), supplying high-quality nitinol to manufacturers.

In February 2009, a personal statement[91] believed to be from Dr. Buehler was prepared and published online in 2006. The statement is interesting, for it now suggests an important variation in the accepted historical view of how nitinol was discovered. If we were to accept this latest statement, it would suggest the inventor already *knew* nitinol was a SMA, but didn't realize how significant its effect would be until after the NOL meeting.

Dr. Buehler also gives a reason for providing his personal statement online. He said:

"I have noticed with great pleasure the rather frequent reports in THE LEAF on NITINOL and its varied applications. The summer issue (Vol. VII, Issue III) emphasizes an overlooked point. That being the RECOGNITION of the inventors, NOL and U.S. Navy in the discovery and early development of NITINOL.

The RECOGNITION aspect along with the frequent misleading reference to its 'ACCIDENTAL DISCOVERY' are the two sensitive areas that need some clarification." (capitalized text as in the original)

In support of his claim, Dr. Buehler stated he observed a martensite structure in NiTi after "...stressing of the specimen's surface during metallographic specimen polishing" as well as a "ringing like a bell" in two warm samples of nitinol, which he thought was caused by a "...change at the atomic or crystalline level..."

Dr. Wang's explanation merely expanded Dr. Buehler's thoughts to state this change at the atomic or crystalline level is probably due to movement of the atoms.

By mentioning a martensite structure and realizing it has to exist in SMAs, Dr. Buehler assumed that nitinol was also a SMA. In fact, he introduces the term "shape-memory" directly into his personal statement, as if the term was known in the 1950s, even though the term was coined after nitinol was *discovered* and sometime after 1963.

But can one make this direct link between martensite structure and SMAs simply from the sounds emitted by an alloy at different temperatures? And if he did know, why didn't he called NiTi a pseudoelastic alloy like the few others found previously which exhibited this shape-memory effect? Or, if he was certain the term *SMAs* was known, why wasn't anyone else in the scientific community using this term?

It is common knowledge for a piece of steel to have a martensite structure (minimum 0.4% weight carbon) when cooled rapidly to help trap the carbon atoms inside the iron crystalline "cages". When heated, the carbon atoms can slip out of the structure, thereby creating a similar solid-to-solid change at the atomic or crystalline level. Yet, steel is not officially classified as a SMA.

Before this time, only two other alloys were *officially* known to behave in this way, namely AuCd and CuZn. These were described as pseudoelastic (to give an indication of how weak the effect was) or thermoelastic (an indication that heat was required to activate the effect) alloys, but no official classification of SMAs was introduced before 1958. Pseudo and thermoelastic alloys were still considered weak elastic alloys and thus scientific oddities, and were essentially ignored by the scientific community. To have said NiTi had martensitic transformation characteristics and, therefore, must be a SMA would have been pushing the boundaries of scientific discourse.

It would probably be more true to say that U.S. metallurgists at NOL probably *did* know something about martensite structures and how the atoms moved slightly with temperature. Certainly a critical idea when understanding a piece of steel making up the structural supports in a large building. However, from the way Dr. Muzzey and others at the NOL meeting were looking at the data on NiTi, they weren't exactly talking about SMAs. Indeed, they acted just as perplexed about the alloy as Dr. Buehler and his assistant Dr. Wiley. It is not likely, therefore, that Dr. Buehler or the others would have known nitinol was a SMA until after the NOL meeting in 1960.

Or was it after 1961? According to Confluent Medical Technologies, Buehler and his colleagues allegedly "failed to even recognise the link to martensite even though Purdy and Parr had previously reported the existence of martensite in Nitinol, dubbing it the 'p phase'."

More specifically, Purdy and Parr stated:

> "TiNi in all alloys containing less than 51 pct Ni decomposes rapidly and reversibly at 36°C to form a previously unreported phase."

The work by Gary Purdy from McMaster University, and J. Gordon Parr, can be found in the article, "A Study of T;E Titanium-Nickel System between Ti_2Ni and TiNi", published in August 1960.

It is starting to look like the people at NOL still did not know what nitinol was or why it showed an elastic response by the end of 1960 (or even if it had a martensite transformation if the work of Purdy and Parr is anything to go by), and definitely no articles were published in the scientific journals by Buehler and his colleagues to support the work into uncovering elastic alloys, let alone any understanding behind the mechanics of this elastic response.

Perhaps Dr. Buehler neglected his duties as a scientist to publish his understanding of what was happening at the crystalline level for nitinol in a scientific journal because of his public duties to answer various questions was taken up most of his time. But to say nitinol was an SMA merely on the basis of how two nitinol bars sounded when tapped together seems a bit far-fetched. The only sensible explanation is to say the discovery was an accident.

This is why virtually every article and reference on the official history of nitinol's shape-memory effect has described it as an "accidental discovery". The observations by Dr. Buehler were building up a picture of a potential SMA. In hindsight, nitinol is a SMA, the most powerful of its type. But we can't call NiTi a SMA until Dr. Buehler carried out the crucial and fundamental test to confirm the shape-memory effect—that is, by heating the deformed alloy (or applying an electrical current). Nor do we see any evidence his colleagues were acting like they knew what type of alloy it was until after it was heated.

Or was it much later?

Whatever happened, a new term was coined by Buehler to classify this new elastic alloy. It was called a *mechanical memory alloy*. After 1963, the term *shape-memory alloys* became the preferred classification (it kind of slipped off the tongue in a nicer way and even captured the imagination of the public when they heard it and realized how cool it sounded). And eventually a number of metallurgists outside NOL checked the scientific literature and realized the term *pseudoelastic* had been used prior to 1958. Until all of this was worked out, NiTi was seen as a total surprise. No one knew what to make of this metal and how to categorize it in the scientific literature. A new branch of science had to be created in order to support the new and unexpected discovery.

Once scientists outside of NOL got hold of NiTi and began the proper study, it was not long before the interest quickly turned to finding ways to change the transformation temperature outside the standard range for NiTi, as well as finding other SMAs (perhaps in the hope of finding something more powerful, as well as to better understand the mechanics behind this remarkable shape-memory effect). In 1965, scientists learned how adding a small amount of a third element such as cobalt (Co) or iron (Fe) to NiTi can retain the strong shape-memory effect, but also dramatically decrease the transformation temperature to well below -50°C. Once these new alloys were created, it inspired the first commercial application of SMAs in a product known as Cryofit—a hydraulic line coupler for properly sealing pipe connections of the U.S. Navy's F-14 fighter aircraft.

Cryofit was a success. The only niggling issue was how to assemble the product before the shape-memory effect gets activated. In the early version of this product, the pipe couplings had to be transported in liquid nitrogen (with a temperature of less than -196°C). After further research, an alternative alloy became available in 1989 known as NiTiNb. With its larger temperature hysteresis, the new alloy made it easier to handle.

Then new SMAs started to come to foray. Most notably, the high transformation temperature alloys of TiPd, TiPt, and TiAu exceeding 100°C. These were developed as early as 1970 and successfully opened up another door of opportunity and new engineering applications.

By January 1979, K. N. Melton of the Department of Metallurgy and Materials Science at the University of Cambridge, and research scientist from Brown Boveri Research, O. Mercier, were looking to understand and find ways to improve the fatigue properties of NiTi.[92] By June 1979, together with assistance from another colleague R. H. Bricknell, they hit upon a new alloy, NiTiCu. The advantage of this alloy was the transformation temperature was similar to NiTi, but with a narrow stress hysteresis.[93] In 1999, Miyazaki showed this alloy has an improved fatigue life.

After 1999, the aerospace and oil industries expressed interest in actuators capable of operating under high temperatures approaching 350°C. From here, a range of new High Temperature SMAs (HTSMAs) were developed and patented, including a few by NASA.

Today, the demand for SMAs has moved into another class of alloys capable of activating its shape-memory effect under the influence of a magnetic field.

Despite Dr. Buehler's important scientific work in discovering the world's most powerful titanium-based SMA in 1958 (even if it was accidental), and opening the doors to serious scientific research into all

SMAs after 1963, there was still another scientific discovery to be made. It has to do with NiTi's superelastic property.

The scientific history of superelastic alloys

The earliest mention of the term *superelastic* for an alloy was made in 1958 by William A. Rachinger while working at the Aeronautical Research Laboratories in Melbourne, Australia. The article, "A super-elastic single crystal calibration bar", was published in the *British Journal of Applied Physics* in June 1958. On closer inspection, it would appear as if Rachinger may have seen superelasticity in the alloy CuAlNi. Is this true?

Here is the abstract provided for this article:

> "Elastic strains of up to 4% can be attained in a copper-aluminium-nickel alloy in the form of a single crystal suitably heat treated. The high elasticity is due to a reversible martensite transformation induced by stressing. The methods of production and heat treatment of large crystals are described together with details of mechanical properties. The alloy can be machined by conventional methods. Suggestions are made for the development of other super-elastic alloys."[94]

On closer analysis of the abstract for Rachinger's article, the elastic strain observed in a copper-aluminium-nickel (CuAlNi) alloy is 4%. This is considered at least half of nitinol. Why Rachinger claimed the alloy he studied was superelastic is unclear. With an elastic strain in the non-superelastic phase for NiTi of between 8% to 10% (depending on purity and for a freshly-made sample straight out of the oven[95]), technically we should call NiTi super-superelastic if Rachinger is correct in his claim. Either that, or it was the first time he had ever observed an elastic alloy of any kind. Certainly the *British Journal of Applied Physics* had not published an article on elastic alloys prior to 1958, which would strongly support this "first time" view. If this is true, it is probably necessary to peg back the superlatives and instead call any alloy over 7% elastic strain as *superelastic*, and anything under this figure should be referred to as *pseudoelastic* (or would *semielastic* be a better term to use?). This recommendation was supported by Dr.

Alberto Teramoto in his White Paper report, "Sentalloy: The Story of Superelasticity":

> "Other shape memory alloys include copper-zinc-aluminium-nickel, and copper-aluminium-nickel, but they do not possess the combined physical and mechanical properties of nickel-titanium alloys."[96]

As elasticity is a physical property, it looks as if Rachinger had never seen a truly elastic alloy like NiTi when he studied CuAlNi.

To add further confusion to this matter, the terms pseudoelasticity and superelasticity get defined in the same way, as we can see in the following 1974 quote by Otsuka et al:

> "The superelasticity, which is caused by a stress-induced martensite transformation upon loading and a reversible reverse transformation upon unloading, has been found in a number of alloys exhibiting thermoelastic martensite transformation, such as Au-Cd, In-Tl, Cu-Al-Ni, Ti-Ni, Cu-Zn-Sn, Cu-Zn-Si, Au-Cu-Zn, and Ag-Cd. The effect in the Cu-Al-Ni alloy was first reported by Rachinger and later investigated rather extensively by Oishi and Brown, although they did not describe how the effect is related with the structure of martensites concerned."[97]

Knowing the definition for pseudoelasticity is exactly the same (see Wikipedia and more reputable sources mentioned earlier), it does not mean superelasticity was discovered by Rachinger. Rather, there is a difference in the strength of the elasticity from an induced-strain to the alloy. Some alloys, such as Cu-Al-Ni, can only be bent so far before elasticity causes the alloy to return to precisely its original shape. Exceed the strain limits, and the alloy fails to get back to the original shape. It will attempt to move back, but never does so precisely. In this circumstance, the alloy studied by Rachinger should be described as having a weak pseudoelastic nature. Superelastic alloys would come much later.

The same "weakness" issue in the shape-memory response can be identified in a 1952 article by J. E. Reynolds and M. B. Bever in their 1952 article, "On the Reversal of the Strain-induced Martensitic Transformation in the Copper-Zinc System". However, the scientists did not use the term "superelasticity" to define this behavior, nor did they recognise any significant elasticity that would warrant such a term at the time. The shape-memory effect of CuZn, although seen as reversible, was, in fact, quite weak.

Even if this is untrue, any alleged observation of a superelastic property in CuAlNi or CuZn still did not capture the interest of other scientists (not even the American medical fraternity or any Japanese metallurgist). Extraordinary. Not even the word "superelastic" in the title of a scientific article was enough to get vast numbers of scientists to read Rachinger's article and make a detailed study of SMAs (or any metal with superelastic capabilities). It seems scientists had to wait for a better shape-memory alloy to come along and see the evidence of a substantial elastic response before anything more dramatic would be done in this area.

That evidence eventually came with the discovery of NiTi. It is clear the shape-memory response displayed by NiTi far exceeded that of CuAlNi or CuZn. Yet still, many scientists were not aware of NiTi's superelastic effect. Firstly, NiTi had to emerge from the laboratories at NOL. That would not happen until 1962. And secondly, a demonstration of the alloy's capabilities was required to excite enough scientists to study the shape-memory effect, but no one was talking about superelasticity of the alloy or any other alloy in the scientific literature.

Today, scientists know of NiTi's superelastic effect, thanks to the work of Dr. Alberto Teramoto and his White Paper report. He was the first scientist to really define superelasticity. He also stated that the first alloy that was officially called superelastic was NiTi in 1986. As his article stated:

> "Dr. Fujiio Miura develops SENTALLOY, the first Super-elastic Nickel-Titanium Alloy."[98]

One could argue that this is in relation to the first superelastic product ever sold commercially. However, Dr. Teramoto's White Paper report made no mention of another superelastic alloy; iif he did, he would have explained why he chose NiTi to develop the SENTALLOY product. The work that would lead to the discovery of superelasticity did not come before the 1980s. Rather, the report makes it clear the work into superelasticity began in 1982 and was completed by 1986, and the alloy that was officially recognised as the world's first superelastic alloy was NiTi. In addition, Dr. Miura demonstrated in that report the difference between the standard shape-memory nitinol wire that was produced at the time by Unitek Corporation and the superelastic form of nitinol called SENTALLOY.

Since then, scientists have officially accepted the term *superelasticity* to describe alloys having this highly pronounced property when the removal of a stress results in a return to their original shape. Anything less pronounced should be called by the traditional name of *pseudoelasticity*.

The present

Since the official discovery of the shape-memory effect in NiTi at NOL and all other fascinating observations such as the superelastic effect, scientists have continued to be left in awe over nitinol's property. As one metallurgist was known to have said about nitinol after spending years studying the alloy:

> "It's a gift from God. Never mind what's going on, just use it."[99]

To this day, the uses for nitinol are endless. New inventions have already sprung up such as new heat engines, artificial hearts, and self-coiling/straightening satellite antennas, to name a few. Nitinol is truly an amazing material.

The big question: Did the USAF already know about these discoveries?

With all this important work into NiTi finally achieved for the scientific community after 1958, how likely is it that nitinol's highly pronounced shape-memory effect was never discovered at any time before 1958? In other words, can we say the USAF had not known about nitinol's interesting shape-memory and superelastic effects?

Already clues have emerged of the USAF"'s interest in NiTi through Dr. Nielsen from late 1947 onwards, not to mention the work into the martensite transformation property of titanium-based binary alloys that somehow grabbed the attention of the U.S. Department of Defense in the early 1950s. Furthermore, it is clear the Roswell witnesses saw a shape-memory (and probably even a superelastic) foil in action in the deserts of New Mexico by early July 1947. With many military witnesses seemingly certain the foil was a metal or alloy of some sort and one General prepared to go as far as to state the foil probably contained titanium to explain the tough physical properties as noted by the witnesses, it seems reasonable to consider the likelihood we have a titanium-based SMA. Combined with its distinctly dark-grey color looking like lead foil, the only naturally dark-grey and tough SMA we know of is nitinol. A coincidence? Well, at the very least, if we are going to explain the Roswell object as a man-made secret military experiment, the military had to know a thing or two about SMAs, including NiTi, if they built the flying object. And if not, someone else did not about SMAs. This means we will need to look more closely at the history of nitinol and SMAs in general from a military point-of-view to find out more.

In the next chapter, we will review the available evidence for the military connection to nitinol and SMAs from scientific literature and any recently released military reports on titanium research conducted by the USAF around the time of the Roswell case. By doing so, we can see how much the USAF had probably known nitinol and SMAs in general, whether the USAF could have had the knowledge and technology before July 1947 to build a titanium-based SMA like nitinol of sufficient purity, and used it in sufficient quantities to construct a 9-meter diameter flying object.

So, what can we learn from recently declassified military reports on titanium research and scientific literature? Did the USAF actually build and test this Roswell object in July 1947?

CHAPTER 8

The Official U.S. Military History of Nitinol

AS THE scientific literature has revealed, the official scientific study into shape-memory alloys (SMAs) began with the discovery of nitinol. The *Encyclopedia of Metallurgy and Materials* states:

> "Originally discovered in the near equi-atomic nickel titanium alloys (nitinols), the [shape-memory] effect has been found also in Au/Cd, Fe/Pt, Cu/Zn, Cu/Al/Ni and In/T alloys."[1]

This is not because nitinol was the world's first shape-memory alloy. Nor did the first sample of nitinol in 1939 reveal a shape-memory property. As the previous chapter has alluded, scientists already knew of two other alloys exhibiting a weak form of this shape-memory effect — AuCd in 1932 and CuZn in 1938—known simply in those days as pseudoelastic alloys. Of course, today we understand them to be "shape-memory alloys".

The words "originally discovered" in the above quote simply refers to the moment when U.S. Navy metallurgists at NOL created a new branch of science for studying all SMAs following the discovery of the most pronounced shape-memory effect ever seen through nitinol after 1958. For once the alloy was discovered, Ryhänen stated:

> "Since that time, intensive investigations have been made to elucidate the mechanics of its [nitinol's] basic behavior."

This is how we should interpret the first quote.

Nevertheless, is it true to say nitinol was never discovered and its shape-memory property recognised prior to 1958? In other words, could the USAF, with possible help from a handful of Battelle scientists have secretly known about nitinol's interesting property before U.S. metallurgists at NOL officially knew about it? Let's see how likely this scenario might be by focusing on the history of nitinol from a military perspective.

As this chapter will present some technical information, a summary of the results can be found at the end.

Air Force Project RAND

In 1946, and backed up with an annual budget to dwarf most other nations, the USAF became aware of titanium developments taking place at the U.S. Bureau of Mines. Seizing on the opportunity to test titanium among other new materials to help give the nation a military edge, the USAF issued a contract with the RAND Corporation to evaluate new materials and determine the feasibility, production methods, and usefulness of the materials for aeronautical and aerospace applications.

Known specifically as Air Force Project RAND, all funding for the project came directly from the USAF by way of the Douglas Aircraft Company. For the study of titanium, the U.S. Bureau of Mines supplied its own samples of the metal directly to Battelle where some selected titanium-based alloys were produced.

Confirmation for the existence of some titanium-based alloys produced by Battelle can be seen in the article *Titanium Binary Alloys*. As Craighead et al. wrote in the introductory paragraph:

> "Early in 1947, as one phase of the evaluation of materials for Air Force Project RAND, Battelle successfully arc

melted Bureau of Mines titanium and obtained a few basic properties of unalloyed titanium and several titanium-based alloys."[2]

According to scientific literature, the only other time titanium was ever alloyed with other elements was around the turn of the 20th century when iron was added to it to see if a harder and tougher material could be produced. As the 1983 edition of the *Encyclopedia of Chemical Technology* stated:

"During this period [early 20th century], titanium was also used as an alloying element in irons and steels."[3]

In 1946, Battelle was asked to produce a selected range of titanium alloys with other elements because the aim was to "study the properties, the welding, and the fabrication characteristics of titanium and its alloys"[4].

With Battelle at the forefront of producing titanium-based alloys, this means that a reputable scientific organization performed what is considered to be the earliest scientific study into titanium alloys. Further evidence of Battelle's involvement in this study is found in *Mert Davies: A RAND Pioneer in Earth Reconnaissance and Planetary Mapping from Spacecraft* by Augenstein and Murray:

"Via a lengthy series of reports, Battelle conducted several dozen studies on rocket, supersonic aircraft, and missile fuels, and on the materials of construction for space vehicles and for very high speed atmospheric vehicles. (It is interesting that the first study to appraise the value of titanium and titanium-based alloys as construction materials was done in this Battelle subcontract, thus suggesting the value of titanium for such vehicles as the later SR-71 and its limited use on even earlier vehicles.)"[5]

However, could the U.S. Bureau of Mines have done this work? In other words, could another organization have been involved in this titanium alloy work?

Partly on the grounds that the Battelle organization provided an independent assessment of the metal's potential, the real reason was that Battelle was the first organization in the U.S. (and the rest of the world) to have developed a new type of *arc furnace* designed specifically to handle the production of titanium-based alloys while minimizing the impurities entering the system. The U.S. Bureau of Mines handled the production of titanium, but it was Battelle that actually created the equipment to combine titanium with other elements in a way that minimised impurities in the final alloyed products.

Does this mean the new equipment was sufficient to make a titanium-based SMA? And if so, did Battelle or the USAF recommend and later manufacture in reasonable quantities a hitherto new titanium-based SMA for use in a secret flying object flown over New Mexico in early July 1947?

If anyone knew of the importance of purity in making a titanium-based SMA, it had to be either the USAF or Battelle, or both. Does this mean the USAF had conducted a secret study into titanium, thus acquiring knowledge of the importance of purity and mentioned this to Battelle? Not according to the work of Dr. Nielsen in the previous chapter. This scientist had to go outside the USAF and enter NYU to use the latest version of Battelle's arc furnace, which used a vacuum chamber. Therefore, the USAF could not have mentioned this "purity issue" to Battelle before July 1947. If anyone knew of the importance of high purity in the production of titanium-based alloys, it had to be Battelle. In fact, the involvement of Battelle in the USAF study to produce titanium-based alloys is clear evidence the USAF did not have their own equipment (secret or otherwise) for rapidly melting and re-solidifying titanium with other elements while minimizing the levels of contamination seeping into the metal. Only Battelle had the technical knowhow to build such equipment by early 1947, and even then it was not sufficient to create a titanium-based SMA. A vacuum chamber had to be added to the next generation of the arc furnace to achieve this important milestone, and that would not come until around mid-1947. And when it did arrive, the USAF appeared not to have known about any titanium-based SMA, let alone one that was made with nickel and titanium, until the decision was made to send one man to NYU to use the equipment in the latter half of 1947.

Craighead et al. officially confirmed this in their article where it is shown how Battelle became the first to develop the "arc furnace for melting titanium and other refractory metals…under Air Force Project RAND".

In other words, neither the USAF nor the U.S. Bureau of Mines built the arc furnace technology, and certainly not one with a vacuum chamber added to it. Only Battelle had the knowhow. And when Battelle developed the technology, it turned out to be the best available piece of equipment Battelle (and any other scientific community in the world) could muster for the task of alloying titanium with other elements.

Yet the technology used in Air Force Project RAND wasn't perfect. In fact, each time titanium was melted and rapidly re-solidified in the arc furnace Battelle scientists noticed that more contaminants would enter the alloy. Fast melting with electricity was not enough to avoid impurities.

Proving their worth once again, Battelle scientists showed exactly why they were the right people for the job. As soon as they noticed this impurity issue, they set out to rectify it. During this period of the study, Battelle made considerable improvements to "the design and operation of this furnace for making small ingots of titanium and titanium alloys"[6]. These improvements included the use of an inert argon atmosphere for melting titanium, as well as methods of cooling the crucible for holding titanium and the tungsten electrode for supplying the electric charge to melt the titanium.

Eventually, a point was reached when the final design for the arc furnace would allow "one man [to] make five to six ½-lb ingots of the easily melted titanium alloys per day"[7].

Impressive. However, did these worthy and necessary improvements help the ingots reach the absolute minimum 99.995 percent purity in order to reveal a potential shape-memory effect in certain titanium-based alloys produced by Battelle, assuming the USAF and/or Battelle were looking for this effect? And were these ingots in sufficient quantity to be manufactured into sheet form for the possible construction of, say, an outer skin component of a secret 9- to 12-meter diameter titanium-based flying machine for the USAF to test by early July 1947?

It all depends on the purity attained by the new arc furnace, on whether a machine existed to create large newspaper-thin sheets out of the tougher titanium-based alloys, and on what type of titanium-based alloy Battelle developed from all this important work. Even if there was by some chance enough titanium to cover the surface of, say, a large flying object of at least 9 meters in diameter and was machined into sheet form, we find a problem on the issue of purity.

As the official history of titanium has shown, there is no evidence in the scientific literature to support a higher purity claim for the 1947 period. There is not even a report from the USAF that they had requested higher purity levels in the metal from Battelle because of some amazing new secret discovery in the titanium alloys made at Wright–Patterson AFB suggesting a possible "shape-memory effect". For if they had, Battelle scientists would have shown their enthusiasm over the metal sometime in early 1947, not 1948 as the *Scientific American* article of April 1949 showed (see the previous chapter for further details).

Instead, the maximum level of purity used by Battelle at the time was equivalent to the fresh ingots coming out of the U.S. Bureau of Mines, which wasn't much better than around 99.95 percent. Of course, Battelle could always ask for better quality titanium if the scientists provided a good reason, but there is no official evidence to show that Battelle or the USAF requested the U.S. Bureau of Mines to improve the quality of the metal.

Purity of titanium alloys was clearly not the aim of Air Force Project RAND.

Even if such a request for higher level of refinement were somehow unofficially made, what would be a good scientific explanation for doing so? The only good scientific explanation for the removal of the remaining levels of impurities in titanium would be to reveal a new or unusual physical property in titanium alloys. The only property of titanium-based alloys we know of that is considered critical in having such high purity work is to reveal a shape-memory effect. This is the only reason we know of. However, without a USAF/Battelle report to confirm this, and so far none exists, we must assume the USAF and Battelle had not requested a higher purity in the metal prior to July 1947, as Craighead et al.'s article shows. Rather, the USAF accepted whatever levels of contamination were present in the titanium coming from the U.S. Bureau of Mines for Battelle to do its job. Battelle may have understood how susceptible titanium was to absorbing impurities from the air, but the amount of contamination still present in the metal was considered "relatively small" and they didn't think it was critical enough to require making the metal purer when investigating the properties of titanium and some of its alloys for the purposes of Air Force Project RAND. As Craighead et al. said:

"The contamination which occurs during the melting operation is relatively small."[8]

How small? If the work involved SMAs, it would be critical for Battelle to mention precisely how *relatively small*. Is it 0.05 percent, or 0.005 percent or less in the level of impurities? Clearly, the article from Craighead and his co-authors didn't say. The most Battelle stated regarding the level of impurities in titanium-based alloys was in terms of the metal's hardness. Generally, the more impurities the metal received, the harder the metal would get. If the hardness was considered not too high, it was assumed sufficiently pure for the purposes Battelle needed the metal for. As Craighead et al. said:

"For example, a button ingot of 50g of Process A titanium was melted and kept molten for 5 min, using the normal operating procedure. The Brinell hardness of the ingot was 130. After remelting, this ingot had a hardness of 143; and after remelting again, the hardness had increased to only 160. These values indicate that only a small amount of contamination takes place during melting."[9]

This must be clear evidence titanium-based SMAs (through a request for extremely high purity titanium) were not on the minds of people at Wright–Patterson AFB or Battelle during the Air Force Project RAND study. The work to build a titanium-based SMA flying object for the military must have happened after the study and before July 1947 assuming the Roswell object was a military experiment. Because to assume some kind of early interest in a titanium-based SMA at around the time of Air Force Project RAND would mean the impurities could not be termed as "relatively small". If one were truly serious about making a titanium-based SMA like NiTi, no scientist would ever accept a statement like that. We are talking about an alloy

291

that needs to be *absolutely* pure, as pure as can be physically achieved. You can't cut corners in something like this.

Therefore, the aim of Air Force Project RAND was to gather scientific data on the essential physical properties of titanium and titanium-based alloys in terms of tensile strength, corrosion resistance, and density to help determine how useful the metal and its alloy would be for building new military aircraft of the future.

Thus, in the end, the conclusion reached in Air Force Project RAND by March 1947 can be seen in the following quote published in the *Encyclopedia Britannica*:

> "A USAF study conducted in 1946 concluded [in March 1947] that titanium-based alloys were engineering materials of potentially great importance, since the emerging need for higher strength-to-weight ratios in jet aircraft structures and engines could not be satisfied efficiently by either steel or aluminium."[10]

Notice the absolute pittance of words indicating Battelle had made an amazing discovery in one or more titanium-based alloys. Not even words to the effect that the USAF had realized the importance that higher purity in titanium would make an enormous difference to the properties of certain titanium-based alloys. The best Battelle (and supported by the USAF) could muster was to say how titanium would have great potential for the aerospace industry.

Well, yes. Anyone can say a piece of metal will have great potential. The question is: what kind of potential? Is the potential to build a flying object containing a SMA? Or to build a flying object that will be lighter and stronger if the costs to manufacture titanium in high quantities is lowered? As far as the Project RAND/USAF report is indicating, it is suggesting the latter, not the former.

The quote is clear. The official conclusion from the report was not a ticket to unfettered research and experimentation into highly pure forms of titanium alloys by the USAF after March 1947, with or without outside scientific assistance. The USAF did not question any aspect of the conclusion. They merely accepted the findings from Battelle, the samples of 99.9 to 99.95 percent pure titanium from the U.S. Bureau of Mines, and virtually everyone else involved in the titanium study. No further questions were asked. If the Roswell object was man-made in nature, other means must have triggered the study into SMAs and ultimately much purer forms of titanium by the USAF.

And it had to be triggered sometime between March and no later than the end of June 1947. The only problem is, Dr. Nielsen's work is not suggesting that the USAF had figured out the importance of highly pure titanium-based alloys by the time the Roswell object crashed in early July 1947.

This has to be a worrying sign for the USAF if it wants the public to believe that the Roswell flying object was a secret military experiment involving a SMA.

When did the U.S. Army show interest in titanium?

Alright then. So could the U.S. Army have been the instigator in the interest into highly pure forms of titanium? And did these people tell the USAF about it?

According to the U.S. Army Material Command Historical Office published in online article titled "The Army Titanium Program at Watertown—A Brief History of a Major R&D Accomplishment", the answer is no, as the following quote shows:

> "In 1948, Watertown [Arsenal] had placed its very first research contract under the titanium program with the engineering school at New York University (NYU) to study phase diagrams in the titanium-carbon and the titanium-nitrogen systems. NYU continued to receive Army support throughout the 1950s and, by 1959, developed and patented Ti-6Al-6V-2Sn, a high-strength alloy still in wide use."[11]

The article also revealed the company responsible for supplying titanium for the U.S. Army in 1948:

> "P.R. Mallory and Sharon Steel joined forces in a venture known as Mallory Sharon Titanium Corp. (later taken over by U.S. Steel and National Distillers) that operates today under the name RMI."[12]

The U.S. Army would not stop there. Watertown Arsenal pursued further titanium research by sponsoring the Armor Research Foundation to look at the titanium-silver (TiAg) system among several other titanium-based alloys.[13] Furthermore, we learn from the previous chapter how the U.S. Army had asked Dr. Nielsen to look at various titanium-based alloys to see which ones would be promising as an effective armour. And the thing that is in common in all of this work is the fact that it all happened after 1947. Again this is not a good sign for the USAF who wants the public to believe the Roswell object was man-made.

Indeed, it is hard to ignore the fact that the U.S. Army's interest in titanium commenced in 1948, not 1947. So if the interest in titanium did not come in 1947 from the Army, or titanium-based SMAs from the USAF before July 1947, this leaves us with the U.S. Navy. But as we shall learn later in this chapter, the U.S. Navy was either preoccupied with other matters or not included in the work. This brings us back to the USAF as the only one to show prime interest in titanium and potentially titanium-based alloys prior to July 1947. But we know the USAF did not have the knowledge or equipment to conduct a secret study into highly pure forms of titanium-based alloys. Yet, unless Battelle had made the discovery about titanium-based SMAs before July 1947 (and we know the enthusiasm by the scientists was definitely not there at the time), the only other people who could raise an interest in the importance of highly pure titanium alloys would have to be the USAF.

As we recall from the previous chapter, according to the *Scientific American* article, no U.S. scientist was expressing interest over the metal anytime in 1947. Why? Because it was not pure enough to reveal important engineering properties. It means no scientist, not even at Battelle, was working on titanium to make any kind of new discovery. So who else could have done this work and knew about titanium-based SMAs?

The U.S. Bureau of Mines? Surely not since this organization was asked to produce titanium to the level of purity capable with its own technology, which was about 99.95 percent. It clearly is not the 99.995 percent or higher needed to reveal a shape-memory effect in a titanium-based alloy.

Not even the quantities of titanium officially produced in 1947 for any commercial or military purpose were available. Unofficially, perhaps something was made. But whatever was made in 1947 was entirely for experimental purposes, namely for Battelle to study a select number of titanium alloys as part of Project Air Force RAND. After that, there is no request from the USAF to mass produce titanium for any secret or otherwise experiment involving a flying object carrying a few people. Even if the request had somehow been made in secret (clearly the request had to be made since the USAF did not have the equipment to make its own titanium alloys), no Battelle scientist was expressing great interest over highly pure titanium in 1947. Only in 1948 do we see this. This strongly points to the fact that a request had not been made, secret or otherwise.

Still not sure? Well, if a titanium-based shape memory alloy was somehow developed and used in a secret flying machine by early July 1947, then the very least the USAF should have done was to figure out how such a scientific feat was accomplished. Surely reports would have been written. And if the USAF could not achieve the work, the USAF would have told Battelle about it between March and July 1947. Then all the equipment would have been manufactured and eventually great quantities of highly pure titanium sheets could have been made to build this Roswell object as part of some hitherto unknown secret military experiment.

Not according to history.

At the very least, if any report exists to support SMA studies by the USAF, they have not come to light to this very day. Certainly the equipment from Battelle to help the USAF conduct its military experiment was not forthcoming by early July 1947, officially or otherwise.

So how did the USAF achieve this remarkable work prior to the Roswell crash?

Did the U.S. Army look at SMAs?

The earliest known official interest by the U.S. Army into SMAs can be found in a document titled, *Martensitic Transformation Induced by Tensile Stress Pulses*. Written in 1985 by Marc André Meyers and Naresh N.

Thadhani of the New Mexico Institute of Mining and Technology in Socorro, the U.S. Army Research Office in North Carolina wanted to know more about the cheaper and easier to manufacture iron-based pseudoelastic alloys such as FeNi, an iron-nickel alloy, and FeNiC, an iron-nickel-carbon alloy, and to understand how fast the martensite phase transformation would occur when the surface of a pseudoelastic alloy reflects a shock wave from a "tensile pulse".

As Meyers and Thadhani stated:

> "This paper describes a systematic analysis of martensitic transformation induced by tensile pulses and proposes that athermal martensitic transformation is an ultra-rapid isothermal transformation."

In plain English, those alloys that do not require to be heated in order to show the martensitic transformation but merely flexed or stressed by a force (known as a *superelastic alloy*) and the speed at which this occurs could be used by the U.S. Army to deflect shock waves as a form of armament. And if so, very hard, thick, and heavy protective metal plates could be replaced with more lightweight, thinner, and highly responsive temperature-independent SMAs for the protection of military personnel.

Unfortunately, this kind of work was not done in the late 1940s. The U.S. Army back then was looking at titanium alloys, but mainly those with vanadium as a tough protective material. The decision to look at SMAs for the same purpose came at least 35 years after the Roswell crash.

A separate version of this document for international scientific review was published in 1986 under the title "Kinetics of Isothermal Martensite Transformations". This paper can be found in *Progress in Material Science* (published by Pergamon Press).

Since the original document was written (and scanned using Xerox Digipath equipment in January 2004, and only recently released online), Meyers continues to receive funding from the U.S. Army Research Office for the last forty years and currently works at the University of California, San Diego, USA.

His colleague, Thadhani, is the Professor of the College of Engineering at the Georgia Institute of Technology, Atlanta, USA, since September 1992. He specializes in studying the response rates of metals and other materials when subjected to shock waves. As the university web site stated:

"His [Thadhani's] research focuses on studies of shock-induced physical, chemical, and mechanical changes for processing of novel materials and for probing the deformation and fracture response of metals, ceramics, polymers, and composites, subjected to high-rate impact loading conditions. He has developed state-of-the-art high-strain-rate laboratory which includes 80-mm and 7.62-mm diameter single-stage gas-guns, and a laser-accelerated thin-foil set-up, to perform impact experiments at velocities of 70 to 1200 m/s."[14]

A renewed USAF interest in titanium in 1948

Then, on May 18, 1948, more than 10 months after the Roswell event and despite the work of Dr. Nielsen at NYU to look at certain titanium-based alloys of interest to the USAF, someone working at Wright–Patterson AFB had to ask Battelle to re-investigate new methods of achieving higher purity titanium and titanium-based alloys, and to study selected alloys, including no less than three notable titanium-based SMAs. Among those SMAs was the all-important NiTi.

As Craighead et al. noted:

"As a result of this early work, Battelle, under contract with the Air Materiel Command, Wright–Patterson Air Force Base, has continued extensive studies of titanium alloys [in 1948]."[15]

When the second call for scientific assistance finally came from the USAF, it didn't take long for Battelle to make further headway on the titanium front. Before the public knew what had happened, Battelle was already producing titanium to at least 99.99 percent purity and as high as the 99.995 percent level[16] for experimental purposes in 1948. Once the new purity level was reached, U.S. scientists became uniformly enthusiastic (an emotion that could not be contained, and, in fact, was allowed to emerge from Battelle and infect the rest of the scientific community) regarding the potential of titanium for various engineering uses, with particular emphasis on the alloy form during that year (hence the article from *Scientific American*). From then on, the status of titanium changed forever. In the case of those older editions of the *Encyclopedia Britannica* after 1947 was when the status of titanium changed, or 1948 to be more precise.

This is why Craighead et al. wrote in his 1950 article:

> "The low density, excellent resistance to corrosion, and high tensile properties of these [titanium-based] materials stimulated a great deal of interest among metallurgists and designers seeking better materials of construction."

This intense interest in titanium among metallurgists and designers outside of the USAF is the same as "uniformly enthusiastic". And that enthusiasm over the metal could only have begun after 1947, specifically 1948.

Now let's look more closely at the evidence supporting the USAF in their renewed aim to bring back Battelle for further titanium research work in 1948.

We will begin by discussing how the formerly classified second progress report on titanium research linking NiTi and Wright–Patterson AFB was obtained under the U.S. Freedom of Information.

Finding the Battelle/USAF Progress Reports

In the 1990s, the internet became accessible to ANU students and staff. In the earliest search for the Second Progress Report using Air Force contract number 33(038)-3736 from Footnote 10, nothing turned up. It suggests the report was probably still classified and held at Battelle, Wright–Patterson AFB, or the U.S. DoD in Washington. Only a U.S. citizen could make the request. However to avoid spooking the DoD by asking a U.S. citizen to request the report and reveal a possible link to this Roswell research (a bit like the situation with the late Steven Schiff), it was better to remain quiet and see what other evidence can be obtained.

On January 9, 2008, another search on Google Scholar revealed additional support for the existence of the Second Progress Report in Volume 25, pp.364-65 of a 1949 metals journal. The article in question discusses the decomposition of the beta-phase in a rapidly quenched titanium-based alloy.

On clicking this link, we learned the article is actually titled "Beta-Eutectoid decomposition in rapidly solidified titanium-nickel alloys" by Krishnamurthy et al. in *Metallurgical and Materials Transactions A*, located in pages 23-33 and published in January 1988.

The Second Progress Report of 1949 is another footnote in this article, reinforcing the view that the report does contain NiTi.

Seeking information from two U.S. Roswell Investigators

During the search, two names cropped up regularly enough as being engaged actively in the latest investigation of the Roswell Incident. Wondering whether these two U.S. investigators[17]—Thomas J. Carey[18] and Donald Schmitt—would have independent information relevant to this research work SUNRISE contacted them by email on January 21, 2008

Nothing more was heard until July 21, 2008 when another U.S. researcher, later turned out to be a close friend of Mr. Carey, named Mr. Anthony "Tony" Bragalia sent an email to SUNRISE enquiring about our work[19].

SUNRISE noticed a considerable effort to learn the name of the Australian researcher who obtained information from the ANU when making the link between the USAF, NiTi and the Roswell case. Strangely, nothing was asked about the nature of the SUNRISE work and anything that could be provided to assist with the SUNRISE research.

The Summary Report of December 31, 1952

Not long after Mr. Bragalia sent his first emails to SUNRISE, we noticed for the first time an online copy of a "Summary Report in the Development of Titanium-base Alloys", published by the Battelle Memorial Institute on December 31, 1952 for the USAF under the same contract number as the unpublished and classified Second Progress Report in Footnote 10. This report never appeared in previous search listings.

On downloading the document[20] on October 9, 2008, we discovered Contract No. AF 33(038)-3736 is a stage 2 work in the study of titanium-based alloys. At least three reports and this 1952 Summary Report covering progress work for Stage 2 are definitely identified and part of the contract as the following image taken from the report shows:

> This is the Final Report on Contract No. AF 33(038)-3736 covering the work done during the period from May 19, 1952, to December 7, 1952. The termination date of this contract was December 31, 1952.
>
> This alloy-development program was started under Contract No. W 33-038 ac-21229, May 18, 1948. The present contract became effective May 18, 1949. In addition to this report, four previous summary reports on alloy development have been issued under these contracts:
>
> Summary Report-Part III, July 30, 1949
> AF Technical Report No. 6218-Part II, June 30, 1950
> AF Technical Report No. 6623, June 18, 1951
> WADC Technical Report No. 52-249, June 18, 1952

These reports were conducted over various dates in the 1950 to 1952 period. A Summary Report Part III also appears to be part of the same contract, but this is not entirely conclusive. It probably refers to work done in Stage 1 prior to May 18, 1949. All the reports were presumably given the same title "Development of Titanium-base Alloys".

Noting that a Summary Report could also mean Progress Report (since we are looking for the Second Progress Report of 1949), and assuming "Summary Report Part III" dated July 30, 1949, is part of AF 33(038)-3736, it suggests the very first summary report for Stage 2 is probably "Summary Report Part I". Seems logical so far. Therefore, could "Summary Report Part II" be the official Second Progress Report of 1949 talked about in the footnotes of two NiTi articles? If so, what about Part I and III, and what do they contain? And do any of the other reports of this contract number contain more information on NiTi? Certainly, the current downloaded 1952 summary report contains no information about NiTi. Therefore, there has to be another report discussing NiTi.

The TiZr (titanium-zirconium) article

About this time, SUNRISE received a scientific article from Mr. Bragalia[21]. This was a TiZr article[22] titled "On the System Titanium-Zirconium" by Paul A. Farrar and Sandford Adler published in the *Transactions of the Metallurgical Society of AIME*, Volume 236, July 1966, pp.1061-1064.

REFERENCES

[1] J. H. deBoer and P. Clausing: *Physica*, 1950, vol. 10, pp. 267-69.
[2] J. D. Fast: *Rec. Trav. Chim.*, 1939, pp. 973-83.
[3] C. M. Craighead, L. Fawn, and L. W. Eastwood: Battelle Memorial Institute, Second Progress Report on Contract #33(038)-3736 to Wright-Patterson Air Force Base, October 31, 1949.
[4] P. Duwez: *J. Inst. Metals*, 1951-52, vol. 80, pp. 525-27.
[5] E. I. Hayes, A. H. Roberson, and O. G. Paasche: *U.S. Bur. Mines, Rep. Invest., no. 4826*, 1951, pp. 246-47.
[6] E. Ence and H. Margolin: *Trans. Met. Soc. AIME*, 1961, vol. 221, pp. 205-06.
[7] W. L. Finlay, J. Resketo, and M. B. Voldahl: *Ind. Eng. Chem.*, 1950, vol. 42, no. 2, p. 218.
[8] E. Ence and H. Margolin: *Trans. Met. Soc. AIME*, 1961, vol. 221, pp. 151-57.
[9] C. J. McHargue, S. E. Adain, and J. P. Hammond: *AIME Trans.*, 1953, vol. 197, p. 1199.

It shows the same footnote linking Battelle and the USAF under the same contract number. More importantly, the footnote gives a specific date for the Second Progress Report: October 31, 1949. It means we have another report under this contract number not mentioned in the final 1952 report. So why isn't it mentioned?

A brief analysis of the TiZr article shows Farrar and Adler have done more than acknowledge the existence of the classified and unpublished Second Progress Report of 1949, they have actually read it by stating in the article "...as by the limited data of Craighead et al". Furthermore, the TiZr alloy was not officially pursued by anyone else since the Second Progress Report was written until the authors decided to reinvestigate TiZr using metallographic and x-ray diffraction techniques.

Why reinvestigate TiZr? It seems the authors found a discrepancy in recent observations made by Mr. E. Ence and Dr. Harold Margolin. The discrepancy is explained as follows:

> "However, a more recent investigation of the titanium-rich region by Ence and Margolin indicated that the solubility of zirconium in α titanium at 500°C is approximately 22 pct with the $\alpha + \beta$ field extending to approximately 47 pct Zr. Therefore in order to resolve this discrepancy the following investigation was initiated."

In other words, Ence and Margolin found new and potentially more accurate data for a portion of the phase diagram for TiZr containing the critical combination of solid alpha (α) and beta (β) crystalline structures. The new data in this specific dual-crystalline region was enough to get the attention of people at Wright–Patterson AFB by assigning Mr. P. L. Hendricks of the Air Force Materials Laboratory to act as the project monitor. The USAF would officially provide sponsorship for this study.

It was interesting to see Ence and Margolin crop up in this article as they worked with Prof. Nielsen in studying NiTi at NYU after the summer of 1947. Now it seems these same people had an interest in TiZr as well. And so too. did the USAF.

Why would the USAF be so interested in this kind of work?

The clue comes when we notice how TiZr possesses some remarkably similar characteristics to TiNi.

In the case of TiZr, the dual-crystalline phase region is said to extend to approximately 47 percent zirconium (according to the work by Ence and Margolin). In NiTi, this phase exists from 45 to 55 percent nickel. And as the literature on SMAs has revealed, a change (or usually called a *decomposition of a phase*) from the alpha into the beta crystalline structure through application of heat (or an electrical current) is something metallurgists have observed during martensite transformation, with the potential to exhibit a shape-memory response as in TiNi.

Are we being led to believe the USAF's interest in Farrar and Adler's work was because the military knew TiZr was a SMA but wouldn't say? Even so, why the interest in TiZr? The article was published in 1966. If it relates to any secret study into SMAs conducted by the USAF in the late 1940s, hasn't the USAF already worked out the scientific explanation behind the shape-memory effect? Or was the USAF still figuring it out in the mid-1960s and was using this as an opportunity to gather the latest and most accurate results from the NYU study to help obtain a possible solution to the problem?

If the USAF had some kind of secret interest in shape memory alloys, there would have to be something in the TiZr article to give credence to this claim.

Looking at the article, we see there are no direct words to the effect that TiZr exhibits a martensite transformation to support a potential shape-memory response. Could this be because the scientists and, therefore, the USAF did not know about SMAs? Or could it be that the USAF already knew but did not want to tell Farrar and Adler about the shape-memory response but wanted to let them study TiZr to understand the alloy better? And if any samples were made, someone from Wright–Patterson AFB could easily come in and test them secretly for their shape-memory response. Possibly. Unfortunately we don't know.

What we do know is that TiZr is indeed a SMA, although not as pronounced as NiTi. Evidence for this can be found on page 3 in the article "Ab initio Ti-Zr-Ni phase diagram predicts stability of icosahedral TiZrNi quasicrystals" by Hennig et al. (published in *Physical Review* B 71, 144103 (2005)):

> "Several experimental studies investigate the properties of ternary Ti-Zr-Ni alloys. Alloys near the composition $Ti_{50}Zr_xNi_{50-x}$ show a reversible martensitic transformation leading to shape-memory behavior.[42]"

Footnote number 42 shows the source for this information, an article written by S. F. Hsieh and S. K. Wu published in 1998 in the *Journal of Alloys and Compounds* 266, 276. It means that when x=50, $Ti_{50}Zr_xNi_{50-x}$ effectively becomes TiZr, thereby, supporting the existence of this behavior by metallurgists.

It is remarkable to see two SMAs present in the Second Progress Report of 1949. Could this be yet another coincidence? Since TiZr is a SMA, does this mean the USAF was interested in SMAs before 1949?

And why the USAF's careful monitoring and sponsoring of NYU scientists in the 1966 work? What was the USAF hoping to discover from this study?

Is TiZr the original composition of the Roswell foil?

The arrival of Farrar and Adler's TiZr article suggests a couple of things:

1. The Second Progress Report (dated October 31, 1949) has information on at least two SMAs: TiZr and NiTi.
2. The Second Progress Report is not listed in the previous downloaded document from December 31, 1952 for contract number AF 33(038)-3736 because the work into NiTi and TiZr (and potentially other alloys) remains classified for some reason and is not meant to be disseminated widely to the scientific community.

Are we on the right track in terms of a link between the USAF, secret SMA research, and the Roswell case? Whatever the truth, it does raise another important question: Is the receiving of this TiZr document from Mr. Bragalia suggesting the interest by the USAF is not

304

in NiTi, but in TiZr, and that NiTi is just an offshoot into what could have been independent scientific research by the military, with no obvious link to anything unusual about the Roswell wreckage? Certainly, Mr. Bragalia's thinking was to move the current research away from NiTi when in his email dated July 25, 2008, he said:

> "...Nitinol is not itself the Roswell material. Rather it was Wright Field/Battelle's attempt at simulating the properties of the debris."

We requested Mr. Bragalia provide us his thoughts and any evidence as to why NiTi is not the alloy of interest to the USAF. On July 27, 2008, Mr. Bragalia gave his opinion why he thought nitinol was probably not the Roswell metallic foil:

> "The Roswell debris itself is material that is future-engineered at the molecular level and is comprised of elements not found on Earth."

What proof does he have to support this claim? In fact, it is not yet a foregone conclusion that the object is exotic in nature. Mr. Bragalia believes the Roswell object is alien and that science can never understand alien technology (a good way to discourage scientists from studying this area). SUNRISE takes on a more balanced position. But if a scientist was to give an answer on the basis of probabilities, then the Roswell object was likely to be a secret military experiment. However, SUNRISE is even more open-minded to other possibilities if the evidence supports it.

With no firm evidence from Mr. Bragalia, and with our aims to be independent in our scientific research work, let's take a closer look at Mr. Bragalia's claim.

Is it TiZr or NiTi?

Mr. Bragalia primary claim is that the alloy making up the Roswell foil contains "elements not found on Earth". Truly a brilliant claim to make if one wanted to stop all further scientific enquiry into the Roswell case (perhaps to avoid further embarrassing the USAF with new evidence).

Fortunately, the available scientific information is sufficient to tackle this claim head on.

Firstly, to say that the metallic foil contains unknown elements (even by 1947 standards) would have to imply a position on the Periodic Table above 120. Unfortunately, we can already see a problem with this claim. Any supposedly unknown elements composing the Roswell metallic foil would increase its mass and make it exceedingly heavy to lift up in the hand while standing on the Earth's surface. You see, the higher you go on the Periodic Table, the more protons, neutrons, and electrons are added to the elements. When atoms combine to form a solid substance, the weight of the substance increases substantially. This does not correlate particularly well with observations of the Roswell metallic foil.

As the witnesses have claimed, the foil was "extremely lightweight" and very hard and strong to resist cutting. For any metal (alloyed or not) to weigh practically nothing and resist cutting would mean the elements comprising the material has to be well below 120 on the Periodic Table with the metal atoms arranged in a close-packing structure and with strong metal bonds holding the crystalline structure to help explain the hardness and strength. It means the Roswell foil was made using the very lightest metal elements but somehow retained the high strength and hardness considered valuable for aerospace applications. Therefore, we can safely say the elements cannot be "unknown" under any circumstances.

Secondly, there is good information from the scientific literature to show why TiZr is unlikely to be the SMA of choice to build a secret titanium-based SMA flying object by the USAF in early July 1947. The major evidence against the TiZr theory lies in the weight, the color and the quality of the shape-memory response, not to mention the temperature of shape-memory activation.

Focusing on the weight issue, we compared the atomic weights of nickel, titanium and zirconium. What we found can be seen in the following table:

Element	Symbol	Atomic Number	Atomic Weight
Nickel	Ni	28	58.71
Titanium	Ti	22	47.90
Zirconium	Zr	40	91.22

Since we are looking for an alloy with the lightest metal elements in its composition. there is no doubt zirconium, the sister element of titanium, has a much higher atomic weight than nickel and titanium. Even if we did replace some or all of nickel with zirconium, the alloy would be appreciably heavier than NiTi. Unless there is a significant benefit in adding zirconium to the NiTi system, such as a much higher melting point or greater hardness and strength, there would be little if any benefit of having zirconium.

Next, we checked the scientific literature for information on the quality of the shape-memory response and other properties considered useful for aerospace applications. We found that NiTi is not only the world's most powerful SMA, but it is also the lightest titanium-based SMA known to science. In addition, NiTi possesses other useful physical characteristics needed to meet the most stringent engineering requirements for aerospace applications, such as corrosion and heat resistance, as well as strength and hardness. As we mentioned to Mr. Bragalia on October 17, 2008:

> "Scientists fully appreciate the importance of not making the alloy too heavy for aerospace applications. This means using the least number of elements and the lightest metal elements possible, and when combined should provide the necessary characteristics for aerospace applications such as high heat tolerance, impact resistance etc."

If this isn't convincing enough, one other piece of information is worth mentioning in regards to TiZr to help quell any thoughts that this could be the alloy found near Roswell.

From the statements we now have, witnesses claim they saw the Roswell foil's remarkable shape-memory behavior at room temperature. As one military witness said, he carried it in his pocket, pulled it out to show to his friends, and showed the shape-memory effect in action. Similarly, the behavior was exhibited to William Brazel early one morning while playing with it under the warm desert sun. For this to occur, the temperature of the martensite-to-austenite transformation must be around 25°C. Already we see, just by looking at the official temperature graph obtained from the *Metal Reference Book* (see the previous chapter), how NiTi can do this at around 51.0 to 51.3 atomic % nickel.

Furthermore we learn from another article that if we add zirconium to replace nickel in NiTi to form the alloy TiNiZr and ultimately TiZr, the temperature for the martensite transformation increases significantly. In the 2003 article titled "Development of a localized Heat Treatment Tool for Shape Memory Alloy Wires using an Ytterbium Fiber Laser" by Ryan Shaun Dennis at Clemson University, on page 6 the author states:

"Another way to change the Ti-Ni balance in a wire is to add some sort of alloying element.... Hsieh and Wu also discover that adding in Zirconium (Zr) will raise the martensite start temperature (M_s) significantly."

How much is *significantly*?

In *The Shape Memory Effect—Phenomenon, Alloys and Applications* by D. Stoeckel published in 1995, on page 4 the author states:

"Some ternary Ni-Ti-Pd [3], Ni-Ti-Hf and Ni-Ti-Zr [4] alloys also are reported to exhibit transformation temperatures over 200°C."

This is not what we want.

In which case, how little zirconium is needed to keep the alloy activating its shape-memory response at room temperature? We are of the firm view that it cannot have too much zirconium for the NiTi system.

Again, the scientific literature reveals the answer.

Out of the scientific woodworks we find a Technical Report titled "Martensite transformation and shape-memory effect on NiTi-Zr high temperature shape-memory alloys" by Pu et al. published on October 17, 1995. In it, we find support for keeping the zirconium quantity to an absolute minimum:

"During martensite transformation of the newly-developed NiTi-Zr high temperature shape-memory alloys (SMAs) the temperature increases along with Zr content when the Zr content is more than 10 at%"

The amount of zirconium must be well below 10 at.%. Precisely how much below this figure has yet to be determined. Even so, the authors clearly indicated how adding the slightest amount of zirconium to the NiTi system poses other problems. Pu et al. has pointed out:

> "NiTi-Zr high temperature alloys possess relatively poor shape-memory properties and ductility in comparison with NiTi-Hf and NiTi-Pd alloys."

In other words, there are better choices for a SMA (that is, NiTi-Hf and NiTi-Pd) than TiNiZr or TiZr. In addition, the authors revealed:

> "As the Zr content increases, the fully reversible strain of the alloys decreases....Stability of the NiTi-Zr alloys during thermal cycling was also tested and results indicate that the NiTi-Zr alloys have poor stability against thermal cycling. The reasons for the deterioration of the shape-memory effect and stability have yet to be determined."

Given the significant problems posed by TiNiZr (let alone TiZr) compared with the much lighter NiTi, one wonders the benefit of TiZr (or TiNiZr) to the USAF in building a flying object when NiTi is clearly the better alloy.[23]

Seriously, given how the USAF were aware of NiTi and TiZr in the Second Progress Report and assuming the military had been deciding which one of these alloys to choose for their secret military experiment allegedly conducted in early July 1947, it would be remarkable not to choose NiTi.

Finally, to put to rest the idea of TiZr being the likely alloy discovered by the USAF in a secret study or examined from debris found near Roswell, we know TiZr is a bright-silver alloy, not a dark-grey alloy as NiTi is and as observed by the witnesses in the Roswell foil. Even with TiNiZr, you have to minimize zirconium well below 10 atomic% to start looking like a dark-grey alloy *reminiscent* of the Roswell foil. Surely, the secret military experiment that went horribly wrong could not have used TiZr. It had to be a NiTi-like alloy, but since we know of no other distinctly dark-grey titanium-based SMA that is lighter in weight than NiTi and we see Battelle and the USAF had studied a couple of SMAs (including NiTi) in the Second Progress

Report as far as we can tell, one would have to choose NiTi as the most likely candidate in explaining the artificially-made shape-memory Roswell foil.

Either that, or the Roswell foil contained a rather significant amount of pure nickel and titanium (to retain the naturally dark-grey color) with just one or more other elements added in small amounts, but not zirconium, as long as there was a significant technological benefit to be had from adding these other trace elements.

All this is suggesting the opposite to what Mr. Bragalia is claiming. In other words, it is more likely the USAF wanted to simulate the much better NiTi or near identical composition by studying TiZr instead. Now here is a more intriguing possibility: could TiZr be a simulation of the NiTi system? In other words, is NiTi the actual composition of the Roswell foil?

The only other possibility is that the USAF or Battelle had indeed invented (probably by accident) TiZr before July 1947, followed by the USAF or Battelle making the discovery with NiTi and using it in their military experiment, such that it could be interpreted as NiTi simulating TiZr. However, once NiTi was discovered, it is unlikely the USAF would go back to TiZr when clearly NiTi is the better of the two for use in any future military aircraft. Certainly, the dark-grey color of the Roswell foil is suggesting this is what happened. Yet for some reason the USAF did go back to TiZr, in 1966 as we are told by the Farrar and Adler article. There is something the USAF still could not figure out about this alloy and potentially NiTi if it related in any way to the shape-memory effect.

However, if the USAF did actually use NiTi or something containing a lot of nickel and titanium in its composition for the foil when building its secret military flying object, we still face a dilemma: the purity issue. We can't get away from this fact. Somehow, the titanium had to be pure for the USAF or Battelle to notice NiTi's shape memory response. But this "purity issue" was not on the radar of both the USAF and Battelle by July 1947. Not even the quantity of titanium needed to build the Roswell object was available by July 1947. So how could the USAF or Battelle have made NiTi (and hence the Roswell foil)?

As we recall, the witnesses observed a shape-memory effect in the Roswell foil. Therefore the USAF had to have known about this too, which means if the flying object is man-made, the USAF must have figured out the purity issue and had some idea of how SMAs worked by mid-1947.

Even if the USAF did somehow use NiTi or something similar of a dark-grey appearance in a secret military experiment and got the purity right as well, the odd thing is the USAF decided not to use NiTi again for the skin and/or structural supports of any future fighter jet because of a lightning strike, which probably caused the deaths of some victims (or were they really plastic dummies?).

Something is telling us this is unlikely. Given how useful the alloy is, the military would not give up on the alloy quite so fast. Something else must have happened.

Analyzing the Summary Report of December 31, 1952

Performing further independent research, we see in this final Summary Report of 1952 for contract number AF 33(038)-3637, three other reports issued under this second stage contract:

1. "AF Technical Report No.6218 Part II" dated June 30, 1950
2. "AF Technical Report No.6623" dated June 18, 1951
3. "WADC Technical Report No.52-249" dated June 18, 1952

together with the final report of December 31, 1952, which is the date the contract officially terminated.

On adding the TiZr article by Farrar and Adler to the mix, it is clear a fifth summary report exists for this second stage contract, dated October 31, 1949. This is the Second Progress Report containing the two SMAs of NiTi and TiZr. For some reason, this report has been omitted.

We can also see contract number AF 33(038)-3637 for this second stage contract into the study of titanium-based alloys commenced on May 18, 1949. What about the first stage contract? When did this commence?

Looking further into the report, we see it states:

311

"In addition to this [final] report, four previous summary reports on alloy development have been issued under these contracts."

Noticing the plural "contracts" and with a total of four reports already identified as part of the second stage contract of AF 33(038)-3736, Summary Report Part III dated July 30, 1949 remains the odd one out, suggesting it should be part of the first contract into the development of titanium and its alloys of W 33-038 ac-21229. Does this mean additional reports titled Summary Report Part I and II not mentioned in this final report exists? And are these missing reports part of W 33-038 ac-21229? If so, why aren't Part I and II mentioned in this final report? Assuming Summary Reports Part I, II and III are part of contract number W 33-038 ac-21229, what is the First Progress Report under contract numbers AF 33(038)-3736 and W 33-038 ac-21229?

Perhaps the First Progress Report could be the one dated June 30, 1950? Except that the Second Progress Report we are looking for under contract number AF 33(038)-3736 is dated October 31, 1949. So there must be another report earlier than this one which must be the official First Progress Report.

Interestingly, a Wright Air Development Center (WADC) Technical Report 53-109 dated April 1953 has been released. The title for this report is "The Tentative Titanium-Silver Binary System" written by Henry K. Adenstedt and First Lieutenant William R. Freeman, Jr. of the Materials Laboratory. In the bibliography section we find the following footnote:

Craighead, C. M., Simmons, O. W., Maddex, P. J., Greenidge, C. T., and Eastwood, L. W. Preparation and Evaluation of Titanium Alloys. Summary Report, Part III. Battelle Memorial Institute, May 1948 - July 1949.

This tells us the Summary Report Part III definitely refers to the first stage contract, and Parts I, II and III cover the period of May 1948 to July 1949. This suggests the first contract number of W 33-038 ac-21229 commenced in May 1948. Confirmation for this can be seen in the final Summary Report of 1952 where it gives a specific date of May 18, 1948. This is the date when all official titanium-based alloy development work had begun for the USAF since Project Air Force RAND ended in March 1947.

The final 1952 report in this contract also states the work into titanium alloys had continued into another contract number, AF 33(636)-384. For the purposes of this Roswell research work, we shall ignore this newer contract number for now.

Beyond that, this 1952 report contains absolutely no information about NiTi or any other SMAs. It is clear the missing Second Progress Report for contract number AF 33(038)-3736 dated October 31, 1949 contains the vital information we need for this research, especially in regards to SMAs. Of course, we cannot ignore the possibility of further information on NiTi to be found in the other reports.

Analyzing the Summary Report of June 18, 1952

Not long after the above report was released, another report[24] under contract number AF 33(038)-3736 suddenly emerged online that had never appeared in previous search listings. We must assume someone made the decision to release these documents for public scrutiny at the right time.

We find this is the second last report before contract AF 33(038)-3736 officially terminated, titled "WADC Technical Report No.52-249", dated June 18, 1952.

What can we learn from this report? On pages 8 and 9, we observe the following statement:

> "This report describes the work done under Contract No. AF 33(038)-3736 during the period May 19, 1951, to May, 1952.
>
> The general objective of this project has been the development of high-strength structural alloys of titanium. At a recent meeting with the Sponsor, however, this objective was expanded to some degree and the required properties were more closely defined.
>
> ...In this report, the emphasis is on the development of the high-strength sheet alloy (Type B-1)."

Reading further into the report we find the alloys of interest to the USAF were titanium-based, and combined with chromium, iron, manganese, vanadium and molybdenum with some minor amounts of iron added to some binary, ternary and more complex alloys. A brief discussion was also made of the Ti-O and Ti-N alloys with additions of phosphorus.

There is still no information on NiTi, TiZr or any other SMA. *Clearly* not the right report.

More importantly, we find the USAF did not see the NiTi-like newspaper thin sheet covering the skin of the secret flying object in mid-1947 to be sufficiently strong enough. It had to look for other high-strength titanium-based alloys in 1952. Or perhaps the USAF still did not know how to harden NiTi in the right way. But this would imply the USAF was not responsible for the object that crashed in New Mexico in early July 1947. Probably not good for the USAF to consider this. Then again, why would the USAF be interested in finding another super-tough titanium-based alloy if NiTi or something similar was already manufactured with such toughness by early July 1947? Very strange.

Or was the USAF trying to avoid using NiTi by finding another high-strength alloy in case any of the Roswell witnesses were to notice the alloy in action and make the link which could prove to be difficult to explain if asked by the media?

If the Roswell case is truly man-made, the difficulty in explaining the situation must come back to the issue of who the "victims" were. These unnamed individuals must be what is keeping the USAF quiet after all this time.

Further confirmation for USAF/Battelle Contract No. W 33-038 ac-21229

Just as things were starting to settle down, we decided to have another check on Google Scholar for Contract No. W 33-038 ac-21229. Miraculously another unclassified report released by the DTIC turned up. The report is titled "Some Mechanical and Ballistic Properties of Titanium and Titanium Alloys" by R. K. Pitler and A. Hurlich, and is dated March 7, 1950.

The report is not part of any USAF/Battelle contract numbers. It was written for the U.S. Army at the Watertown Arsenal Laboratory (WAL). Our fortuitous timing to gain access to this report simply helps us to provide further confirmation for the existence of the first USAF/Battelle contact on titanium via number W 33(038) ac-21229 in the References section as shown below:

7. Bimonthly Progress Report for March and April 1949 on Development of Titanium Base Alloys and Processes for their Commercial Production.   Contract W-33-038-ac-21229 to Wright Patterson Air Force Base.   Batelle Memorial Institute.   30 April 1949.

However, reference number 7 as we see here does give further insights into the first USAF/Battelle contract. It can be noted that a series of bimonthly progress reports (rather than yearly reports) was produced under this first contract, all of which appear to have the general title of "Development of Titanium Base Alloys and Processes for their Commercial Production". The report shown in reference number 7 is dated April 30, 1949 and it appears to refer to work done by Battelle from March to April 1949. The date of April 30, 1949 also suggests that this is when the contract officially terminated and Stage 2 commenced.

Did the report by Pitler and Hurlich contain anything interesting by way of SMAs and their link with the USAF and Battelle? Unfortunately no, but it did mention the USAF's interest in finding impact-resistant titanium-based alloys which, by implications, would make one question why the Roswell foil was not sufficiently tough enough to meet this demand.

On closer inspection of the report, we see the interest is in the benefits of titanium and titanium alloys as a form of armour and a

315

structural material for equipment in the Ordnance Department, revealing such information as:

1. Titanium alloys have higher strength-weight ratios beyond those possible with aluminium (Al) and iron (Fe) alloys.
2. Tests on the ballistic impact resistance of thin sheet titanium alloys showed some promise, but thicker plates of unalloyed titanium were superior to heat-treated alloy steel of equivalent weights.
3. The tensile properties of titanium can be varied by alloying and heat treatment.

The report also recommended further investigation of titanium alloys for armour and as a structural material because of "limited ballistic and mechanical tests" on available titanium and titanium alloys manufactured by different commercial companies with different production methods.

While the report may not have discussed anything about SMAs such as NiTi, it does suggest the hardness exhibited by the Roswell foil when it resisted dents and cutting was still on the USAF's mind by April 1949 when the report mentioned "ballistic impact resistance" and the development of new "armor" using titanium alloys. Why? Hadn't the USAF made a metal hard enough for the job since July 1947? Something is just not right.

Looking further, we see that when the report was completed, it was distributed to various interested parties including Mr. J. B. Johnson at Wright–Patterson AFB. And as we know, Wright–Patterson AFB was the place where the interest in NiTi (and now TiZr) originated soon after the Roswell foil was sent there for analysis. It is starting to look like whatever restarted the military's interest in titanium and its alloys in 1948 had all stemmed from whatever work the people at Wright–Patterson AFB had been doing and possibly helped along by some materials found near Roswell.

Would this, by any chance, have anything to do with the Roswell foil? Or was the foil just another alloy in a long list of other ordinary titanium-based alloys that the USAF had discovered without any outside scientific assistance?

Two more scientists have seen the classified USAF/Battelle Second Progress Report

On October 20, 2009 Mr. Bragalia provided SUNRISE with a copy of one report and a scientific article for our analysis:

1. "On the NiTi (Nitinol) Martensitic Transition, Part I" by Dr. Frederick E. Wang of the U.S. Naval Ordnance Laboratory dated January 1, 1972.
2. "Beta-Eutectoid Decomposition in rapidly solidified Titanium-Nickel Alloys" by Krishnamurthy et al. dated October 23, 1984.

The first document comes from the scientist who first explained the idea of solid-to-solid phase transformation in the crystalline structure of NiTi during shape-memory recovery to Dr. Buehler. He would join Dr. Buehler as a co-inventor of nitinol and later set up two U.S. companies specializing in the manufacture of nitinol products.

The second is the scientific article we found in January 2007. One of the authors, S. Krishnamurthy, is described in the report as the co-principal investigator at Metcut-Materials Research Group at Wright–Patterson AFB. Another author linked to Wright–Patterson AFB is F. H. Froes, chief of the Air Force Wright Aeronautical Laboratories and Materials Laboratory. An overseas senior lecturer from the University of Sheffield in England named H. Jones also participated in the work.

In both the report by Dr. Wang and the article by Krishnamurthy et al. we find the all-important Second Progress Report of October 31, 1949 as shown below:

Document 1 - Dr. Frederick E. Wang

```
6.  C.M. Craighead, F. Fawn and L.W. Eastwood; Battelle Memorial Inst.,
Second Progress Report on Contract AF 33(038)-3736 to Wright
Patterson Air Force Base(1949).
```

Document 2 - Krishnamurthy et al.

13. C. M. Craighead, F. Fawn, and L. W. Eastwood: Second Progress Report, Contract AF33 (038)-3736, Battelle Memorial Institute, Columbus, OH, 1949.

Beginning with the article by Krishnamurthy et al., we see it is focused on "rapid solidification (RS) processing of titanium alloys"[25] with emphasis on NiTi because the process provided "increased homogeneity, microstructural refinement, and formation of unique structures". As stated on page 1:

> "The work reported here on titanium-nickel alloys is part of an exploratory investigation on rapidly solidified eutectoid-forming titanium alloys."

It is interesting to see how the authors noticed microstructural improvements during this rapid solidification process. As we have noted before, minimizing microstructural defects are known to improve the shape-memory effect of NiTi.

Also, realizing the rapid solidification process is a common method to create an amorphous metal, one can't help wondering whether someone at Wright–Patterson AFB could be interested in developing amorphous NiTi to build something more cheaply and easily without the need for expensive equipment to melt titanium in a vacuum or inert gas, while possibly enhancing the shape-memory effect of NiTi at the same time. The question is, is this all related to the Roswell foil that was analyzed in 1947?

As for the report by Dr. Wang, he must know the USAF has studied at least two SMAs—NiTi and TiZr—from the Second Progress Report. Does this mean Dr. Wang knew the USAF had studied SMAs? He would not say. When SUNRISE found his email online and asked for his view on the footnote linking the USAF and nitinol, Dr. Wang never replied. Is this because he is protecting his position as official inventor of nitinol in case he agrees the USAF had probably known about it too?

On reading the report, we find the part where the Second Progress Report of 1949 is mentioned within the historical background section of Dr. Wang's report:

> "Craighead, Fawn and Eastwood (1949) carried out a limited study of the Ti-Ni phase diagram up to approximately 11.6 at.% nickel within a limited temperature range but did not define the eutectic or eutectoid temperatures."[26]

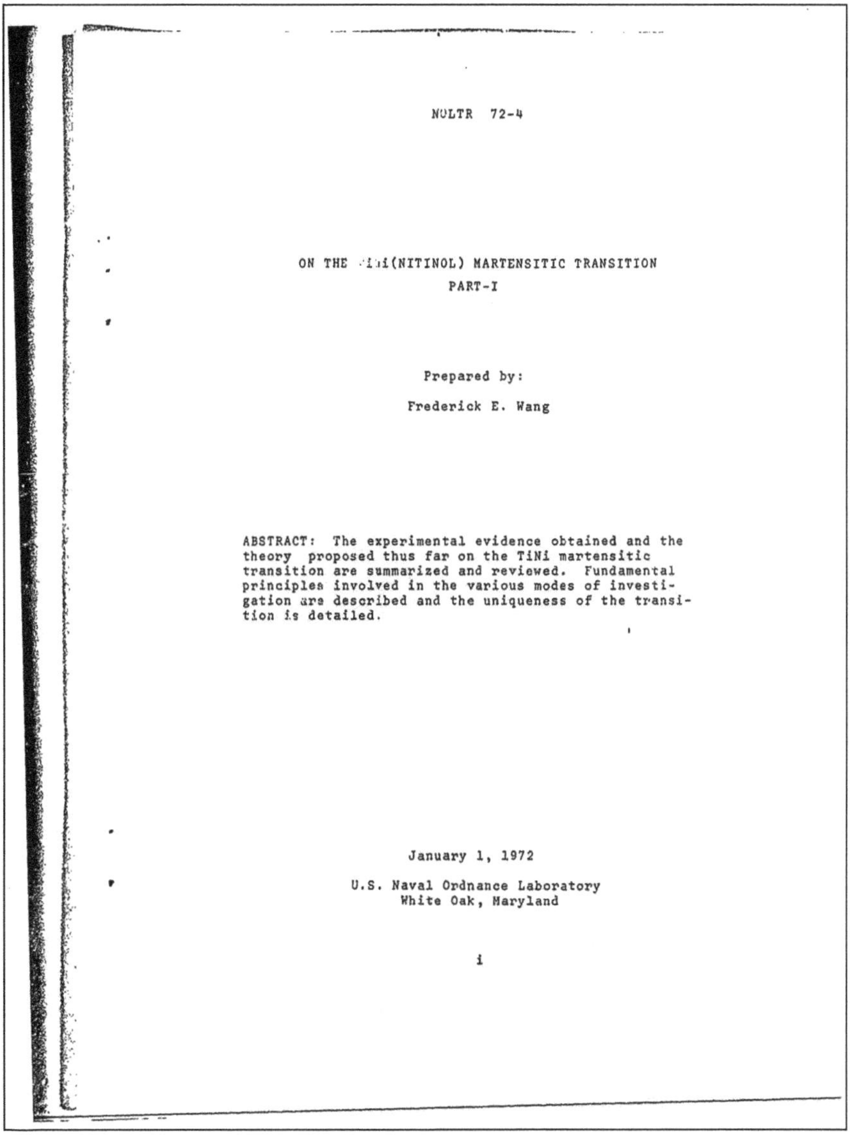

NOLTR 72-4

ON THE TiNi(NITINOL) MARTENSITIC TRANSITION
PART-I

Prepared by:

Frederick E. Wang

ABSTRACT: The experimental evidence obtained and the
theory proposed thus far on the TiNi martensitic
transition are summarized and reviewed. Fundamental
principles involved in the various modes of investigation are described and the uniqueness of the transition is detailed.

January 1, 1972

U.S. Naval Ordnance Laboratory
White Oak, Maryland

i

Dr. Frederick E. Wang's 1972 Report.

While it does give us further details into the classified report, we
now see the focus is not in the equi-atomic range where the shape-
memory effect is most notable, but in the titanium-rich end of the
alloy's composition. Dr. Wang also noted the aim of the study by
Craighead et al. was not to find the eutectic or eutectoid temperatures[27].

319

The focus on the titanium-rich end is confirmed in the WADC Technical Report 53-109 "The Tentative Titanium-Silver Binary System" dated April 1953:

> "The binary equilibrium diagrams of titanium and its alloying elements can be considered absolute prerequisites for the understanding of all technical titanium-base alloys.
>
> In the course of the alloy development work, which the Materials Laboratory has been sponsoring for several years at Battelle Memorial Institute, the first, very preliminary knowledge of the phase relationships at the titanium-rich end of such diagrams was obtained. This investigation has also furnished a broader knowledge of the applicability of the different alloying elements for practical alloys.
>
> In view of the aforementioned necessity for a knowledge of the binary equilibrium diagrams, and the paucity of information on this subject in the literature, it was decided to start at an early sate, a broader research project on binary diagrams of titanium and the most promising alloying elements."[28]

Does this mean that Battelle scientists had no knowledge of shape-memory effects in NiTi? According to the 1953 report, Battelle was among five research institutions studying titanium alloys. As the report stated:

> "Investigations was started on ten systems, the work being divided among five research institutions."

Each of the institutions provided a summary report for their own particular area of titanium alloy research:

> "After the first year of research, a summary report was prepared by each contractor and these reports have been published as [Air Force] Technical Reports [AFTR]."

Then the report gives a list of titanium-based alloys studied by each of the five research institutions. These are:

320

Armor Research Foundation
Battelle Memorial Institute
Massachusetts Institute of Technology (MIT)
New York University
University of Notre Dame

For Battelle, it suggests the only alloys the USAF was interested and had asked Battelle to investigate were:

titanium-manganese (Ti-Mn)
titanium-tantalum (Ti-Ta)
titanium-tungsten (Ti-W)

Whereas in the same table we see the New York University appears to be the only research institution to be investigating NiTi with report numbers AFTR 6596, Parts 1 & 2 covering this work by NYU scientists.

Below is the table as shown in the 1953 report:

Contractor	Systems Investigated	Report No.
Armour Research Foundation	Ti-Si, Ti-Cb, Ti-Mo	AFTR 6225
	Ti-O, Ti-Al, Ti-Fe-Cr	WADC TR 52-16
Battelle Memorial Institute	Ti-Mn, Ti-Ta, Ti-W	AFTR 6516, Parts 1 & 2
Massachusetts Institute	Ti-Cr, Ti-Cu,	AFTR 6595, Parts 1 & 2
of Technology	Ti-Cr-O	WADC TR 52-255
New York University	Ti-Ni	AFTR 6596, Parts 1 & 2
University of Notre Dame	Ti-Fe	AFTR 6597, Parts 1 & 2

However, we know Battelle had been studying NiTi thanks to the existence of the Second Progress Report under contract number AF33(038)-3736 dated October 31, 1949. If the 1953 report cannot reveal this NiTi work conducted by Battelle, then it raises the possibility that perhaps one or more scientists at Battelle could have known something about the alloy's interesting shape-memory effect and kept it secret.

Whether or not Battelle knew, we certainly cannot put the USAF in the same position of not knowing about shape memory alloys since the witnesses made it abundantly clear they had observed the effect and commented on how interesting it was. With numerous military witnesses willing to state on the record that the Roswell foil is a metal or alloy (only an alloy can exhibit the pronounced shape-memory effect seen in the foil), the USAF had to have known something about SMAs. But why a study in the titanium-rich end of the alloy's composition? One possibility is that the USAF had wanted to see how far in the titanium-rich end of NiTi this shape-memory effect may extend and then compare this with other titanium-based alloys having a similar phase relationship diagram and so perhaps determine if the scientific knowledge the USAF had acquired about the effect is correct. Or it could merely be interested in a general study of titanium-based alloys.

Is there a way we can determine which one is the true situation for the USAF?

Beyond that, we find Dr. Wang's declassified report provided useful information about NiTi, including a brief historical background of the studies carried out on the alloy going right back to the world's first NiTi sample produced by Wallbaum et al. in 1939. As Dr. Wang noted, these early studies showed "the difficulties associated with the understanding of the TiNi transition", or shape-memory effect to put it another way, because of a variety of reasons. As a result, through his report, Dr. Wang wanted to inform the reader of the different interpretations offered by people investigating the alloy for the TiNi transition and gave his own insights of why there are differences based on his knowledge. The aim of writing his report can be seen on page ii when Dr. Wang said:

> "...the primary intention of this writing is to challenge and to stimulate further research and eventually derive a true understanding of not only the TiNi transition but also the martensitic transition as a whole."[29]

Further confirmation for the existence of the Second Progress Report of 1949

Despite being an unpublished and classified USAF/Battelle document, another individual—H. O. Teeple—witnessed and probably read the Second Progress Report of 1949.

Teeple was conducting his own research into nickel-based alloys at the time for use in paper manufacturing. He had close contacts with Battelle scientists who gave him some of the latest developments at the institute. It is here where he learned of the Second Progress Report containing NiTi and the apparently high interest to the USAF. He successfully gained access to the classified report. He noted the work on NiTi and later wrote the footnote confirming he has seen it in his 1950 article "Nickel and High-Nickel Alloys"[30] in the *Industrial and Engineering Chemistry* publication.

Beyond that, only a handful of scientists from NYU, those at Wright–Patterson AFB, and the official co-inventor of NiTi, Dr. Wang, had the privilege of viewing the unpublished and classified USAF/Battelle Second Progress Report.

The Second Progress Report of 1949 is released

On August 12, 2009, the U.S. DoD public information branch known as the Defense Technical Information Center (DTIC) made the official Second Progress Report of October 31, 1949 available to the public.

At last, we got the report.

A quick look at the report shows NiTi is not mentioned in the contents page (only TiZr). However, on running through each page of the report, we find NiTi and TiZr is mentioned in a section titled "Evaluation of Experimental Titanium-base Alloys". In this section, we find the all-important authors names: C.M. Craighead, F. Fawn, and L.W. Eastwood.

The Second Progress Report is divided into two parts: pages 1 to 59, which are missing, and pages 60 to 119, which were declassified and released by DTIC.

According to footnote number 99 in H.O. Teeple's 195 article, "Nickel and High-Nickel Alloys", the missing pages *should* contain a third SMA studied by Battelle called TiNiCo. As Teeple stated on page 1990:

> "Craighead et al. (99) reported binary alloys of titanium with nickel and cobalt and ternary alloys of titanium-chromium alloys with nickel and cobalt among others. Charts on properties and diagrams are included."

Does this mean the USAF also studied NiTi-Cr and NiTi-Co among other alloys? Scientific literature has indicated NiTi-Co and NiTi-Cr are SMAs. In fact, the literature reveals how the addition of cobalt (Co) helps to extend the martensite transformation temperature range of the shape-memory effect in NiTi for cobalt amounts not exceeding around 1.5 at.%. It means the report contains no less than *two* confirmed SMAs (i.e., TiZr and NiTi) with a potential for a further two more to come out of the bowels of this report. Unfortunately, further details for these two alloys happen to be located in the missing pages, just out of our reach.

It means the full report is not exactly fully declassified.

Are we to assume this is because half the report is illegible to read and DTIC has decided to leave it out? Or does it contain more sensitive information?

Focusing on the section crucial to this research, we find the first half of Craighead et al's work involves identifying the extent of the dual alpha and beta crystalline phase field at the titanium-rich end of the composition for several binary titanium-based alloys, such as NiTi and TiZr, and in determining the solvus lines (the boundary between the phase fields). Quenching the alloys rapidly in cold water at different compositions and temperatures is used to lock in the crystalline structures (p.65, PDF p.15).

Now this is interesting. Isn't this dual crystalline alpha and beta crystalline structures in a given solid phase of an alloy crucial in understanding the theory behind the shape-memory effect of alloys? At present, scientists believe a solid-to-solid crystalline change, where a decomposition of the alpha crystalline structure to a beta crystalline structure when heated within a certain phase field and vice versa when cooled is currently the leading explanation for how SMAs work. Even if the explanation is not known, this dual-crystalline region is where the shape-memory effect is most likely to be found. If this is true, then it seems we cannot ignore the possibility that the USAF were fully aware of the shape-memory effect in NiTi (and TiZr). It is just that the USAF wanted to see how far this dual-crystalline phase region extended. And the part of the phase diagram for NiTi with the least data was in the titanium-rich end of the spectrum. That is what this report focussed on.

What was the composition range of the alloys studied by Craighead et al.? According to the report:

> "A nickel-rich phase appeared in the microstructure of the [NiTi] alloy containing about 7.5 percent nickel when the specimen was quenched from 1450°F, but is absent…in the specimen quenched from 1550°F. From the available data, the eutectoid composition [in the binary titanium-nickel system] is placed between 6 and 7 percent nickel." (p.68, PDF p.18)

The report mentions the "additions of 1 to 10 percent zirconium were also studied" (p.60, PDF p.10) or "[a]dditions of 1.0, 2.5, 3.5, 5.0 and 10.0 percent zirconium were made to Process A titanium" (p.80, PDF p.32) but claim "[n]o alloys of interest were noted" (p.60, PDF p.10).

All this may suggest the USAF/Battelle were not involved in any secret SMA research by not focusing in the equi-atomic range for NiTi or wherever the region in the phase diagram for dual crystalline structures may exist for TiZr when the Battelle scientists claimed no alloys of interest were noted. However, this is not proof they didn't know about SMAs. What we find clearly from this section of the report is the focus on determining how far the dual crystalline structure region extended into the titanium-rich end of several alloys, including NiTi. We know the region exists in the equi-atomic range for NiTi, which is where the shape-memory effect occurs. But precisely how far either side of the equi-atomic range does this region go? Clearly a logical question to ask if a shape-memory effect had first been observed in the equi-atomic range.

Also, it is possible Battelle scientists could have been asked to find other alloys having this dual crystalline region, which might explain how TiZr came to exist in the report. If so, and apart from raising the likelihood that certain people at Battelle were aware of the shape-memory effect of certain alloys, on testing these alloys at Wright–Patterson AFB, the USAF could have easily discovered the shape-memory effect of TiZr and kept it quiet, which would explain why the USAF wanted to monitor NYU scientists in the study of the alloy in 1966 (and Prof. Nielsen through Dr. Margolin after 1947). Whatever the truth, as far as the composition range is concerned, it appears the titanium-rich end is where more data were needed for this 1949 report.

As for the titanium-germanium (TiGe) and titanium-silver (TiAg) alloys analyzed in this section of the report, these may not be considered SMAs by scientists, but their inclusion would suggest they were needed for comparing the results with the NiTi and TiZr systems when we see the same dual-crystalline regions mentioned as well. Any differences, such as not showing a shape-memory effect despite having the dual-crystalline structure, might help to explain how shape-memory alloys work.

For example, in Table 16 (p.66, PDF p.16), we see heat treatment of the TiGe alloy ranging from 1450°F to 1750°F together with variation in the germanium composition of between 0 and 1.0 percent. From this table, there is an obvious transition from the alpha crystalline phase at 1450°F to the beta crystalline phase above 1700°F, with a combination of the two crystalline structures appearing around the mid-temperature range of between 1550°F and 1650°F. Within this dual crystalline phase region, a percentage of how much of the alpha and beta crystalline structures were identified is shown. The percentages do not stay the same. The amounts vary with changes in the temperature, as well as the amount of germanium added to titanium.

Likewise, a similar table is presented for the titanium-silver alloy[31]. In fact, all the alloys studied had a table developed to show this same dual crystalline phase region.

Naturally, anyone with an interest in SMAs would ask: what happens if the temperature is increased? Would a change in the crystalline structure from the alpha to the beta form lead to a shape-memory response?

Indeed, one can't help realize the possibility that this dual crystalline phase region was a critical aspect to the USAF/Battelle work at the time. Why such emphasis in this phase region? Is this because the USAF and Battelle had known something about the shape-memory effect of alloys such as NiTi and TiZr? Certainly we cannot say the work by Craighead et al. categorically proves no SMA work had been carried out in secret by 1949.

Or maybe it was just a coincidence? Perhaps this is the only report where NiTi was studied, in which case, the work could be safely ignored and, therefore, probably has no bearing to the Roswell case. Well, maybe. Except further information on NiTi does exist in the previous bimonthly report.

According to Craighead et al.:

> "In the previous bimonthly report, data on the mechanical properties and the response to heat treatment and aging of titanium-germanium and titanium-nickel alloys were listed in Tables 1 and 2" (p.65, PDF p.15)

If NiTi was not meant to be of any major interest to the USAF and Battelle, it was mentioned again in the previous report. So maybe it is just these two reports and nowhere else? The work might still be a coincidence.

There is no indication in this report to suggest further study of NiTi would take place in the future.

Supporting this view is the way the authors concluded the study in this section of the report by stating "...the data do not justify further investigation of binary titanium-germanium or titanium-nickel alloys." (p.68, PDF p.18).

Yet Dr. John P. Nielsen and other NYU scientists continued to study NiTi and TiZr after this Second Progress Report was written, all the while still garnering the auspicious interest of the USAF at Wright–Patterson AFB.

Continuing with the analysis, we see the same heat treatment and rapid quenching were applied to NiTi to help reveal the extent of the combined alpha and beta crystalline phase fields. We see this in Table 17 (p.69, PDF p.19). A third gamma phase[32] is also included because a nickel-rich phase is formed within the titanium beta phase matrix as the amount of nickel added to titanium exceeds 7.3 percent. A tentative phase diagram for nickel-titanium showing the different phase fields is revealed in Figure 20 (p.70, PDF p.20).

We also see the same combined alpha and beta crystalline structures in a given phase region locked in through rapid quenching for the titanium-silver (TiAg) alloys between 1550°F and 1650°C (see Table 20 on p.76, PDF p.28). Initially the work into this alloy system involved a composition range of 1 to 2 percent silver according to the Summary Report, Part III (p.68, PDF p.18). It was extended for this report in the range 2.5 to 5.0 percent silver. This system displays more similar crystalline structure phase results to NiTi, including not just the simultaneous presence of alpha and beta structures but also a gamma phase of silver-rich regions within the titanium matrix as seen in Table 17 (p.69, PDF p.19) together with a tentative phase diagram in Figure 20 (p.70, PDF p.20).

Again, we find the same alpha and beta crystalline phase region present in various percentages when the composition exceeds 5 percent zirconium and when the TiZr alloy is rapidly quenched from a temperature between 1450°C and 1650°C as revealed in Table 21 (p.83, PDF p.35). Figure 29 (p.84, PDF p.36) shows a tentative phase diagram for the titanium-zirconium system.

Somehow, we cannot escape this dual-crystalline region issue from this section of the report.

With the first crucial section out of the way, we move onto the second section. In it, we find considerable effort by the authors to observe the mechanical properties of various titanium-based binary, and more complex ternary alloys except for NiTi, whose data had already been published in the previous bimonthly report. Mechanical properties included measuring hardness, tensile strength, and ductility.

When testing for hardness, the phrase "fabrication to sheet" form for the alloys has cropped up as being integral when testing for mechanical properties (p.61, PDF p.11). No doubt, the Roswell foil was fabricated to sheet form. Except for some reason the USAF was not quite happy with the hardness characteristics of the foil it had allegedly manufactured in early July 1947 when in this Second Progress Report we see further testing for hardness was needed. A very tough client to please it would appear.

For this section, the titanium-silver (TiAg) (pp.68-77, PDF pp.18-29), titanium-beryllium (TiBe) (pp.78-80, PDF pp.30-32) and titanium-zirconium (TiZr) (pp.80-84, PDF pp.32-36) alloys were also measured for their tensile strength, hardness and ductility. Additional hardness data were also obtained for titanium-tantalum (TiTa) and titanium-columbian[33] (TiNb) and are presented in Table 19 (PDF p.22).

There was considerable effort to gather hardness data for these selected alloys, possibly for comparison with similar data obtained for NiTi in the previous bimonthly report. For example, the alloys were either heat treated to specific temperatures ranging from 1450°F to 1750°F, hot rolled, or hot rolled for four hours to simulate aging to help determine any variation in the hardness data. From this, the authors learned TiAg did not increase the hardness (results revealed in Figure 25) when hot-rolled for 4 hours at 750°C (p.76, PF p.26).

In the case of the titanium-zirconium alloy system, the authors "concluded that titanium-zirconium alloys in the range of composition investigated do not show any significant response to heat treatment or aging". In other words, these alloys would not show significant hardening with heat treatment and aging.

As for the other alloys, they did not show appreciable differences in the hardness data compared with TiZr or TiAg.

One must ask, how would the hardness data of these alloys compare with NiTi? More importantly, would NiTi develop similar hardness results to the Roswell foil? And why persist with obtaining hardness data if the nitinol-like Roswell-foil already possessed unusually high hardness characteristics?

As we recall, observations of the Roswell foil suggest it had a high level of hardness in the shape-memory activated phase, making it impossible for witnesses to cut, pierce or rip the foil with scissors and other tools. To achieve something similar with NiTi, you have to regularly cold work the alloy, such as bending it and letting it return to its original shape. By doing this over time, incredible hardness is attained, which is how U.S. scientists eventually patented the observation in 1967[34]. However, when you get a very pure sample of NiTi out of the oven and play with it, the alloy is remarkably pliable and soft (i.e., easily bent).

Could the USAF had been confused by the hardness data they got with NiTi from Battelle after testing a fresh sample from the oven and realizing the hardness level was not the same as the Roswell foil? If so, was gathering this hardness data important to determining how to make NiTi reach the same hardness level as the original Roswell foil? However, all this would imply the military had not manufactured the Roswell foil under any circumstances. Not good for the USAF if it is trying to hide something from the public. But if the military had manufactured it, there would be no need to gather hardness data. The USAF would already know the level of hardness achievable with NiTi and applied it as we see in the Roswell object. And then we would have seen this dark-grey Roswell foil put onto the skin of fighter jets immediately after 1947 to show the advanced knowledge of the USAF in developing the world's best super-tough titanium-based alloy. Unfortunately, this is not how it happened. Even when the first official titanium aircraft was built for the USAF in 1952-53, no dark grey alloy was used. If this is true, then the only other possibility is that the USAF had discovered something else in the desert and this was enough to avoid using the alloy in their military arsenal and choose another to keep their work on NiTi and other aspects of the Roswell wreckage secret.

Whatever the reason for the USAF to ignore the Roswell foil it had allegedly created and the aim of gathering this data (yet again if the Roswell foil had already attained the required hardness), it seems the work would eventually pay off with the discovery of at least one other high-strength titanium-based alloy. As stated in the report:

> "...steps have been taken to re-evaluate the more promising high-strength alloys. This work is directed toward selecting an alloy composition on which extensive engineering data will be obtained." (p.61, PDF p.11).

Alloys showing greatest promise in terms of high-tensile strength and hardness turned out to be those containing vanadium. Apparently, adding vanadium to a titanium-based alloy would be better for achieving the required hardness than using NiTi or the nitinol-like Roswell foil.

How odd?

With the analysis of Craighead et al.'s work out of the way, let's look at the rest of the report.

In the next section titled "Investigation of Refractories for Melting Titanium", by P. D. Maddex and L. W. Eastwood, there was interest in working out how susceptible titanium tetraiodide (TiI_4), written in this report as iodide titanium, for creating pure titanium is to absorbing oxygen by adding "known amounts of oxygen...as TiO_2" (p.62, PDF p.12). Also included in the study for this section was a search for new materials to make high temperature crucibles for holding molten titanium and its alloys.

In the case of the impurities issue, we see Maddex and Eastwood chose TiI_4 instead of titanium tetrachloride ($TiCl_4$). As we recall, the Kroll method relies on chlorine to react and isolate titanium from the naturally occurring TiO_2 to form a relatively stable structure known as $TiCl_4$. However, scientists have understood that iodide provides a much stronger intermolecular van der Waals bonding than chlorine (as can be seen from the higher melting point of TiI_4 which is 150°C, compared with -24°C for $TiCl_4$), making it less likely for the titanium to react to oxygen.

Compared to the Kroll method using chlorine, the use of iodide is commonly known as the Van Arkel process, and is used to make highly pure forms of titanium metal.

To produce TiI_4, scientists combine raw titanium ore where titanium is in the form of TiO_2 and some aluminium iodide (AlI_3) leading to the following oxide-iodide exchange reaction:

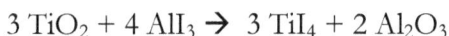

$$3\ TiO_2 + 4\ AlI_3 \rightarrow 3\ TiI_4 + 2\ Al_2O_3$$

Or if titanium is already in the form of a piece of metal and a higher purity of the metal is required, scientists simply add pure iodide inside a furnace heated to 425°C leading to the following reaction:

$$Ti + 2\ I_2 \rightarrow TiI_4$$

In the case of titanium already locked in by chlorine to form $TiCl_4$, the iodine can replace the chlorine by using hydrogen iodide (HI) through the reaction:

$$TiCl_4 + 4\ HI \rightarrow TiI_4 + 4\ HCl$$

Whatever the reaction used, all this information is telling us one thing: by 1949, the USAF wanted high purity titanium and its alloys and neither the military nor Battelle were entirely confident of the stability of the TiI_4 structure to attain the purity level required. This is rather unusual considering the USAF or Battelle had allegedly produced a highly pure form of a titanium-based nitinol-like (because of its dark-grey color) SMA by early July 1947. For some reason, this latest study had to focus on TiI_4 to see how well it could resist impurities, primarily oxygen in the air. Therefore, the authors wanted to test the theory inside a vacuum furnace at elevated temperatures with a known amount of oxygen added to the system via TiO_2 to determine if any titanium in TiI_4 was reacting to the oxygen. If so, there could be impurities seeping into various titanium alloys, let alone titanium itself.

Amazing for the military to still be worried about this impurity issue after 1947.

As revealed in the Summary section of the report:

> "Standard specimens of iodide titanium, containing known amounts of oxygen added as TiO_2, were prepared and submitted to Dr. G. Derge, of the Carnegie Institute of Technology, for vacuum-fusion analysis. The results reported from this laboratory indicate that the vacuum-fusion technique, as it presently exists, yields fairly reliable results. Relatively minor inconsistencies in the analytical results were obtained. At the present time, it is not known whether this is inherent in the analytical technique, or merely reflects slight non-uniformity in the composition of the sample ingots."

When it came to the issue of crucibles, the Summary section of the report stated:

> "Tests were completed on the evaluation of 'hot-pressed' titanium carbide and graphite crucibles lined with tantalum carbide and tungsten boride as refractories for molten titanium. Melts were prepared in crucibles made of zirconium oxide (stabilised with CaO), calcium oxide, calcium oxide fluxed with TiO_2, and aluminium oxide. The stabilized zirconium oxide crucible was the first refractory tested which had areas not wet by the molten titanium. Therefore, additional melts in this type of crucible are planned to evaluate this refractory further. None of the

other refractory materials tested appear to be useful." (p.61, PDF p.11)

Further confirmation is seen in the next section titled "Studies on the Chemical Analysis of Oxygen in Titanium by the Chlorine-Carbon Tetrachloride Method", by E. J. Center and A. C. Eckert, where the authors claim more work on test crucibles and investigation of refractory materials will continue (p.119, PDF p.77).

It means the quantities of these highly pure titanium samples were also not exactly riveting after 1947 in the sense one could build vast sheets of the titanium-based alloys to help build a large flying object. Furthermore, being made inside crucibles whose materials were yet to be determined to handle the high temperatures suggests very little confidence in making large amounts of titanium-based alloys by April 1949. So naturally, one must ask, how much less confident were the USAF and Battelle in making a titanium-based SMA in the sort of quantities seen in the Roswell foil by early July 1947? If the foil was so important at the time and the confidence was so low, wouldn't the question of the right materials to make the crucibles have already been determined in early 1947?

Indeed, if such concerns about choice of materials to build crucibles and the stability of the $TiCl_4$ and TiI_4 forms had already been made by early July 1947, then these should have been resolved by 1949. Apparently, the report suggests they weren't.

Therefore, how likely was it that the USAF or Battelle had the capabilities to produce a nitinol-like Roswell foil in large quantities and at high purity to build at least the skin of the large flying object by early July 1947?

The First Progress Report of 1949 is released

Not long after the official release of the Second Progress Report, another important document would be declassified. This time it was the official USAF/Battelle First Progress Report of contract number AF 33(038)-3736 dated August 31, 1949. The report was declassified and released by DTIC on February 11, 2010 after Billy Cox placed a U.S. FoI request.

The date also shows it is another example of a report not mentioned in the final progress report of AF 33(038)-3736 dated December 31, 1952. It makes one wonder how many other hidden reports one can expect to find under AF 38(038)-3736?

Superficially, we see the First Progress Report is structured in the same way as the Second Progress Report. For example, there is nothing in the contents page to suggest NiTi had been studied. Similarly, the authors' names of Eastwood, Fawn, and Craighead do not appear in the contents for a specific section. It is almost as if the report is written to avoid highlighting the importance of the NiTi study to the USAF. As for the sections of the report, these are essentially the same as in the Second Progress Report.

On closer inspection, the report mentions NiTi under the section titled "Evaluation of Experimental Titanium-base Alloys" by authors Craighead, Fawn, and Eastwood. While the Second Progress Report focused on developing a phase diagram for NiTi, the main aim of this section is to measure and uncover a pattern in the data gathered for tensile strength, hardness levels and elongation of NiTi with different percentages of the nickel composition, as well as the methods of hot rolling and aging of the alloy. It also compares that data with a selection of other titanium alloys.

This report proves hardness was important to the USAF for NiTi by August 1949. Yet, for some reason, this data did not exist prior to manufacturing the nitinol-like Roswell foil in early July 1947. Why the hardness data so late in the game?

As in the Second Progress Report, the titanium-germanium (TiGe) alloy is used for comparison purposes. The data is later compared with two ternary titanium alloys comprising molybdenum (Mo) and manganese (Mn) combined with a small amount of a third element from the following list: nitrogen (N), copper (Cu), chromium (Cr), iron (Fe), cobalt (Co) and carbon (C). The alloys developed for this report were tested in "both the as-hot-rolled temper and after aging the hot-rolled sheet four hours at 750°F" with the aim of measuring the tensile strength, hardness and elongation.

Germanium was added to titanium in the range 0.01 to 1.0 percent. Nickel was added to titanium in the range 1.75 to 15.0 percent (p.6,

PDF p.14). This again re-emphasizes the titanium-rich end of the composition.

Table 1 suggests the USAF was seeking a pattern in the tensile strength, elongation, and hardness data as more nickel was added to the titanium in NiTi (p.6, PDF p.14). This may be used to give an estimate via extrapolation of the tensile strength up to the 50 percent nickel range of the original shape-memory NiTi alloy, assuming there is a pattern in the data.

The tensile strength pattern for NiTi indicates an increase in the strength from 80,000 pounds per square inch (psi) but leveling off slightly above 120,000 psi with increasing nickel (p.9, PDF p.17). NiTi has much higher tensile strength results than TiGe. In fact, this section of the report suggests TiGe showed no improvement to the mechanical properties by adding extra germanium or aging the heat rolled alloy at 750°F.

There is a trend toward increasing hardness under the Vickers Hardness (VHW) column for NiTi and TiGe. The hardness for NiTi does not level off—it simply goes up with increasing percentage of the nickel (p.9, PDF p.17).

The elongation in one inch of NiTi and other titanium alloy samples also shows a pattern, with significant changes taking place from 1.75 to 15.0 percent nickel in NiTi with the percentage decreasing from 22.5 percent to 2.0 percent, respectively (p.6, PDF p.14).

There is no obvious improvement to the tensile strength and hardness of NiTi and TiGe when aged for 4 hours at 750°F after hot-rolling at 1450°F.

Table 2 attempts to measure changes to the hardness of NiTi and TiGe when hot-rolled at different temperatures and with varying composition amounts of the nickel or germanium elements. Again, the results appear inconclusive at different temperatures—only the composition range of the elements added to Process A titanium (the name given to the pure titanium for use in alloying elements) affects hardness.

Again, similar extrapolations can be made on the hardness data and other mechanical properties for NiTi approaching the equi-atomic range if the USAF had wanted to do this.

Funnily enough, in order to avoid any further interest in the alloys NiTi and TiGe by other scientists, the authors decided to write the following statement: "The present data do not justify further investigation of binary titanium-germanium or titanium-nickel alloys" (p.5, PDF p.13). This is *exactly* what the Second Progress Report said. So why keep on studying NiTi in different reports? In addition, in his accompanying letter to the report, Eastwood indicated how the alloys studied were also included in the Summary Report (part I, II or III) of the USAF Contract Number W 33-038 ac-21229. As Craighead et al. stated:

> "Data obtained during this period on alloys already under study on May 18, 1949, are included in the above-mentioned Summary Report."

If the work on NiTi is really meant to be so unimportant or uninteresting to these scientists (and presumably for the USAF as well), why continue studying the alloy again and again and again? Why NiTi at all? Who exactly is making the decision to combine nickel and titanium and study its properties with the help of Battelle scientists and then later claim it is not important? A funny way of not showing interest in NiTi. Whoever was making the decision, it was most probably the USAF. But who at Wright-Patterson would be showing such great interest in gathering data over a seemingly uninteresting and ordinary alloy and yet try his hardest to discourage others to look at the alloy in every report?

And why specifically NiTi?

In Table 3 (pp.10-12, PDF pp.18-20), tensile strength and hardness data are included for the ternary titanium-based alloys of interest. These may be used as further comparisons with NiTi for any secret study on the alloy at Wright–Patterson AFB.

Figures 3-9 (pp.12-18, PDF pp.24-30) indicate greater hardness and tensile strength for the titanium-molybdenum (Ti-Mo) binary alloy compared with NiTi. Adding small amounts of carbon, copper, iron, cobalt, nickel or chromium to TiMo can further improve these mechanical properties (either to a minor or more significant degree depending on composition range).

A similar result can be found when adding 0.1 percent nitrogen to the TiMn alloy without sacrificing the elongation of the alloy, but the ductility of the alloy decreases dramatically when 0.2 percent nitrogen is added (p.20, PDF p.32).

The "as-hot-rolled temper" gives better tensile strength and hardness for ternary TiMn-based alloys compared with "a solution heat treatment at 1600°F". However, the aging method shows inconsistent results (pp.27-28, PDF pp.41-42).

Moving away from the work of Craighead et al., in other sections of the report we notice a continued interest in finding ways to measure how chemically inert and pure the titanium is in the tetrachloride solution at various temperatures, and in finding materials suitable for holding molten titanium and its alloys.

USAF/Battelle Summary Report Part III dated July 30, 1949

The accompanying letter for the First Progress Report dated September 16, 1949 written by Battelle scientist L. W. Eastwood (a member of Dr. Craighead's team in analyzing NiTi and TiZr) provided other useful details into the content of an earlier report, called the "Summary Report Part III". In the letter, he stated:

> "Part III of the Summary Report, dated July 30, 1949, under Contract W-33-038 ac-21229, describes the development of analytical methods and the study of refractories carried out during the period May 18, 1948, to May 18, 1949. In addition, it contains data obtained during the interval May 18, 1949, to July 30, 1949, on alloys which were in the process at the expiration of the preceding contract, May 18, 1949. At the request of Mr. J. B. Johnson, this latter information obtained during the first two and a half months of the present contract was submitted in lieu of the first regular bimonthly progress report."

It is good to see Eastwood is able to confirm Summary Report Part III (and hence Parts I and II) is definitely part of contract number W-33-038 ac-21229.

With further research, we also discover J. B. Johnson is the Chief of the Metallurgy Division at Wright–Patterson AFB. It is likely Dr. Craighead and his team of Battelle scientists were under instructions from Mr. Johnson to analyze certain alloys of interest to the USAF, as well as deciding which alloy data should be included in the various progress reports under these USAF contracts.

If anyone at Wright–Patterson AFB had intimate knowledge of the relationship between NiTi or some other kind of dark-grey titanium-based SMA and the Roswell foil, Mr. Johnson would be the person to ask.

The U.S. Navy's involvement

Leaving aside the analysis of the USAF/Battelle progress reports into titanium and its alloys for the second stage contract, there is further evidence to show NiTi was an alloy of intense and somewhat secretive interest to the USAF as early as 1948.

Firstly, it is clear the titanium industry officially began on American soil by E.I. du Pont de Nemours & Co. Inc, suggesting the U.S. Bureau of Mines had finalized its titanium commercialization pilot plant and sold a license to this U.S. company to use the technology. And secondly, the USAF decided it would commence the first stage of its renewed interest into titanium and its alloys on May 18, 1948 with the help of Battelle through contract number W-33-038 ac-21229.

During the period when the first stage into the study of titanium and its alloys had commenced, Battelle had asked the U.S. Bureau of Mines (perhaps without consultation with the USAF) to produce the world's first tentative NiTi phase diagram. Long et al.

A tentative titanium-nickel diagram. J. R. Long, E. T. Hayes, D. C. Root, and C. E. Armantrout. *U.S. Bur. Mines, Rept. Invest.* No. 4463, 13 pp.(1949).—Powders of 98.66% pure Ti, minus 35 mesh, and 98.93% pure Ni, minus 200 mesh, were pressed at 50 tons/sq. in. to form compacts which were welded in Fe sheaths and rolled at 700–1000°, followed by H_2O quenching after soaking 1 to 48 hrs. at the rolling temp. For subsequent heat-treatment, samples from the rolled compacts were sealed in quartz tubes which were evacuated and back filled with He. Because of impurities the alloys behaved like complex ternary alloys. The addn. of each 1% Ni up to about 11% lowers the m.p. of Ti approx. 50°. At this compn. the m.p. is 960° and there is a eutectic at about 33% Ni. Ni is sol. in the high temp. form of Ti, showing max. soly. of 11% at 960°. This solid soln. is stable down to 890°. At lower temps. it begins to break down forming 2 or 3-phase structures. One of the new phases that appears is the hexagonal low-temp. form of Ti contg. less than 0.5% Ni in soln., and is termed α. The γ-phase is a body-centered constituent contg. approx. 41% Ni. Decompn. of β is complete at about 765°; below this the alloys consist entirely of α and γ. The diagram suggests that the most useful alloys may be found in the range of 0.5–10% Ni.
C. W. Schuck

338

would eventually publish the work in February 1949 under the title *A Tentative Titanium-Nickel Diagram.*

The work was outsourced probably because it was safe to do so knowing a number of Battelle members worked at the U.S. Bureau of Mines. At any rate, the nickel range for this work was limited to below 41 percent, and the titanium being produced at the U.S. Bureau of Mines didn't appear to be sufficiently pure enough to reveal any unusual shape-memory properties. Either that or the composition range was kept well away from the equi-atomic range of 47 to 52 percent nickel for NiTi.

We see evidence of titanium not being manufactured in a sufficiently pure form in the *Chemical Abstracts* of 1949, where it is stated how the U.S. Bureau of Mines was producing NiTi in sealed "quartz tubes which were evacuated and back filled with He [helium]"[35], but later in the same abstract we see the statement:

> "Because of impurities the alloys behaved like complex ternary alloys."

Such concerns of impurities in the titanium-based alloys, such as NiTi, can only reflect the quality of the titanium being produced from this organization at that time.

While Battelle may have succeeded in not linking this work to the USAF nor reveal the possibility of a shape-memory effect in NiTi in the equi-atomic range with anyone at the U.S. Bureau of Mines, the U.S. Navy got a whiff of what was happening, probably from a Battelle insider working at the U.S. Bureau of Mines or at the institute itself. Together with high-level discussions at the U.S. DoD in Washington, the U.S. Navy learned there was considerable interest in titanium and its alloys from the USAF and later the U.S. Army, as well as a strong interest in NiTi.

Much wrangling must have taken place in the top military ranks as the U.S. Navy tried every method possible to figure out the mystery of NiTi. Unfortunately, it got the U.S. Navy nowhere.

In the end, the U.S. Navy got fed-up with the continued stone-walling by the USAF and probably from the head of the Battelle team involved in the NiTi work, Dr. Craighead, to the point where top Navy brass decided it was time to set up their own independent and open symposium on titanium on December 16, 1948. This was to gather more details from experts in the field, including Dr. Craighead. Rear Admiral Thorwald A. Solberg, the Chief of Naval Research, approved the symposium. As the ONR stated in the Foreword of their report on the symposium, page ii:

> "The purpose of this meeting was to bring together those government agencies and industrial organizations actively engaged in the research and development of titanium and its alloys so as to provide, for the first time, a comprehensive review of the titanium research effort and the progress thus far attained."

At the conclusion of the symposium, Mr. Harold C. Cross from Battelle prepared the final report for the ONR dated March 1949. At the request of the U.S. Navy, the work on NiTi by the U.S. Bureau of Mines had to be included (page 27). Any other information on titanium was kept at a general level, such as finding suitable crucible materials to handle melted titanium and other elements as discussed by Dr. Craighead in his section of the report. None of the discussion papers included in the report revealed the reason for the USAF's secretive interest in NiTi.

Before the ONR symposium report was published, the U.S. Bureau of Mines had to prepare and write a manuscript containing the NiTi results obtained for Battelle. According to the manuscript, it was completed sometime during December 1948, but was probably not intended to be published. However, it would be published as a 13-page soft cover report in February 1949, just a month before the Symposium Report came out.

The official title of the ONR symposium report is "Titanium— Report of Symposium on Titanium, Sponsored by Office of Naval Research"[36]. Its existence has been previously confirmed in the unclassified U.S. military document titled "Some Mechanical and Ballistic Properties of Titanium and Titanium Alloys" by R.KK. Pitler and A. Hurlich as seen below:

REFERENCES

1. "Titanium, Report of Symposium on Titanium, Sponsored by Office of Naval Research, 16 December 1948." Office of Naval Research, Dept. of the Navy, Washington, D. C., March 1949.

The ONR symposium report has been available for download since March 2010.

Since the report failed to give a clear and unequivocal explanation for the USAF's unusually high secret interest in NiTi, the U.S. Navy went quiet on the alloy until, in 1958, Dr. Buehler at the Naval Ordnance Laboratory (NOL) decided to combine pure titanium and nickel in the equi-atomic range and study its properties. From there, the *official* scientific history of SMAs began.

Interest in super tough and extremely lightweight alloys by U.S. military

On January 29, 1949, Clyde E. Williams (1893–1988), chairman on the Panel of Metallurgy (and head of the Battelle Memorial Institute in Columbus, Ohio, between 1934 and 1953), wrote a confidential memorandum[37] to the head chairman of the Committee on Basic Physical Sciences at the Research and Development Board about military requirements for new materials:

> "The organization of sub-panels supporting this Panel was begun in November [1948]….Accordingly, this statement cannot undertake to deal with the metallurgical programs in detail. The Panel has considered only the most urgent problems and the programs directed to their solutions.
>
> The following sub-panels have submitted reports of initial surveys of work underway in their respective fields:
>
> Super High Temperature Materials
> Ferrous-Base High Temperature Materials
> Ceramic Coatings for Metals
> Ceramics
> Aluminium
> Magnesium

Titanium and Zirconium
Carbon and Alloy Steel
Welding and Joining
Corrosion and Surface Treatment
Magnetic Materials

Those reports, some of which are quite detailed, are being circulated among all interested committees of the Board, as well as among the Technical Services of the Departments."

There was particular interest in "metals and alloys of increased strength-weight ratios and toughness", "super high temperature materials for use in the temperature range, 1800-3000°F", "Nickel-base and cobalt-base alloys for use at temperatures up to 1800°F", and "methods of producing titanium on a commercial scale..."

Apparently, all this interest was in 1949, not 1947.

The entrance way to the Battelle Memorial Institute.
Mr. Clyde E. Williams was also the director of this institute in the late 1940s and early 1950s.

In the light of USAF/Battelle studies into TiZr, it is also interesting to see titanium and zirconium as two metals mentioned in the document.

There is also a mention of cobalt as being useful in some alloys of interest to the military. Earlier we asked why zirconium would be added to NiTi unless there was a technological benefit in doing so. Scientific literature tells us there is no benefit and, in fact, adding zirconium to NiTi would reduce the shape-memory effect, as well as result in the loss of other useful physical properties. However, if no zirconium is added and only a small amount of cobalt replaced some of the nickel and/or titanium, we would have a different situation. As long as we keep the amount of cobalt to a minimum in order to retain the lightweight characteristics of NiTi, adding around 0.8 to 1.2 percent cobalt can actually extend the shape-memory response of NiTi to a much wider temperature range without loss of other physical properties, including the shape memory effect.

Could NiTi-Co be the alloy used in the Roswell foil?

Come to think of it, we have to ask why the continued interest in any new super-tough materials assuming the Roswell foil was already manufactured by the USAF and/or Battelle? The best that Williams could fathom can be seen in the following quote taken from the document:

"The performance of weapons and carriers under development depends largely on the materials used in their construction, and better materials are essential to meet the new requirements."

Either a natural progression by the military to look at new materials, or something was found and analyzed that sparked the interest into new types of materials, but for some reason the USAF did not want to make it obvious the composition of the alloy that was found near Roswell in early July 1947 in case the media and other scientists were to ask some difficult questions about the incident. Why? Indeed, what is so special about this Roswell object and its "victims" that the composition of the Roswell foil cannot be known to the public after all this time even when we are confident it must contain Ni and Ti? And if it isn't, what other possible titanium-based SMA could it be? And why choose not to use this amazing alloy again?

Somehow we cannot ignore the possibility that this document could be related to the Roswell case.

When did the U.S. military titanium industry begin?

Yet the quantities of pure titanium in 1948 still had not reached the crucial levels needed to be siphoned off to the USAF where a large titanium flying object could be built. The U.S. DoD had first to encourage U.S. companies to manufacture titanium in large quantities before enough of the metal would be available for the USAF to build a flying machine. The magic year for this to eventually happen for the military was 1950.

The *Encyclopedia Britannica* stated:

> "As a result, the [US] Department of Defense provided production incentives to start the titanium industry in 1950."[38]

It should be noted that 1950 was not the year when titanium was in ample supply for military use. The year 1950 was merely to encourage manufacturers to start mass-producing titanium for the military. The point at which adequate quantities of titanium to build a decent-sized military flying object would come after 1950.

It was only when a number of American companies began working with the U.S. government on titanium (with the promise of lavish financial government support) that the quantities of titanium reached the thousands of tons by around 1955, making it feasible to build at least *one* large military titanium flying machine. As the 1987 edition of the *McGraw-Hill Encyclopedia of Science & Technology*, Volume 18, p. 380, has indicated:

> "By the mid 1950s a number of Japanese and American companies were producing many thousands of tons of metal using the Kroll method, a practice which still dominates the industry."

The world's first official titanium military aircraft

Luckily, the U.S. military didn't need to wait until 1955. In fact, not much longer than 1951 as the scientific literature tells us when J. B. Johnson (a familiar name from previous studies into NiTi) and E. J. Hassell at Wright–Patterson AFB expressed interest in titanium for building military aircraft because of its useful aerospace characteristics as the following abstract[39] shows:

> **Titanium in aircraft.** J. B. Johnson and E. J. Hassell (Wright-Patterson Air Base, Dayton, O.). *Metal Progress* 60, No. 3, 51-5(1951).—Strength-to-wt. ratios, in tension, are given of 304 and 347 stainless steels, 2 Al alloys, one Mg alloy, and Ti and one exptl. Ti alloy at 70, 300, and 700°F. Ti and some of its alloys are inherently stable, have good mech. and corrosion-resisting properties, and are useful up to 800°F. W. A. Mudge

From 1952-55, the USAF began implementing titanium as a structural and external skin component in their experimental jet aircraft known as the X-3, where it remained in the unadulterated light-gray color of either alloyed or unalloyed titanium. In other words, no dark-grey shape-memory titanium-based metallic foil was ever employed in the making of this experimental plane.

The Douglas Aircraft Company manufactured the aircraft—with its two Westinghouse XJ-35 turbjets—and only one model was ever produced.

As John Powell wrote:

> "[The] X-3 had a slender fuselage and a long tapered nose. Its primary mission was to investigate the design features of an aircraft suitable for sustained supersonic speeds.
>
> The X-3 was manufactured by the Douglas Aircraft Company. There was only one model produced. It was 66 ft long, 12 ft high and had a wingspan of 22 ft. Two Westinghouse XJ-35 turbojets, equipped with afterburners, powered the X-3. It was capable of takeoff and landing under its own power.
>
> The top speed of the aircraft was just over the speed of sound. It reached and altitude of 41,318 ft.

A secondary purpose of the X-3 was to test new materials such as titanium."[40]

It makes one wonder why the Roswell foil was not used again in the X-3, using the same technology for which we are led to believe by the USAF and/or its contractual affiliates (e.g. Battelle) had already existed unofficially before July 1947.

Is this a question of cost? Was the alloy too expensive to re-use again? Not according to a study by Emmanuel G. Mesthene of The RAND Corporation.

How much was the U.S. military willing to invest in titanium?

In a report written by Mesthene titled "The Titanium Decade", he noticed a considerable amount of money was put into the establishment of the titanium industry after 1947. In fact, Mesthene sees the spending as being highly inefficient. Anyone who wanted to work with titanium could virtually get funding from the government without question. As Mesthene said in his Summary:

> "The titanium development program illustrates inefficiencies inherent in Government-industry contractual arrangements aimed at rapid advances in basic technology. More than half the total cost of the program to the Government was the result of subsidizing the creation of a titanium metal industry. It is argued that such Government programs can be more efficient in the future if they recognize more specifically and are aimed more directly at the technological objective, and if the contracts with industry contain more direct rewards for research and development work as such (rather than indirect rewards in the form of production orders)."[41]

Mesthene estimated the total amount spent by the U.S. Bureau of Mines in the research and development of a new titanium extraction process developed on the Kroll method and "establishing the pilot-plant feasibility of the Kroll process" between 1942 and 1948 compared with the total amount spent by the U.S. Government and industry for the period 1942 to 1958 was roughly "half of 1 percent of

346

the total cost". However, after 1947, the money spent by the U.S. government reached US$200 million, with a further US$200 million from private industry. As Mesthene writes:

"The decade 1948-1958 involved virtually all the costs."[42]

In another quote, Mesthene said:

"It took sixteen years—1942 to 1958—to bring titanium from a paint pigment to a useful structural metal. As nearly as I can figure it out, the cost of doing so was about $200 million to the Government and almost as much more to private industry, for a total not far short of a half a billion dollars."[43]

Mesthene also indicated how the period from 1948 to 1958 involved "the scaling up of the Bureau of Mines' methods to production-plant size", and the development of new fabrication procedures to handle titanium and its alloys, such as the "sheet-rolling program". As Mesthene said:

"It was not until the sheet-rolling program got underway that melting and fabrication began to get the full attention they deserved."[44]

It is interesting to see the "sheet rolling program" mentioned. Considering the Roswell metallic foil was fabricated in sheet form to the thickness of a typical newspaper or cigarette foil, it seems such a program came after the Roswell event, not before. Something is suggesting the USAF and Battelle had no mass-scale fabrication program established for rolling titanium into large sheets anytime in 1947. Yet somehow the USAF has managed to create the outer hull of a 9-meter diameter flying disc in one single sheet according to the witnesses.

Leaving aside the technology for fabricating titanium sheets, this does not look like an issue of cost. Given how keen the military was to see titanium manufactured at any cost, there would be no reason why the Roswell foil could not have been re-manufactured and used to build their prototype X-3. Nor is it a question of the Roswell foil not having the necessary aerospace attributes for a top-quality flying object. The witnesses made it clear the Roswell foil was super lightweight, extremely strong, very hard, resistant to high temperatures, and had the ability to return to its original shape, all in a paper-thin dimension. Truly awesome! We are effectively talking about the Rolls Royce of aerospace materials.

There is absolutely no reason why the Roswell foil could not have been reused to form the outer skin of the X-3. Every effort, with no expense spared, would have been attempted to reuse or create more of this Roswell foil to build more military flying machines. Unfortunately, there is no evidence to support this.

So what happened here?

Did the USAF really invent a dark-grey nitinol-like SMA in a so-called secret military flying object by July 1947? Or did something else happen?

The U.S. military's history of superelasticity

Then in 1986, the scientific community had a new word in its vocabulary to describe any SMA that can return to its original shape without a change in temperature, and do it in a highly elastic way. It was *superelasticity*.

In 1947, the USAF recovered in the New Mexico desert pieces of a dark-grey, paper-thin foil with the ability to return to its original shape. The shape recovery occurred at room (or air) temperature (with no apparent signs of extra heating required), although the witnesses' quotes suggest that differences in temperature under the desert sun, body temperature, and higher temperatures using a blow torch seemed to affect how elastic the alloy was.

For example Bill Brazel recalled playing with the foil. He described it in the following words (taken from Chapter 3):

> "...The only reason I noticed the tinfoil (I'm gonna call it tinfoil), was that I picked this stuff up and put it in my chaps pocket. Might be two or three days or a week before I took it out and put it in a cigar box. I happened to notice when I put that piece of foil in that box, and the damn thing just started unfolding and just flattened out. Then I

got to playing with it. I'd fold it, crease it, lay it down and it'd unfold. It was kind of weird. I couldn't tear it."

Furthermore, the military witnesses were confident the foil was definitely a metal, with one witness being certain it was an alloy, and another witness claiming that it probably contained titanium. Since we are dealing with a shape-memory effect, it most certainly had to be an alloy. Whatever this alloy was, it was very tough, and the behavior of the foil was considered unusually elastic at around body temperature (or 37°C). If this is true, it looks as if the USAF already knew something about superelastic alloys by 1947 and had realized that at least one particularly tough alloy with this superelastic property already existed at body temperature.[45]

Here is another quote, this time from Sergeant Robert Smith, showing the importance of body temperature in creating what appears to be a reasonably elastic response from the foil:

"...It was just a little piece of metal, or foil, or whatever it was. Just small enough to be slipped into a pocket. I think he just picked it up for a souvenir. It was foil-like, but it was stiffer than foil that we have now. In fact, being a sheet metal man, it kind of intrigued me, being that you could crumple it and it would flatten back out again without any wrinkles showing up in it."

The words "crumple" and "unfold" are quite intriguing. They suggest a certain high level of flexibility such that the foil could sustain significant deformation and yet return to its original shape very quickly without a change to its temperature. This is essentially what it means by a superelastic property. Also, keeping the foil in a pocket against the human body somehow made the foil incredibly flexible. So flexible, in fact, that it attracted the attention of various witnesses, both civilian and military alike.

Assuming the Roswell foil is made of NiTi or some other titanium-based alloy with a body-temperature superelastic property, this is rather curious. Because, for some reason, the USAF needed Battelle to help them look at NiTi and other titanium-based alloys after 1947. So, how could the USAF know about this effect? Are we to presume the people at Battelle had known about this? Unfortunately no scientist at Battelle

349

or anyone else outside of this establishment had made an official discovery of a superelastic alloy in 1947. Even if someone could do it unofficially, the technology was not available before July 1947 to make the alloy pure enough to reveal this superelastic property. Neither could scientists predict the discovery of superelasticity because not even the early few scientists who saw a weaker form of the shape-memory property called the pseudoelastic effect in a couple of alloys were stating they were superelastic in any way. So how could the USAF know that such an alloy existed, and how did it make the stuff in large enough quantities for witnesses to observe this type of foil in action by July 1947?

It is clear someone had managed to create what appeared to be a superelastic foil long before the scientific community found out. The question is, who? And how was it possible to make such an alloy in 1947?

Whatever the truth, it seems the actual official history of superelastic alloys may have started much earlier than we have been led to believe by the scientific literature.

Summary

As at time of writing, it is fortunate to see the timely declassification of at least two crucial USAF/Battelle reports.

Looking at the formerly classified USAF/Battelle First and Second Progress Reports in the second stage development of titanium-based alloys, we see no use of the term *martensite transformations, pseudoelastic alloys,* or other similar technical terms. Even the composition range chosen for NiTi suggests they were not interested in the equi-atomic region where the shape-memory effect is most noticeable at room temperature. This may lead us to suspect the USAF and Battelle were probably not aware of SMAs at the time. Yet no amount of careful positioning of the final composition range for this study—in this case at the titanium-rich end of the composition range for NiTi—or lack of specific terms to suggest a shape-memory effect can deny the possibility they did know. The work of Craighead, Eastwood, and Fawn in the report clearly emphasizes the establishment of a dual crystalline phase region for various alloys, including nitinol. This is crucial information. The existence of a dual crystalline phase is paramount to determining a potential shape-memory effect in any alloy.

When combined with witness claims of a shape-memory nitinol-like Roswell foil that ended up at Wright–Patterson AFB for analysis, it becomes more than just a possibility...it is highly probable the USAF and Battelle did know.

If this does not seem sufficiently compelling, the reports show Battelle had been requested, under this second stage contract with USAF, to perform alloy research on no less than two known SMAs— NiTi and TiZr. Moreover another SMA—NiTiCo—is believed to have been mentioned but was removed from the Second Progress report (i.e., missing pages), with the potential for NiTiCr to be another. In all, we have no less than three, and potentially up to four, SMAs in the one report—specifically the Second Progress Report of October 31, 1949. And it isn't the only report to reveal a study into NiTi and other SMAs. Earlier reports indicate the interest in these alloys went back for some time. Apparently as early as when the study into titanium-based alloys commenced on May 18, 1948.

For what is supposed to be an ordinary alloy of presumably no real interest to the USAF (and kept repeating this mantra in different reports), NiTi was given more than just a cursory examination. The First Progress Report for contract AF 33(038)-3736 claims the work into the alloy went back to the previous contract number W 33-038 ac-21229 (the first contract on titanium to be issued by the USAF with Battelle after 1947), as if suggesting the alloy was the principal reason for issuing the contract with Battelle. And all the while, in both the First and Second Progress Reports, the USAF and Battelle tried to claim that no further study of NiTi was warranted or required as if to discourage other scientists from learning more about the alloy, but they still studied it a number of times for some reason.

It makes one wonder how many more titanium-based SMAs, including NiTi, we are going to find if all Battelle/USAF Progress Reports for the period from 1948 to 1949 are released for public scrutiny for both USAF/Battelle contracts. If all this work was meant to be a coincidence, it is a remarkable one indeed given what we have seen so far.

The First and Second Progress Reports may not prove or disprove conclusively the USAF had direct knowledge of SMAs. Nor does it prove NiTi is or is not the alloy the witnesses observed in the Roswell foil. However, on the basis of probabilities, and the fact that too many witnesses reported a shape-memory effect in the nitinol-like Roswell-foil that ended up for analysis at Wright-Patterson AFB, as well as the fact that enough military witnesses are adamant the foil was indeed a metal or alloy, it is highly unlikely the USAF were not aware of SMAs at the time.

In addition, it would be astonishing if NiTi was not somehow linked in some way to the Roswell foil, given the considerable attention given to it by the USAF and Battelle at the time. Furthermore, the dark-grey color of the Roswell foil is virtually the same as NiTi. The fact that we cannot identify another dark-grey SMA containing titanium can only raise the prospect that NiTi or a similar alloy containing significant amounts of nickel and titanium could well be the foil the witnesses observed.

Finally, from the reports dated after 1947, one can see the scientific concerns Battelle and the USAF expressed in the refinement of titanium using TiI_4 and in building suitable crucibles for holding the high-temperature molten metal. This might be considered a rather crucial observation, as it would support strongly the view that the USAF and Battelle did not have the technology to produce large quantities of a pure titanium-based shape-memory metallic foil by early July 1947.

CHAPTER 9

Are there alternative materials?

MUCH HAS been said in this book about shape-memory alloys (SMAs), with particular emphasis on NiTi. Not surprising considering that NiTi (or some NiTi-X variant where X is another element of less than 1.5 wt%) can support virtually all the physical characteristics of the Roswell foil in its natural form. For example, the Roswell foil is almost certainly an alloy considering enough military witnesses claimed it was a metal of some sort. It most likely contained titanium to explain the toughness and high temperature resistance shown when subjected to a blowtorch test. There is no need to add other materials by way of a pigment or an oxide on the surface of NiTi to give it its dark-grey color, nor is there a need to laminate other materials to NiTi to help improve its physical characteristics, including its shape-memory effect. The only things that are required to match the observations of the Roswell foil is to produce extremely pure NiTi in newspaper-thin sheet form and to significantly cold-work it and activate its shape-memory response for the hardness to increase appreciably.

Despite the great promise shown by NiTi, we still need to be conservative by asking if there could have been another material made in 1947 to match these observations. In other words, was there a polymer (plastic or rubber) or some other SMA that could show the same pronounced shape-memory effect; dark-grey color; high-temperature resistance; high strength and hardness to resist tearing, cutting, and piercing while at the paper-thin dimension; and extremely light weight?

Here is a list of the best alternative materials known at the present time.

Kapton

Kapton is a polyimide sheet developed by DuPont in the late 1960s. It is commonly used inside computers and to encase spacecraft instrumentation as a means of electrically and thermally insulating wires. The material has a high tensile strength to prevent ripping and breaking should it need to bend significantly, such as what you would find wrapped around laptop hinges to allow conducting wires to transfer electricity between the screen and the main body of the computer.

The material has a stable crystalline structure in the range from −269 to 400°C. The top end of this temperature range is considered exceptional for a polymer. Because of this, DuPont has stated in its brochure:

> "Kapton® does not melt or burn as it has the highest UL-94 flammability rating: V-0. The outstanding properties of Kapton® permit it to be used at both high and low temperature extremes where other organic polymeric materials would not be functional."[1]

Despite the glowing recommendations provided by the company for Kapton's high temperature resistance the maximum rated temperature is certainly not enough to resist the extreme heat of a blowtorch. In fact, using this material as electrical wire insulation has been implicated in some aircraft crashes caused by electrical fires, such as the crash of Swissair Flight 111 over Nova Scotia in September 1998. While this is rare, care must be taken not to exceed the recommended temperature range.

Among other known physical properties, Kapton is chemically resistant. However, its ability to resist cutting and piercing is not particularly good.

On the question of flexibility, Kapton is outstanding in this regard, and there is even a sense of the material being able to return to its original shape so long as the bending is not too significant. Unfortunately, this elastic response is imperfect. Furthermore, it is still possible to put a permanent crease in Kapton with enough pressure, unlike the Roswell foil where regular pounding with a sledgehammer would not leave a crease or dent of any kind. If we are to address these physical limitations:

1. A true shape-memory material needs to be laminated to Kapton (e.g., rubber); and
2. The entire material must be sandwiched between two layers of a much harder and higher temperature resistant material to prevent it from creasing, denting and burning. However, this hardened material must somehow be flexible enough to allow the shape-memory material to return to its original shape with relative ease.

3. The material must look like a metal or alloy, which means a metallic paint must be applied, or metal particles should be impregnated into the surface of the material.

Unfortunately, such a material will not be as thin as a sheet of paper especially if we add other materials available from the 1947 period. And there are still the questions of whether it would be flexible enough to show a shape-memory property as well as resist the high temperatures of a blow torch.

On the positive side, Kapton can be metallised. While this metal look may increase the chances of Kapton matching the appearance of the Roswell foil, it is best to discount this material on the grounds of all the known limitations with Kapton as already discussed. Therefore, it is unlikely Kapton can come close to matching the witnesses' observations.

Mylar

Also known under the trade names Melinex and Hostaphan, Mylar is more generally described by its chemical name as BoPET or biaxially-oriented PET. PET stands for polyethylene terephthalate, and biaxially-oriented means the polyester film is:

1) rendered in the amorphous state by quickly cooling it onto a chilled surface, such as a large metal roller, and
2) heated with additional rollers and stretched in two directions using special machinery

in order to create the final toughened product.

As a result of this manufacturing process, the following physical properties[2] can be achieved with BoPET:

- Strong adhesive strength and brightness.
- Excellent puncture resistance.
- Outstanding mechanical strength.
- Excellent electrical insulation.
- Good durability and dimensional stability.
- High resistance to temperature and chemicals.
- Low water absorption and gas permeability.
- Superior gloss and transparency.

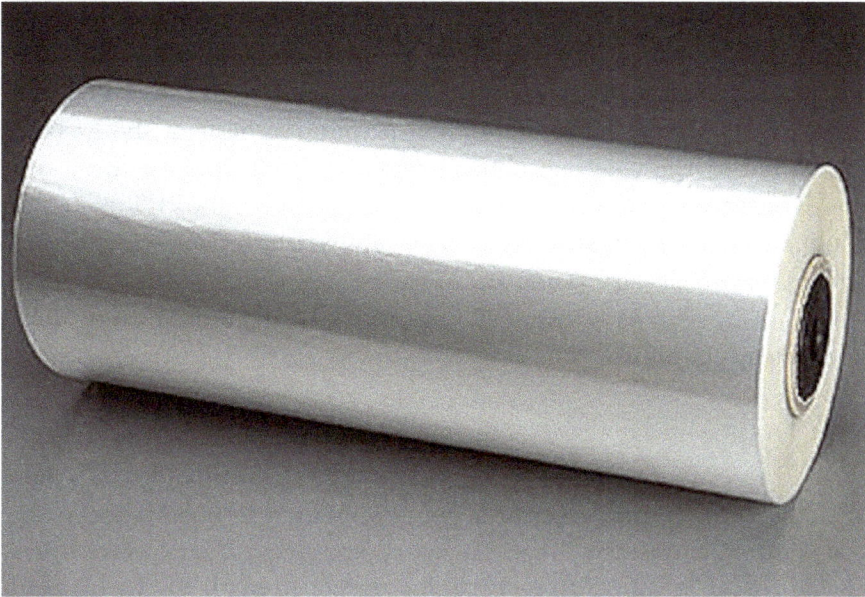

A metallised version of BoPET does exist (classified under BT1031, BT1031C, PETMET BT1031CY, PETMET BT1032 etc.). It is used in tough, bright, silver-colored packaging for holding a variety of foods, such as potato chips, and milk powders. When sealed, it is hard to break apart (even with teeth or a blunt instrument), but can be cut with scissors. Apart from packaging, other uses for this material include the shiny, helium-filled balloons used for carrying meteorological instruments.

In terms of whether Mylar can match the properties of the Roswell foil, the thermoelastic polymer nature of this material means the application of heat can yield some semblance of an elastic response. However, don't get too excited. This shape-recovery property is imperfect. To get the shape to return to a smooth state (i.e., without creases) in a paper-thin form, it is necessary to laminate it to a proper shape-memory polymer or alloy of similar thickness. Even with lamination, a paper-thin form of Mylar would certainly not have resisted the impacts of sledgehammer blows (it would have torn, or left behind permanent marks).

Despite the relative toughness exhibited by Mylar and other similar BoPET materials for general packaging purposes, these materials are still nowhere close to the Roswell foil. Indeed, the biggest problem in the case of Mylar is not just that its hardness is insufficient to resist cutting, but that its melting point of around 254°F (or 200°C)[3] is too low. Just on the last point, knowing the Roswell foil was subjected to a blowtorch by USAF personnel during preliminary tests at the Roswell AAF without showing signs of melting (and felt cool to the touch quickly after removal of the heat), Mylar and other BoPET materials would struggle to withstand such high temperatures. It is for this very fact that a metallised version of Mylar or any other BoPET material would not be seen as a reasonable material to explain the observations of the Roswell foil.

Kevlar

Also known as Aramid, Kevlar is commonly used in bulletproof vests. It is made of numerous tough carbon-coated nylon-like polymer fibres woven in such a way as to increase the tensile strength of the material when enough fibres are bent by a projectile (e.g., a bullet), thereby providing enough force from the material to counteract the impact of the projectile.

Despite this high strength capability, the factor that eliminates Kevlar from the present discussion is the ease by which it can be cut or pierced with a sharp instrument. When using a pair of scissors or throwing a knife or spear at a piece of Kevlar, the incredibly tiny area impacted by the projectile means that not enough fibres can work together to provide enough opposing force to counteract the energy of the projectile. Unless the fibres have increased hardness, the sharp projectile will tear through Kevlar. To solve this problem, Kevlar needs another material to be laminated onto it to give it the required hardness, but then this would reduce any ability to return to its original shape. Unless the material providing the hardness is broken into tiny pieces and joined together with flexible linking such as chain mail, the final product would definitely not have the paper-thin dimensions necessary to match the Roswell foil. And given that Kevlar has limited shape-memory ability in the first place unless a shape-memory polymer is laminated to its surface, such a process would only increase its thickness even more, making it less likely to fulfill the requirements needed to be comparable to the Roswell foil.

Kevlar is an understandable material to consider as a possible contender for the Roswell foil because of its perceived toughness, but such consideration is limited to special conditions.

Graphene

Graphene is an alternative material that has the necessary thickness, lightweight, and strength to match the Roswell foil at the paper-thin dimension. In fact, graphene is described as a single atomic layer of graphite. With a thickness of around 0.345Nm, this is certainly much thinner than any ordinary paper.

Graphene is composed entirely of carbon atoms bonded together in a "honeycomb" hexagonal lattice where each carbon atom is connected to three others. Since each atom carries six electrons—two in the inner shell and four in the outer shell—and the chemical bonds use up three of the outer electrons, the fourth electron is free to move around on the surface of the graphene sheet, allowing graphene to behave like a metal in conducting electricity and heat. However, it is in the hexagonal structure formed by these three chemical bonds and the incredibly short chemical bond lengths of approximately 0.142 Nm which is what gives graphene its tremendous strength. This ensures that there can be no instability in the structure due to thermal fluctuations at the single-layer dimension. In multilayer graphene, the strength can only increase, making it tougher to rip apart. Not only that, but the strong and short chemical bonds also helps to increase the hardness of graphene. This means for a multilayered graphene, a material can be created of sufficient hardness to prevent cutting and piercing. As a comparison, graphene is the hardest and strongest substance when compared to any other material of the same thickness. If multilayered graphene is used to create a 3-dimensional object, it would be almost as hard as diamond. In fact, very few other materials would have greater hardness than this multilayered graphene.

Because of the thinness of graphene, the weight of one square meter of graphene is roughly 0.77 milligrams. Compared to ordinary paper, this is almost 1,000 times lighter.

The only drawback of graphene is its lack of a shape-memory response. To its credit, graphene is quite flexible, which means that it is possible to laminate graphene onto the surface of another polymer containing a shape-memory response. Still, given graphene's remarkable thinness, a paper-thin shape-memory metallic-looking polymer material with graphene covering its surface could be created to approach the kind of properties seen in the Roswell foil. However, for this to be possible, there must be absolutely no imperfections in the crystalline structure of graphene to get the best flexibility needed for the shape-memory effect to work best. More easily said than done considering the removal of all imperfections is extremely expensive using current technology due to the great difficultly in manufacturing to this level of quality.

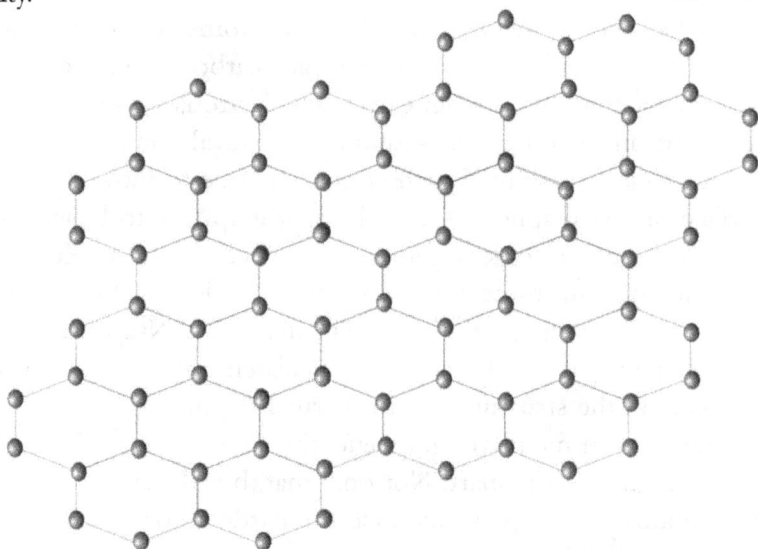

Bearing in mind that the first monolayer graphene ever produced was in 2004 and that manufacturing with no imperfections in the crystalline structure remains an extremely difficult affair at the present time, it is highly unlikely that the USAF could have achieved this feat in the late 1940s. Of course, the USAF is strongly encouraged to prove us wrong by releasing a report to explain how it managed to make graphene with or without assistance from the scientific community prior to the Roswell event.

Hytrel

Invented by the American conglomerate, DuPont de Nemours and Company, in the 1980s, Hytrel is commonly touted as a high performance engineering copolyester elastomer that combines the flexibility of rubber, the strength of plastics, and the processibility of thermoplastics.

Among the claimed benefits of Hytrel include:
- Excellent strength
- Heat resistance
- Low temperature properties
- Chemical resistance
- Good processibility
- Rubber elasticity.

It should be noted that heat resistance according to DuPont's technical specifications is limited to long term application of heat under 150°C where the physical properties of the material does not appreciably breakdown. As for low temperature properties, this is limited to -50°C for the elastomer component of the material to retain its elasticity.

Hytrel is commonly used as cable insulation, and can be used to create molded products that often end up in car interior surfaces (such as the rubber and plastic like materials covering the steering wheel) because of its good retention of physical properties at elevated temperatures as you would find on a summer day inside a car with the windows wound up.

How it is made

Copolyester elastomers are composed of two different types of chemical building blocks, or repeating units, making up the chainlike molecules. Each repeating unit is described as a long sequence, or block. Hence the use of the name *block copolymers* in scientific literature to describe this type of material. One chemical repeating unit is a hard substance. It usually consists of polybutylene terephthalate (PBT), a polyester resin that gives Hytrel its stiffness (and hardness). The other chemical repeating unit is a soft substance made of any one of a number of polyethers or flexible polyesters. A typical example could be polytetramethylene ether glycol.

When combining these two different substances, temperatures of just over 200°C are needed to melt them. Upon cooling, the PBT sequences appear as clusters of hard crystalline regions. It is through these regions where a combination of mechanical interlocking and strong intermolecular attractions are formed to hold the softer elastomeric sequences. However, this holding together is not chemically interlinked into anything like a permanent crystalline structure as occurs in vulcanized rubber. Because of this subtle difference, Hytrel has the advantage of being able to be re-heated to just above the melting point of the crystalline domains (about 200°C) and re-cooled without losing the properties Hytrel is renowned for. It means Hytrel is a recyclable product. Furthermore, it can be processed using standard thermoelastic processes, such as injection molding, blow molding, calendaring, rotational molding, extrusion and meltcasting. And if that is not enough, different coloring agents can be added to give it the right appearance, even of the metallic kind. However, Hytrel often requires a discoloring stabiliser to avoid losing its color over time, less so with metallic additives.

In fact, the hardness of Hytrel can also be easily adjusted. The Shore D (Durometer) hardness scale is commonly used in the polymer manufacturing industry to specify how hard certain materials are. The range is from 30D to 82D, with higher values indicating a harder material. For example 50D is equivalent to the hardness found in solid truck tires, whereas 75D is found in those hard plastic hats used by miners (typically made of a material called HDPE). Therefore, it is technically possible to make a harder substance out of Hytrel. The process merely involves adding more of the PBT to the softer elastomer substance.

In relation to the Roswell foil, Hytrel won't meet the required physical properties. Despite the promise of greater hardness through the addition of PBT, a sledgehammer blow can still crack this material. Of even greater concern has to be how well Hytrel can withstand high temperatures of a blowtorch. We know subjecting the Roswell foil to a blowtorch showed it did not melt. Hytrel, on the other hand, will melt under the tremendous heat of a blowtorch. Also, the removal of heat to the Roswell foil resulted in a number of witnesses claiming it felt unusually cool when almost immediately touched. We know nitinol can do this in a highly pure sample without any problems. Hytrel can't do anything like this.

On the positive side, Hytrel can be made to look somewhat dark grey and even metallic with the right metallic additives or paint coating. That's the only advantage this material brings to this discussion.

Vulcanised rubber

Or why not a piece of toughened synthetic rubber that has gone through a process of vulcanization? Here, the rubber is heated with sulfur to create an improved and tougher form of rubber[4] with higher tensile strength while maintaining its elasticity (assuming no more than 10% sulfur was added by weight of the mixture) over a greater temperature range. In fact, this material is reminiscent of SMAs in the sense that it will:

1) contract when heated;
2) gives up heat when stretched;
3) become less elastic as it is cooled; and
4) retain its original shape.

All that is required is to spray the rubber with metallic paint and one could almost get away with describing it as an SMA (maybe even, dare we say it, nitinol as well). Furthermore, the elasticity of vulcanized rubber is effective at room temperature just like the Roswell foil. So naturally this raises the question as to whether the USAF could have created a metallized version of vulcanized rubber in 1947 to help explain the Roswell foil's interesting metallic and pronounced shape-memory properties.

Before we can answer this, let us look more closely at the vulcanization process and the known physical properties of the rubber formed from it.

In terms of the process itself, the exact chemical reactions that take place between rubber and sulphur are not fully understood. Chemists are settling on the idea that the two substances combine chemically to form what are called "bridges" (or cross-links) made of chains of sulfur atoms designed to hold together the long-chain elastic molecules of rubber.

Once the new solid piece of rubber is formed from this vulcanization process, scientists have noted the following physical properties:

- Higher tensile strength
- Able to resist abrasion and swelling.
- Able to tolerate large deformations under load.
- Can quickly recover its original shape once the load is removed.
- A non-sticky material compared to natural rubber.
- Good durability.
- Can be manufactured into highly precise shapes and dimensions.

Already an exceptionally tough and elastic material thanks to the sulfur, the quality of vulcanized rubber can be further improved with minor addition of either zinc oxide or the amorphous carbon as commonly found in soot. The fine powdered form of these additives are combined at temperatures between 140 to 180°C. Their benefit is primarily to increase the hardness and durability of the final vulcanized product.

As for the history of vulcanization, the process had its origins between 1932 and 1934 when Nathaniel Hayward and Friedrich Ludersdorf observed a reduction in the viscosity of natural rubber when sulfur was added under heat. Unfortunately, for whatever inexplicable reason, they did not pursue the discovery further to see if a solidified material would form and whether the physical properties achieved with this new rubber material would be something worth commercializing.

Whether inspired by this early work or more of an independent effort, it seemed Charles Goodyear (1800-1860) looked at this process more closely in 1939. Not long after, he moved to Woburn, Massachusetts, USA, where he carried out a series of experiments to improve the curing of rubber. Together with his collaborator, Nathaniel Hayward (the same guy who worked with Mr. Ludersdorf), they eventually noticed how a tough new rubber material was formed. However, since Mr. Goodyear did not immediately patent his invention (or probably came a bit late), Thomas Hancock (1786-1865) was the first to provide a patent for the volcanization of rubber on May 21,

1845 in England. Quickly following suit came Mr. Goodyear with his own patent in the United States on June 15, 1844. While there continues to be a long-running dispute as to who invented the process, various manufacturers have seized on the opportunity to sell this new tougher rubber compound under different brand names, such as Ebonite and Volcanite to name a few.

Charles Goodyear

Apart from vastly improved cure times and a reduction in the amount of energy required to create the new rubber compound thanks to the work of George Oenslager in 1905 when he discovered how thiocarbanilide helped to accelerate the reaction of sulfur with rubber, the new material has pretty much revolutionised the industrial world to this day. Prior to this time, the only material people could use to seal small gaps between moving machine parts was to grab some leather and soak it in oil and somehow ram it inside the gaps. Now precise shapes and dimensions can be manufactured out of this improved rubber. More importantly, the machines can operate at high temperatures and pressures due to the relative toughness of this new rubber material, as well as its ability to return to its original shape once the pressures are reduced.

Today, volcanised rubber can be found in those hard-wearing rubber shoe soles, hoses and conveyor belts. As for rubber tires, the Goodyear Tire and Rubber Company was posthumously named after Mr. Goodyear and founded in 1898 where it specialized in this product and the rubber-sulfur chemical process thanks to the emerging car industry in the United States.

In terms of using this material to explain the Roswell foil, the only two significant drawbacks is that the material will neither resist heat from a blow torch, nor resist cutting and piercing at the paper thin dimensions despite its perceived toughness. Certainly the rubber is highly flexible to the point where it can avoid cracking or breaking at extreme deformations and even show a certain level of shape recoverability at room temperature. However, without the necessary toughness in the high temperature aspect and hardness to match the Roswell foil, we can effectively rule out vulcanised rubber from this discussion.

Non-NiTi annodized SMAs

Unless there is a secret exotic polymer that could have been made by the USAF in the late 1940s and can match all the observations of the Roswell foil, we are not left with many options to consider, except for the possibility that we might be dealing with another titanium-based SMA. The only niggling problem is that NiTi (or NiTi-X) was the only alloy in the titanium family at the time of the Roswell crash to have a shape-memory effect at room temperature (or lower), and is the one that the USAF studied very soon after the Roswell event according to the progress reports on titanium alloys.

Or perhaps there was another equally tough and more secret non-titanium-based SMA invented by the USAF at the time of the Roswell crash that could explain the Roswell foil. If this is possible, it must have been anodized[5] or given an oxide layer to help bring out a dark grey color. Unfortunately, the problem with this idea is that the USAF had chosen, for some inexplicable reason, not to reuse this alloy in the development of new dark-grey colored fighter jets after 1947. And it would be a long time (not until the 1960s) before the first few highly secret U.S. spy planes would come out with a substantial amount of this dark-grey alloy to match those of the Roswell foil. Why did it take so long to reuse this alloy? Or if the USAF did not use the Roswell foil again, why not? Certainly it is not because the alloy is terrible for aerospace applications. Far from it. It is the best alloy in existence at that time and even to this day, judging from the properties of the Roswell foil. Furthermore, it seems to be one where the USAF could manufacture it with remarkable ease given the considerable quantities involved.

However, if the Roswell foil wasn't made of NiTi, what other SMA could it be?

Relying on the available scientific literature, the only structural metals of interest to the USAF in early 1947 for building aircraft components were iron, aluminium and titanium. If there was an alternative SMA, it would have to be composed principally of iron, aluminium or titanium.

For titanium we know NiTi is the best candidate. So let us not pursue this area.

Aluminium-based SMAs

In the case of aluminium, we know CuAl is a SMA that could have been manufactured by the USAF in 1947. Its transformation temperature is a little too high for practical use (and certainly not at room temperature), but by adding zinc to this alloy to form the ternary system CuZnAl, the transformation temperature drops dramatically. With the right amount of zinc (in the range of 15 to 30 wt.%) and 3 to 7 wt.% aluminium with the balance being copper, the transformation temperature can range from -100 to 100°C. More importantly, the alloy uses relatively inexpensive metals and relies on conventional metallurgical processes for its manufacture. This makes it the cheapest SMA in existence, and the first copper-based SMA to be commercially exploited. Nevertheless, even if one could anodise this alloy to have the distinctive dark-grey color of the Roswell foil, the biggest problem lies with the way the transformation temperature changes over time with continuous use (i.e., aging) and actually increases when the alloy is heated in excess of 100°C (the alloy's high-temperature crystalline structure essentially decomposes to the more stable martensite form), thereby quickly losing its room-temperature elastic effect. Also, compared to other SMAs, the elastic effect is modest at best. A maximum recoverability strain of around 5% is reported for this alloy compared to nitinol with its much higher recoverability strain of between 8 to 10%. Finally, the grain size of this alloy is large and gets larger with each transformation, making it vulnerable to brittleness. Additions of grain-growth control additives (mainly from boron, cerium, cobalt, iron, titanium, vanadium and zirconium) of less than 1 wt.% are essential to minimizing this brittleness problem. Because of these disadvantages, CuZnAl is rarely used today.

A more common and relatively cheap alternative SMA is CuAlNi. Usually containing 11 to 14.5 wt.% aluminium and 3 to 5 wt.% nickel with the balance being copper, this alloy is easy to make using conventional metallurgical techniques. The only significant problems with using this alloy lie in the fact that it shows transformation temperatures in the range 80-200°C depending on its composition, and that its mechanical properties are rather poor unless the aluminium content is kept below 12 wt.%. Furthermore, the transformation temperature and quality of the shape-memory response is particularly sensitive to the aluminium content. Therefore, it is imperative to always keep the aluminium within the 3 to 5 wt.% range. As for reducing the transformation temperature, approximately 2 wt.% manganese can be added. And in terms of the brittleness of the alloy due to the large grain size, the same grain growth control additive used in CuZnAl is required. However, these elements need to be carefully added to avoid upsetting the stability of the shape-memory crystalline structure, which in turn can affect the quality of the shape-memory response.

Al-Ni has been mentioned in some sources as a SMA, but it is not a particularly good example. Depending on composition, the alloy has a tendency to exhibit brittle fractures, and has low ductility at ambient temperatures. It only improves at higher temperatures where the strength increases. Together with its low density and lightweight, it is often used in special applications like coating blades in jet engines and gas turbines. The only other way to improve the properties of Al-Ni is to alloy it to other elements. For example, small additions of boron will increase the ductility of the alloy. In the case of the shape-memory effect, copper will have to be added to form Cu-Al-Ni to help improve the effect.

In more recent times, nickel has been replaced by less than 0.5 weight % beryllium (Be) to extend the transformation temperature range from -200°C to 100°C. It is reported the CuAlBe alloy system shows a superelastic property. Other than that, the alloy is not noted as a particularly tough alloy, and certainly not in the range we know titanium alloys to be famous for.

Here is the full range of known aluminium-based SMAs:

Cu-Al (copper-aluminium)
Cu-Al-Ni (copper-aluminium-nickel)
Cu-Al-Be (copper-aluminium-beryllium)
Cu-Zn-Al (copper-zinc-aluminium)
Ni-Al (nickel-aluminium)
Ni-Al-Re (nickel-aluminium-rhenium)
Ni-Co-Al (nickel-cobalt-aluminium)
Ti-Nb-Al (titanium-niobium-aluminium)
Ti-Ni-Cu-Al-Mn (titanium-nickel-copper-aluminium-manganese)

As there are very few SMAs containing aluminium (we have covered practically all of them here), this leaves us with those SMAs made of iron. So let us see if there is a tough enough iron-based SMA to closely match the observations of the Roswell foil.

Iron-based SMAs

The earliest example in scientific literature of an iron-based SMA is FeMnSi, studied in the 1980s. Since then, other iron-based SMAs have emerged based on that structure, such as FeMnSiCrNi, the corrosion-resistant FeMnSiCrNiCu, and FeMnSiCrCu. Not long after this, scientists started to take notice of some shape-memory effects in FeNi, FePt, FePd, FeNiC, FeNiTiCo, and FeNiCoAl. And from there it was learned that German chemist Erich Scheil did look at the martensite transformation in FeNi as early as 1928 without realizing its shape-memory property. At any rate, iron-based SMAs are not as cheap to manufacture as their aluminium-based and copper-based counterparts. If there were a way to reduce the costs, iron-based SMAs would have the potential to transform civil engineering. Attempts to achieve this can be seen in the 1996 thesis by Huijun Li from the Department of Materials Engineering at the University of Wollongong titled, *The Development of New Iron-based Shape-Memory Alloys*. Here the author has suggested the replacement of up to 10wt% Ni with 1wt% Cu produces a new class of SMAs such as FeMnSiCrCu that "are much less expensive"[6] to manufacture.

Leaving aside the cost issue, iron-based SMAs (also called Shape-Memory Steels, or SMSs) have an advantage over those made of aluminium: the shape-memory response is slightly better. Indeed, a maximum recovery strain of around 5.4% has been achieved for FeMnSiCrCu. Not quite as spectacular as nitinol, but at least it is better than nothing.

A much better iron-based SMA to emerge from further study has to be $Fe_{20.2}Mn_{5.6}Si_{8.9}Cr_{5.0}Ni$ (wt.%). According to the article[7] published on September 14, 2014 by Y. H. Wen and his colleagues, the addition of nickel and chromium helped to suppress "the formation of twin boundaries" considered the stumbling block of many iron-based SMAs in not achieving the large recovery strains that nitinol is famous for. A standard Fe-Mn-Si-based alloy has a low recovery strain of less than 3%. Reducing the density of the twin boundaries by adding two other elements, on the other hand, has managed to attain a tensile recovery strain of 7.6% in the new annealed cast alloy. That is almost as good as nitinol.

Recently, a major breakthrough in superelasticity has been made in iron-based SMAs with the discovery of $Fe_{28}Ni_{17}Co_{11.5}Al_{2.5}Ta_{0.05}B$ (at. %). It has been reported by Tanaka et al. that this alloy can show a massive superelastic strain recovery of greater than 13% at room temperature. If there are any disadvantages in this alloy, it would have to be in the weight and toughness: it is not the lightest SMA, nor is it the toughest. Still, this is considered a major development in SMAs since Buehler first set his eyes on the spectacular shape-memory effect of nitinol after 1958.

Why a titanium-based SMA is best

Yet despite the availability of something as powerful as NiTi for its shape-memory effect, the time period in which scientists officially looked at iron-based SMAs and finally made the discovery of an ultra superelastic alloy makes it seem unlikely the USAF would have studied these iron-based SMAs in the late 1940s. However, even if we entertained ourselves on the thought that the military had studied them, at least on an unofficial level (perhaps via a random "trial-and-error" method of combining a bit of iron and nickel at the right composition ratio and then crossing their fingers with the hope of finding something interesting in its properties, and later adding a few more elements to the mix to create the deluxe iron-based SMA edition), more significant work would still need to have been done prior to July 1947 in order to develop a theory of how SMAs work. Talk of "twin boundaries" and ways to reduce them to maximise the shape-recovery effect would have had to be high on the list of things for the USAF to look at. Unfortunately, there is no evidence to support this in any of the U.S. DoD reports that we have.

But even if this is untrue, at some point, the USAF must have realized that a powerful SMA had to exist in the titanium-based family, one that could explain the incredible toughness and lightweight nature of the Roswell foil. Iron-based SMAs are not noted for having extreme hardness to resist cutting and piercing at the newspaper-thin dimensions of a sheet of this stuff, and do not have the same ability to withstand the high temperatures of a blowtorch as a titanium-based SMA does, especially NiTi. If the USAF was truly interested in finding the toughest SMA, it would have been best to go for a titanium-based SMA. Forget the rest. In that case, which titanium-based SMA would be best? The only titanium-based alloy to have the level of shape-recoverability seen in the Roswell foil at around room temperature; to have the other desirable physical properties of the Roswell foil, such as hardness and high-temperature resistance; and to grab the attention of the USAF so soon after the Roswell crash, according to the reports available to us, had to have been NiTi. If it was not NiTi, some other unique and yet unidentified and very powerful titanium-based SMA must have been created by the USAF.

Yet any decision to use titanium to make an SMA is subject to one important consideration: it must be exceptionally pure to provide a shape-memory effect. It means the work into developing high-purity titanium-based alloys and the theory to predict the existence of a titanium-based SMA must have come before July 1947. This would have been essential if the USAF was to have any hope of predicting (and manufacturing) a more powerful titanium-based SMA, via NiTi or some other titanium-based alloy. Otherwise, the USAF would have been (and still will be) forced to rely on an annodised "dark-grey" form of the FeNiCoAlTaB alloy as the best option (even if it is not quite as tough as NiTi in certain respects).

Clearly sticking to a titanium-based SMA would be the last thing the USAF would have wanted to do, especially if the aim is to hide from the public something significant, because of the risk that people might find out and ask questions later about how the Roswell foil was manufactured.

Well, what if it was this iron-based superelastic alloy and not NiTi? It does not make a difference; we still have the problem of why the USAF decided against using it again after July 1947. As we have seen from history, the USAF chose to look at a titanium-based SMA through NiTi. And as if the USAF was still not happy with NiTi, it had to look at TiZr, and potentially a couple of more NiTi-X alloys containing a little bit of cobalt and chromium. Not exactly a ringing endorsement for the iron-based SMA it had allegedly made for the Roswell foil despite its amazing toughness at the newspaper-thin dimensions (which must take us back to the question of what is so special about the "victims" that the public cannot be told about and learn their names assuming this was a military experiment that was conducted by the USAF? The special and highly sensitive nature of these "victims" to the USAF is the only explanation one can fathom for keeping the alloy secret to this day).

Furthermore, no report has ever been released by the USAF to confirm the work it did in 1947 on SMAs. Not even a basic theory to help the military to make a scientific prediction of a tougher and more lightweight titanium-based SMA leading up to the manufacture of the Roswell foil. This is odd, considering that whoever built the Roswell object must have known a lot about shape-memory effects and realized that a nitinol-like SMA (most probably containing titanium as the primary constituent) was the best material to use for aeronautical and aerospace applications and must have wanted to test it by early July 1947.

However, even if we were to entertain ourselves once again, this time with the possibility that the USAF does have a highly detailed report to explain the theory behind SMAs prior to July 1947, there would be yet another problem. As we have seen in a previous chapter, the knowledge behind valency electrons and the formation of chemical bonds for properly understanding the properties of elements and the crystalline structures of compounds was not finalized until 1948, together with the special case into the chemical bonds of metals and alloys by 1949 from the world's top expert in chemistry at the time, Dr. Linus Pauling. Furthermore, the way Battelle scientists got excited by his work into metals and alloys in the same year (note earlier in this book the help needed by the USAF from Battelle to study titanium-based alloys as if implying no knowledge of how SMAs work), it would be impossible for the USAF to have developed a theory on its own or received help from Battelle to predict the existence of NiTi or some other super-tough titanium-based (or a superelastic iron-based) SMA in 1947 without this crucial knowledge.

Not even Dr. Pauling himself was involved in any contract with the USAF to study SMAs.

And if we could somehow assume we are dealing with a superelastic iron-based SMA in 1947 (highly unlikely for the reasons discussed earlier), it has taken scientists between the 1980s and 2010 to finally discover Fe-Ni-Co-Al-Ta-B. To be conservative, we are talking about more than 20 years of work. For the USAF, it only began to look at new metals by late 1946, and the main element they were interested in at the time was titanium, not iron. Even if we could accept the interest as lying predominantly on the iron side of things, to achieve more than 20 years of work into iron-based SMAs in less than a year would be an extraordinary feat of genius. And after all of that hypothetical effort, the USAF would suddenly give up on the alloy (presumably because of the deaths of a few victims), in favor of the traditional stainless steel and aluminium for aircraft construction, and after 1947 decided it would look at new titanium alloys, including NiTi, leading up to the world's first titanium aircraft in the early 1950s? It simply does not make sense.

It is now starting to look like the USAF did not look at iron-based SMAs. In fact, it is unlikely that any report exists to support any kind of SMA work by the USAF dated prior to July 1947 to explain its knowledge of SMAs, if there ever was any such knowledge. Therefore, whatever material was used on the Roswell object appears to be one that the USAF had discovered entirely by accident, rather than having been invented based on a careful study of known *pseudoelastic alloys*, followed by development of a theory for the shape-memory effect, predicting a more powerful titanium-based (or iron-based) SMA, and later manufacturing this new, highly pure alloy in large quantities to build a flying object. Of course, in what sense was this a "discovery"? Was it a random combination of elements and careful analysis of the physical properties resulting in the discovery of a shape-memory effect? Or was it a discovery of a genuinely foreign object with highly advanced engineering features not manufactured by anyone on Earth? Given the amount of time it has taken scientists to discover a powerful iron-based SMA, and the level of purity required to make a titanium-based SMA, it seems we are led to believe it is the latter.

It does not matter if there had been a purely random combination of elements made at the time and at the right composition. The purity and quantity issue makes a mockery out of any suggestion that the USAF did make a titanium-based SMA. Likewise, one can argue that it would take too much time to discover a powerful enough iron-based SMA.

However, even if we could accept the "foreign object" theory as the most probable explanation, the biggest problem we now face is the fact that no one else has come forth to claim responsibility for the object that crashed in the New Mexico desert. Clearly someone must have made this Roswell object. It does not get made on its own through natural forces. The shape-memory effect and other highly advanced features of the materials show that this had to have been an artificially-made flying object.

So who really built this flying object? Perhaps the answer to this final and most troubling question may involve something far more disturbing than we had dared to imagine.

CHAPTER 10

Was the Object Alien- or Man-Made?

A STUDY OF the witnesses' statements has revealed two conflicting accounts of the nature of the lightweight metallic foil found near Roswell. Depending on who did the observing —the witnesses or the USAF—it seems the foil is:

1. A flimsy, bright silver foil made of aluminium; or
2. A tough, dark-grey, most probably titanium-based, shape-memory foil.

Whatever this foil is, it is almost certainly an SMA considering enough well-trained military witnesses reported seeing a metal of some sort. But irrespective of what this foil is, this does bring us to the question of the object itself. More specifically, what kind of flying object crashed near Roswell in early July 1947? Not surprisingly, like the Roswell foil, we have two distinct possibilities for the type of object allegedly found:

1. A secret weather balloon from Project Mogul; or
2. A secret high-speed disc-shaped flying object.

The USAF naturally supports the former. Witnesses, including a number of outspoken military personnel in their later years, are willing to place their bets on the latter.

Likewise, one could argue the same for the bodies allegedly found with the flying object. They could either be:

1. Dummies; or

2. Alien bodies.

Again, the USAF supports the former, and the witnesses are more inclined to support the latter.

Or, to boil down all the differences we see from the opposing views to their very essence, the USAF alleges that the object is man-made, while the witnesses are adamant that the object is exotic, in the sense that it represents something extraterrestrial.

Is there a way we can determine which of these two options is closer to the truth? And most importantly, can we determine what type of object had crashed near Roswell in early July 1947?

How scientists presently view the Roswell case

When scientists are faced with a controversial issue such as the Roswell case, they can only work on the basis of probabilities that are based on the available evidence.

Furthermore, how probable the explanations might be for this case depends on the question being asked. For example, if the question is whether the Roswell object is man-made or alien-made, the titanium-based alloy NiTi (or an iron-based SMA) of the Roswell metallic foil would seem, at first glance, not to be so alien in its composition. Furthermore, any talk of extremely lightweight metallic foil, plastic sheets and plastic-like structural beams resembling balsa wood are reminiscent of man-made materials and has persuaded most scientists to think this is how a weather balloon is constructed. After all, the USAF was using weather balloons in July 1947, and it only seems fair for scientists to view the USAF's explanation as reasonable and probable. Also, any talk of alien bodies and a crashed disc-shaped spaceship has not been backed up with a *bona fide* piece of alien hardware. Even indirectly, there appears to be no evidence of a new technology based on the reports of UFO sightings that scientists have examined. And since UFOs usually have a prosaic "man-made" or "natural" explanation according to the statistics[1], it is much safer for the scientists to stick with the man-made explanation, whatever that might be in precise terms. With these factors in mind, one can see why there are scientists who are willing to support the USAF's explanation of a weather balloon.

For example, the resident physicist and moderator of the MadSci Network, John Link, was asked, "How do scientists view the Roswell incident in 1947?" He said:

> "My opinion is that most scientists agree with the assessment that the material is from a reconnaissance balloon."[2]

Of course, this is not to say that the Roswell object could not be alien. As Carl Sagan once said, "Absence of evidence is not evidence of absence". In other words, scientists cannot, as yet, disprove the alien explanation.

Furthermore, since the USAF is still unable to provide direct evidence, such as remains original weather balloon for scientists to analyze, it remains reasonable to keep a scientific mind open to the number of witnesses who consistently reported seeing super-tough and somewhat unusual materials. In addition to this, the USAF still cannot explain who the "victims" are in this case. And why a weather balloon? Indeed, what were the victims doing inside the balloon, and what was the aim in having these "victims" participate in the flight? Even more odd is the fact that the shape-memory metallic foil found near Roswell, especially considering its toughness, would be more useful in a high-speed flying object such as an aircraft or missile. Unless the balloon had an attached jet engine (which beggars belief how the balloon can more quickly through the air given the enormous drag it would pose), why the continued inability of the USAF to properly explain the case? Is the military really that incompetent in conducting thorough investigations on the matter? Or are they hiding something far more devastating from the public?

Consequently, it is perfectly reasonable for some scientists to be open-minded on the Roswell case.

If the question being asked is nothing more than whether the Roswell object could be alien or man-made, the majority of scientists will understandably favor the man-made explanation, although the more open-minded scientists who acknowledge the likelihood of finding alien life existing in the universe, of which some intelligent creatures may have solved the problem of interstellar travel, will at least acknowledge the possibility of the object being alien-made considering the USAF has still not provided all the information on the case as it should and, thus, will keep an open-mind should new evidence be found. Otherwise, without evidence, the default position is to remain skeptical of any explanation—whether it be from the witnesses, or the USAF.

For example, Michael D. Swords of Science Studies at Western Michigan University wrote an article in the *Journal of Scientific Exploration* titled "A Different View of 'Roswell — Anatomy of a Myth'". Based on the information and evidence he had received by 1998, he concluded:

> "Roswell: What was it? Without the physical specimens to test, it is impossible to say. At this moment, it seems to me, that a crashed piece of non-terrestrial technology is "the least unsatisfactory hypothesis". But I am ready to learn."[3]

On the other hand, if the question is framed differently by focussing on whether the found Roswell object materials could have been manufactured at the time of the crash, the probabilities can change in ways we might not expect. It can potentially make the difference between whether scientists consider the Roswell case more closely, or to ignore the evidence in favor of other available evidence if scientists believe the probability of another explanation is higher based on this other evidence.

So, how likely is it that the USAF could have manufactured an NiTi-like SMA and used it to build a flying object of a reasonable size to carry several "victims" by early July 1947, thereby lending probability to whether the object is man-made or alien-made?

The nine requirements

To determine the likely nature of the object that crashed, it is clear from the evidence that the object must satisfy the following nine requirements:

1. *There must be at least two wreckage sites.*
   We know that one wreckage site has been proven to exist beyond a reasonable doubt, as revealed by the official USAF Reports of 1994 and 1997, various newspapers articles, several radio broadcasts and the large body of eyewitness testimony. What is not clear, and has never been mentioned by the USAF, is the location of the second wreckage site and what was found there (with the strong possibility of a third site at which a few of the bodies were found, located somewhere between the first wreckage site and the final resting place of the main body of the object). The significant loss of materials blown out and no complete machine remaining after the explosion, the fan-shaped distribution of debris on the ground, and the USAF's continuing claim in a secret memo to General Ramey that a disc (and victims) had been picked up, all suggest that the main body of the object and its occupants remained in the air for some time before crashing elsewhere. Whether the second wreckage site was as close as 40 kilometers, or up to 200 kilometers, away (suggesting the object was moving at high speed) is irrelevant. What is relevant to this discussion is the fact that there must be at least one other wreckage site containing the bulk of the materials, which should give some indication of the size and shape of the object and whether any victims had been present inside it (which there had to have been, according to the secret memo held by Ramey).

2. *The flying object must be large.*
   The most abundant material found at the first wreckage site was the metallic foil, together with smaller amounts of the other extremely tough and lightweight plastic-like sheets and structural beams. This debris was scattered in a clearly discernible fan shape, between 200 (narrowest point with the highest concentration of materials) to 400 feet wide, stretching

no less than "three-quarters of a mile" (and up to 1 mile with the likelihood of a few more pieces lying scattered farther away along the line of travel by the object). As there is no way of knowing the shape (or exact size) of the object from what was found at the first wreckage site, and the USAF insists that the object is a disc in two memos and a news release, a greater quantity of the materials must have remained in the air and landed elsewhere. It is here, at the second crash site, where witnesses allegedly saw a 9-meter metallic disc ripped open by an explosion or impact, and where most (if not all) of the small bodies (or dummies) were found. All of this suggests that the object was relatively large.

3. *The Roswell foil is a metal of some sort*
The Roswell foil was almost certainly a metal, given the high temperature resistance shown by the foil using a blowtorch. Furthermore, it must be an alloy (i.e., one metal element combined with one or more other elements) to explain the shape-memory effect as noted by the witnesses. No metal made of a single element can display the kind of pronounced shape-memory effect seen in the foil.

4. *The Roswell foil contains titanium*
Given that one high-ranking military official believed the alloy probably contained titanium, it is likely to be a titanium-based SMA. And it must have a significant amount of titanium to show exceptionally high temperature resistance, great hardness, high strength, and extremely lightweight. But even if we are dealing with an iron-based SMA, the principal metals used for military aircraft in 1947 were steel and aluminium, with smaller amounts of other elements to help increase strength and reduce weight. Neither steel nor any other known alloys of aluminium and iron manufactured at the time could be made hard enough, in newspaper-thin sheet form, to prevent denting and cutting, or to resist the temperature of a blowtorch. A metal of a much higher melting point, together with other favorable engineering properties, is required. This leaves us with the only other principal metal to gain the official attention of the USAF at the time: titanium.

5. *The Roswell foil must possess a strong shape-memory response at room temperature.*
   The Roswell foil displayed a strong shape-memory effect. Witnesses reported the effect occurring at room temperature.

6. *The Roswell foil was distinctly dark-grey in color.*
   Witnesses reported seeing a "dirty stainless steel" or dark-grey metallic foil. Perhaps the closest witnesses could describe the color of this foil is by referring it to a piece of lead (or sometimes tinfoil), except that lead is much heavier and does not exhibit a shape-memory response. From a scientific perspective, only NiTi, perhaps with small amounts of other elements added to it (e.g., cobalt), is the closest alloy to match the color, shape-memory behavior and other physical properties of the Roswell foil in its natural form. Unless the USAF can show us another dark-grey alloy (either anodised or not) that possesses a similarly powerful shape-memory effect and has the same super-tough properties as the Roswell foil, the best candidate is NiTi (or NiTi-X).

7. *The purity of the titanium-based Roswell foil must exceed 99.995 percent in order to exhibit a shape-memory response.*
   If we are dealing with a titanium-based SMA, the purity of the Roswell foil must be extremely high, since any titanium-based alloy displaying a shape-memory effect like NiTi must meet or exceed a purity of 99.995 percent.

8. *There must have been an available man-made technology to make this SMA by no later than the end of June 1947.*
   The purity and quantity of the titanium-based SMA used in this mysterious flying object that crashed near Roswell must have been attained and manufactured no later than the end of June 1947.

9. *There must be a report written by the USAF or its contracted associates to support the knowledge of SMAs.*
   If the Roswell foil is an iron-based (or even a titanium-based) SMA, the USAF must show evidence by way of a theory

written into a detailed report dated prior to July 1947 about how SMAs work. The theory must also show how the USAF managed to predict a titanium-based SMA using NiTi, TiZr and a couple of NiTi-X alloys (containing chromium or cobalt).

Now, assuming that the USAF had carried out a secret military experiment, what kind of man-made flying object could the military have used to test this SMA and/or anything else it was trying to figure out at the time? Let us consider the possibilities that it was a weather balloon, a military aircraft, a missile, a high-speed airship, or something else entirely.

Could the object have been a weather balloon?

In the case of the USAF's Project Mogul balloon explanation for the Roswell incident, we immediately have a problem. Given what we now know about Project Mogul, the metal used to construct the balloons was aluminium. There was absolutely no evidence presented by the USAF to show the balloons had used expensive, high-tech, and potentially secret materials, such as a new type of powerful and tough SMA. Even Dr. Charles B. Moore, the head scientist of Project Mogul, never asserted to investigators that SMAs, or even just a tiny bit of expensively pure titanium (or a sophisticated combination of elements added to iron to create a SMA), were used in the construction of his secret weather balloons. No titanium, no iron, and definitely no SMAs? How odd. Yet, the witnesses seem certain about the shape-memory effect exhibited by the Roswell foil. And the scientific evidence for the extraordinary toughness of the foil, and the USAF's focus on titanium at the time, tells us that this foil has to contain titanium in a reasonable proportion of the alloy's composition. Even if it wasn't titanium, there is no mention of an iron-based alloy, not even one with a special shape-memory property. Dr. Moore clearly does not recall ever seeing or hearing of any SMA being used in his balloons.

Perhaps the foil's composition was kept secret from Dr. Moore, or he had forgotten about the special nature of the foil he used? Assuming the foil had been made by the military or Battelle in quantities vast enough to build Dr. Moore's balloons and he was not told about the foil's composition, Dr. Moore could have assumed the foil was aluminium. Even so, a shape-memory effect in the foil was not something the local rancher, Major Marcel, and many other witnesses could easily forget or assume was ordinary. Yet for some reason the foil was not considered by Dr. Moore or his other team members to be sufficiently unusual or worth remembering in terms of its shape-memory effect after being given clearance to discuss Project Mogul openly with the USAF. Very strange. Or did the USAF secretly ask Dr. Moore and others to keep quiet about this aspect during the interviews? Again it would not make sense. No reason to maintain secrecy if this object was meant to be a weather balloon. Furthermore, we already know about SMAs, so it really should be no big deal. Unless there is something special about the "victims", we have to assume that the object itself was man-made and made of fairly ordinary materials (well, at least we know about NiTi and one powerful new iron-based SMA in recent times).

Or, it is possible that Dr. Moore could be hiding the use of a titanium-based SMA from the public and the USAF by claiming the material was aluminium, because the latter was part of the composition of the foil used in the Mogul balloons, but he could not (or would not) give the exact composition.

If so, the only known SMAs containing aluminium (Al) are:

Cu-Al (copper-aluminium)
Cu-Al-Ni (copper-aluminium-nickel)
Cu-Al-Be (copper-aluminium-beryllium)
Cu-Zn-Al (copper-zinc-aluminium)
Ni-Al (nickel-aluminium)
Ni-Al-Re (nickel-aluminium-rhenium)
Ni-Co-Al (nickel-cobalt-aluminium)
Ti-Nb-Al (titanium-niobium-aluminium)
Ti-Ni-Cu-Al-Mn (titanium-nickel-copper-aluminium-manganese)

We can discount Cu-Al, Cu-Al-Ni, Cu-Al-Be, Cu-Zn-Al, Ni-Al, Ni-Al-Re, and Ni-Co-Al, as they cannot resist the high temperatures of a blowtorch; you need significant amounts of titanium to be present to provide a high level of heat resistance. This, then, leaves the titanium-based alloys Ti-Nb-Al and Ti-Ni-Cu-Al-Mn. The alloy Ti-Ni-Cu-Al-Mn is the only one to look vaguely dark-grey, but when it comes to the extreme lightweight nature of the Roswell foil, only NiTi is lighter in weight than Ti-Ni-Cu-Al-Mn. If we are to accept the USAF's position that there was a secret military experiment, the enormous effort to make the materials extremely lightweight would have required the use of the lightest titanium-based SMA, which is NiTi. We can safety ignore any iron-based SMA because they do not exist in this list of aluminium-based alloys. The best one to use has to be NiTi. If it wasn't NiTi, it certainly could not have been Ti-Nb-Al or TiZr (one of the alloys looked at by the USAF after 1947). These two alloys are bright-silvery in color. Then again, if Dr. Moore was trying to hide a titanium-based SMA and was confident of the color, the Roswell foil could have been Ti-Nb-Al. It contains the aluminium that Dr. Moore mentioned to investigators as well as titanium. Unfortunately, the color of this alloy would contradict too many other witnesses who claimed the foil was dark-grey in color. Yet, for some reason, Dr. Moore maintains that no dark-grey foil was observed, just aluminium, or a silvery aluminium-like foil. Either the foil contained very little titanium and nickel (or just mostly titanium), or Dr. Moore does not correctly recall (or does not want to reveal) the color. Otherwise, the more likely explanation is that Dr. Moore is correct in seeing a bright-silvery foil in Project Mogul, it is just that the dark-grey foil mentioned by the witnesses must refer to a completely different project. This seems to be borne out by the fact that Dr. Moore does not mention the existence of a shape-memory effect in the foil he observed. No shape-memory effect means Dr. Moore must be talking about a different project not related to the Roswell case.

If not, then perhaps Dr. Moore was not privy to certain secret tests involving the use of a new alloy in a Project Mogul balloon. Yet, incredibly, not one NYU scientist would leak details of an SMA used in Project Mogul. The fact that certain Battelle scientists could not contain their excitement in 1948 after making titanium at an unprecedented level of purity and seeing its properties, it seems highly unlikely that a top secret project on SMAs could be kept secret by the NYU scientists. In fact, Dr. Nielsen had been working at NYU to study NiTi. We all know about this alloy from the scientific literature. Why couldn't NYU scientists get excited by the prospect of discussing Dr. Nielsen's work on NiTi for the USAF investigators or some type of aluminium-based SMA used in Project Mogul?

Given the lack of excitement from the scientists or any mention of a shape-memory effect means it is highly unlikely that Project Mogul is related to the Roswell case.

Leaving aside the composition of the foil for the moment, another discrepancy is the size of the object compared to a typical Project Mogul balloon. We are talking about a large balloon with numerous reflectors dropping a lot of debris over the first wreckage site, and an even larger balloon and many more reflectors to explain the 9-meter-diameter disc later found in the Plains of San Agustin. No single Project Mogul balloon was large enough to account for all of this wreckage. Even if it was, this does not explain the USAF's description of the object as a "disc" in its secret memo held by General Ramey. A torn-up balloon on the ground would not be described as "disc-shaped".

And why use a SMA in a balloon, secret or otherwise? The only good scientific reason for having used a hefty amount of a super-tough, most probably titanium-based, SMA would be to provide significant structural support and a strong outer skin to help withstand heavy impacts as well as friction with the air at high speeds. Since a tough plastic-like material, however, was used for the structural beams, the alloy was likely used principally for the skin. So, why the skin? And why in such large quantities? A balloon does not move fast enough to warrant the use of so much of this expensive, high-temperature-resistant, super-tough alloy, given the level of purity required to display its famous shape-memory response, unless it was in order for the payload to survive a high-speed impact with the ground after the balloon is destroyed. Yet, given the amount of metallic foil found near Roswell, and later at the second crash site, the question arises as to what it was in the payload that needed such protection. With one secret memo suggesting that victims were found, would it seem reasonable to suggest that humans were having, dare we say it, a joy flight across the desert in one of these expensive balloons? Well, the experiment could not have been that important since the USAF is not able to tell us who the victims were. In that case, it would make more sense if the container holding the payload had had a fridge and a bar for drinks to account for the metallic foil found on the ground.

Project Mogul or not, it is unlikely that the object was a weather balloon.

Could the object have been a military aircraft?

A faster man-made flying machine that would be more likely to benefit from the use of a tough SMA like NiTi is the military aircraft. Could the object that crashed near Roswell have been a military aircraft, perhaps of a new "disc-shaped" design? It all depends on how much material was used to construct the aircraft. Since we know more than two individuals were on board at the time of the crash, in accordance with the "victims" in Ramey's secret memo, the aircraft must have carried no fewer than two individuals. However, since the witnesses suggest that a number of bodies were found, including more than one that had allegedly fallen out of the stricken object as it traveled through the air just prior to crashing some distance away, let us be conservative and say we have at least four crew members. This should give us a reasonable indication of the likely size of the aircraft.

To determine the feasibility of an aircraft having the required size and the probability it could have been flown by July 1947, shown below is a table of the empty weights (i.e., without passengers, fuel, and other payload) of a number of U.S. military aircraft available in 1947:[4]

Name	Year	Crew	Empty Weight (kg)
Vultee V-72 (A-31/A-35) Vengeance	1940	2	4,672
Brewster A-32/SB2A Buccaneer	1939-43	2	4,501
Douglas A-33	1935-37	2	2,436
Beechcraft UC-43 Traveler	1934	5	1,399
Douglas A-20 Havoc	1944-45	3	7,706
Douglas DC-3/C-47	1935-46	3	8,300
Douglas A-1 Skyraider	1945-47	1	5,429

In the official *Roswell Report*, the USAF conducted a search of air crashes during the period from June 24, 1947 to July 28, 1947 and came up with the following:

Name	Year	Crew	Empty Weight (kg)
Douglas A-26C Invader	1945-47	3	10,385
Martin P-5 IN (closest is a prototype version of P5M Marlin)	1947-62	11	22,900
Fairchild C-82A Packet	1946-59	3	14,739
Lockheed P-80A	1944-52	1	3,592
Culver PQ-14B	1942-50	1	366

Interestingly, the USAF claims that none of these aircraft were near the area where the first debris field was found during the required timeframe of the night of July 2, 1947:

> "USAF records showed that between June 24, 1947, and July 28, 1947, there were five crashes in New Mexico alone, involving A-26C, P-5 IN, C-82A, P-80A and PQ-14B aircraft; however, none of these were on the date(s) in question nor in the area(s) in question."[5]

Does this mean that it was a foreign aircraft? Either that, or the USAF is still hiding the facts of the Roswell case in terms of another military experiment involving a secret aircraft of its own.

If it was a foreign aircraft, it would have to have been the Russians who built it. No other nation was close to realizing the importance of titanium for building an aircraft and developing the technology to mass-produce the metal. But did the Russians have the knowledge and technology to achieve it? And what kind of aircraft are we talking about?

Surely it could not have been one of the Russian fighter jets of the late 1940s. They were too small[6] to fly all the way to the U.S. from Russia (unless the Russians had a secret aircraft carrier sitting right off the coast of California or in the Gulf of New Mexico that was unknown to the Pentagon). It would be more logical to have deployed a large airship with some kind of motorized control and enough food for half a dozen pilots to reach New Mexico. However, like weather balloons, airships are slow-moving objects compared to military jets. So, it isn't clear why the Russians would need so much of an SMA for the airship's outer skin. As for the speed of an airship, the U.S. Army would surely have detected an airship moving over New Mexico and shot it down with remarkable ease. Yet, the U.S. Army and the USAF were not aware of anything crashing near Roswell by early July 1947. The one possibility remaining is that it was a new technologically-advanced Russian airship capable of flying at high enough speed to avoid American detection (or had special stealth capabilities), which would explain the need to incorporate a significant amount of a super-tough, titanium-based SMA in its manufacture.

Even if this was possible, we still have the significant problem of how the Russians could have built a titanium-based SMA in vast quantities. In fact, the Russians were not looking at building vast amounts of any SMA by July 1947. As far as the evidence in the scientific literature shows, the Russians had no official knowledge of titanium-based SMAs. We only start to see a fleeting interest when the Russians decided to officially look at a pseudoelastic alloy known as AuCd in 1950, and NiTi in 1951. Not exactly in the timeframe to account for the Roswell crash.

Even if, somehow, the Russians did have the knowledge and the interest, it is not likely that they would give away a high-tech SMA to the Americans.

Also, there is no evidence to suggest that a commercially sized titanium production plant existed in Russia in 1947. The lack of dark-grey fighter jets emerging from that nation would be a pretty good testament to this.

With the Russians not at the forefront of this kind of advanced titanium or any super tough SMA technology, we are left with a couple of other "man-made" possibilities: an American fighter jet, or an American missile fired over the New Mexico desert. Or did the USAF develop a new type of fast-moving, symmetrically shaped metallic airship of its own with a new type of engine? Whatever the object was, only the Americans had the potential to achieve something like this. But did the USAF actually do it?

Going back to the military aircraft explanation, the first evidence against a possible U.S. military jet being involved in the Roswell case has to be the metals that were used to build it. According to "Evolution of U.S. Military Aircraft Structures Technology" by Paul et al., Volume 39, *Journal of Aircraft*, 2002:

"For 60 years, aluminium alloys have been the primary materials for airframes. No other single material has played as major a role in aircraft production as aluminium: specifically, the 2000 and 7000 series ingot alloys in various heat treatments.

In 1944, Alcoa developed 75S (Al–Zn–Mg–Cu) expressly to meet the aircraft industry's need for higher strength. This alloy first saw service on the B-29 bomber, and according to the 1945 aircraft yearbook, 'Practically all the new war planes were utilizing high strength 75S.'

The principal aluminium alloy since 1945, however, has been Alcoa's 24S (Al–Cu–Mg–Mn) that contained the same alloying elements as the older 17S but in different proportions for greater strength. Produced in 1917, 17S was Alcoa's version of Duralumin (from the French word dur meaning hard) patented by Alfred Wilm in Germany in 1908. Wilm's Dural aluminium alloy was used for the Zeppelin structure in 1911. Alcoa's 17S was used in construction of the U.S. Navy airship Shenandoah, which first flew in 1923. This alloy began the U.S. stressed metal skin semimonocoque structure revolution."[7]

This quote suggests that titanium was not the favored metal for use in a military aircraft in the late 1940s, at least not officially speaking. Unofficially, it is always possible, but how probable? Did the USAF use titanium to secretly build an unusual "disc-shaped" flying object? It all depends on whether the USAF did have the technology to reach the required purity to display the shape-memory effect in a titanium-based alloy like NiTi.

Or maybe the USAF was interested in an iron-based SMA? If so, it would have been a strange decision by the USAF to stop using it. We know the Roswell foil did work as required (i.e., it displayed its shape-memory effect with such an unmistakable form) and was tough enough too. Even if the unthinkable had happened and the object had been hit by lightning, the military aircraft industry would not have stopped using this SMA.

As a case in point, the civilian aircraft industry has continued to grow despite a few planes getting hit by lightning and a few even crashing. But this has not caused civilian aircraft engineers to stop using a metal for the outer skin. Instead, the engineers developed new technology to handle lightning strikes, which is, clearly, a far better solution. Yet, incredibly, we find the USAF did not do this despite allegedly having manufactured a highly expensive, titanium-based (or cheaper iron-based) SMA that did its job very well by mid-1947. In fact, this remarkable alloy was never used again as far as we can tell, assuming this was an ordinary secret military test of a new SMA. Not even the world's first official titanium military jet, known as the X-3 in 1951-52, would get the SMA treatment, and certainly not of the dark-grey variety. Too expensive? Surely not, since the USAF was prepared to pay a prodigious amount of money later to use titanium on the X-3 as Emmanuel G. Mesthene of the RAND Corporation noted in his 1962 document titled *The Titanium Decade*. And even if it wasn't to build a titanium-based SMA, there is no reason to stop manufacturing an iron-based SMA. Clearly the iron-based variety would be cheaper and easier to manufacture. To choose not to use the iron-based SMA, it can only mean one thing: the USAF did not create an iron-based SMA. However, since the USAF did look at titanium-based alloys including NiTi, it must mean the SMA found near Roswell must be a titanium-based alloy.

We should not be surprised by this choice. Titanium has to be the best material yet for the outer skin of a craft for which extreme lightweight characteristics and toughness is required. Therefore, the most logical course of action would be for the military to have developed new technology to handle lightning strikes. That would make absolutely perfect sense. Yet, as history has shown, this is not the case. Why?

Leaving that aside, let us return to the issue of the amount of material used to build the Roswell flying object.

Noting the extremely lightweight nature of the Roswell materials (and size of the bodies involved) and assuming we are dealing with a titanium-based SMA, it is best to choose from the two tables the lowest figure for the empty weight of a craft with a crew of more than four (to account for the number of victims). This would be the Beechcraft

UC-43 Traveler, at an empty weight of 1,399kg, with its ability to carry five passengers (probably the number of bodies found at the second crash site near the Plains of San Agustin). Despite being the lowest figure in terms of weight, this is still a lot of titanium[8]. In 1948, that would be nearly half of the officially-known titanium manufactured that year with a bit left over for testing new titanium alloys until a discovery was made in the SMA front. In 1947, there are no official data on the amount of titanium produced.

Even if we assume that the USAF had unofficially produced this amount of titanium by mid-1947, the manufacture of a titanium-based SMA like NiTi required titanium of exceedingly high purity, which, as far as the available evidence tells us, wasn't produced. Without the crucial technology from Battelle for a vacuum arc furnace (the world's first of its type was installed at NYU, not at any military installation site in the United States or abroad). The only time we hear from the scientific community of sufficiently pure titanium (to the level needed to make a SMA) was after the USAF had requested Battelle to study titanium alloys and investigate methods to improve the purity after May 18, 1948. We must presume this was because none of the scientists working at Wright-Patterson AFB had succeeded in making pure titanium. Or if they did, with the help of Dr. Nielsen at NYU, it wasn't in the quantities required to build a large aircraft carrying at least four crew members. After 1947, titanium of sufficient purity was limited in quantity and restricted to experimental purposes as the *Scientific American* article of April 1949 has indicated. But in 1947, there was no talk by any scientist outside of Wright-Patterson AFB (the ones who never worked for the USAF) of highly pure titanium (reaching the 99.995 percent purity mark), not even for experimental purposes. The required purity was simply not yet achieved, since not one U.S. scientist officially expressed any excitement over titanium in alloy form. Only in 1948 do we see this excitement.

And if the USAF needed outside scientific assistance from Battelle and the equipment at NYU to do its work after July 1947, it is unlikely the USAF could have achieved any titanium-based SMA before July 1947. In fact, we would have to describe this decision as evidence that the USAF could not make a titanium-based SMA. It would have been impossible.

Even if we assume the purity and quantity issue had been solved by the USAF, the military would still make the strange decision of not using the titanium-based SMA to build new fighter jets. At the same time, the USAF had to request outside scientific assistance after 1947 to study titanium alloys and to query methods of improving levels of purity. Extraordinary. Then we have the request from the U.S. DoD from 1950 onwards for U.S. companies to mass produce titanium. Why the request at such a late stage? It makes the possible scenario of the military manufacturing its own pure titanium-based SMA by mid-July 1947 seem ludicrous. If this is untrue, where is the evidence of a large-scale titanium manufacturing and purification plant at Battelle or at some U.S. military installation to build an object the size of the one that crashed near Roswell in mid-1947? And where is the report to show that the USAF had acquired the knowledge to attain the level of purity necessary?

Additionally, if the USAF could have used its own secret titanium manufacturing and purification plant at one of its own military installations to make all the titanium-based alloys it wanted and to experiment with it, why would the military ask Battelle for help, or to use the equipment at NYU to do the testing? Surely, the military knew everything there was to make these alloys with ease and in any quantity it liked for as long as it wanted. It makes no sense to ask anyone else for help. Unless, of course, the military had no titanium manufacturing and purification plant of its own.

Even if this is untrue, there is also another argument against the idea of the USAF unofficially having made its own titanium-based SMA: the number of people involved in the mining, extraction and manufacture of titanium and whether such a large number of people could all keep a secret on this work.

According to official data, the production of titanium in 1948 reached three tons[9]. In later years, the production did increase dramatically. But in 1947, the production of titanium was non-existent. Actually, for some reason, no one bothered to specify a figure for that year. Are we to assume the amount of titanium produced in 1947 was really that abysmal? Or is the amount produced being kept a secret to this day just to keep everyone else off the track of the USAF having conducted an unofficial experiment in the New Mexico desert? But if that were the case, why would there be so much secrecy over titanium in 1947 and not in later years? Because it is obvious that the first data on titanium quantities (i.e., three tons) came out in 1948. Soon after that occurred, another information leak would make the rounds, this time in relation to the discovery of more refined samples of titanium and the production of selected alloys for experimental purposes, and immediately U.S. scientists expressed amazing enthusiasm over the metal. So, if the amount of titanium produced in 1947 and its purity were meant to be a secret, some scientists at Battelle were not doing a good job of keeping quiet after 1947.

Furthermore, this secrecy would have had to have extended to those people mining the ore, extracting the metal out of the ore, and refining it to the minimum 99.995 percent purity. We could be talking about hundreds, if not thousands, of people. Yet, remarkably, not one detail leaked about the extra titanium being sent to the USAF or Battelle for a secret military experiment conducted by early July 1947. These ordinary miners, truck drivers, executives of the mining companies, and others who may have been involved in refining the titanium must have done a better job of keeping quiet than the Battelle scientists, or even the military personnel who saw the original Roswell foil and mentioned its shape-memory effect.

Does the USAF still want to maintain an unofficial stance on the production of titanium in 1947? Then it will need to provide a report to explain the knowledge acquired and the technology available by mid-1947 to build a relatively large flying object made of a highly secret, and highly pure titanium-based SMA. And even if we are dealing with an iron-based SMA, we still need the report in order to see how the USAF knew about SMAs and what made it realize something important would exist in the titanium-based (or iron-based) alloys. This is the absolute minimum. But even with this report, another report needs to be released to explain why the USAF pursued titanium-based alloys when, in fact, the Roswell foil (if made of iron, but it would have to be one with a unique composition that even scientists today do not know about) is clearly the most powerful SMA and a particularly tough one for that matter. When combined with its extremely lightweight nature, it must surely be the best SMA for aeronautical and aerospace applications. So why stop making this iron-based SMA in favor of a much more expensive titanium-based SMA when the perfect alloy had already been made?

Leaving the purity issues of the titanium-based alloys aside, as well as whether the Roswell foil contained principally iron in its composition, there is yet another problem to contend with regarding this military aircraft explanation for the Roswell case. In particular, there is the issue of the shape of the object. Two secret official memos and an official news release unequivocally mentioned the shape of the flying object as a disc. Was it a new military jet without wings? Well, this isn't a good shape for a craft moving at high speeds through the air. It would face considerable instability problems during the flight, unless the USAF had developed another radical new technology (following the radical new technology of a titanium-based SMA) to reduce air pressure near the surface of the disc-shaped object. Now that would be a truly useful piece of technology. Yet we find no evidence of this technology being used in other aircraft (just like the alloy), and certainly not by the late 1940s. And why was the shape not retained if it was a successful test flight? We know this object must have lifted off the ground and flown a considerable distance over the New Mexico desert before it was, presumably, struck by lightning. Creating a flying disc should not have posed any sort of problem for the USAF at the time, if the technology existed. Yet we never see this disc-shaped object manufactured ever again.

In the absence of yet another USAF report, this time explaining the shape of the secret military aircraft design, it is unlikely that the Roswell object was a military aircraft.

Could the object have been a missile?

A missile is another object designed to fly at high speeds. The use of extremely lightweight, yet tough, titanium-based alloys in the missile's construction would have been a consideration for the USAF when investigating the Roswell case.

So what has the USAF discovered?

Interestingly, the USAF considered this possibility and stated in the *Roswell Report*:

"[USAF] researchers did not...go to the U.S. Army to review historical records in areas such as missile launches from White Sands, or to the Department of Energy to determine if its forerunner, the Atomic Energy Commission, had any records of nuclear-related incidents that might have occurred at or near Roswell in 1947. To do so would have encroached on GAO's charter in this matter. What Air Force researchers did do, however, was to search for records still under Air Force control pertaining to these subject areas."[10]

It would appear that the USAF not only ruled out the likelihood of a missile test of its own, but also had not considered a titanium-based alloy as a potentially worthy material for the construction of a missile. The USAF believes that if it was a missile, someone else must have done the work on titanium-based alloys and the missile test. Even if such a mysterious entity responsible for this missile test could be tracked down and identified, the USAF is not mentioning the possibility of having discovered an SMA, even an iron-based alloy, during the recovery operation.

So, who could have conducted a missile test if it wasn't the USAF? Knowing the level of sophistication and advanced technology needed to manufacture the SMA used in the Roswell object, it would have to be either the U.S. Army or the Atomic Energy Commission. Oddly enough, the USAF did not check with the U.S. Army or the Department of Energy. And without a word from either establishment about such a test, it seems like these entities are doing a fine job of keeping secret the missile test, or they simply did not carry out the testing.

Well, how likely is it that the U.S. Army or the Department of Energy could have achieved this kind of work?

The answer is, *very unlikely!*

For a start, the Department of Energy did not request Battelle to study titanium-based alloys. Nor did it have a titanium manufacturing and purification plant of its own in 1947, or at any time after 1947, to develop its own SMA. If that is not enough, the Department of Energy has not come forth to claim responsibility for the Roswell object. In conclusion: it is unlikely this department could have manufactured it.

As for the U.S. Army, we discussed in a previous chapter how this military establishment became interested in titanium in 1948—after the USAF had approached Battelle for the second time in its history and expressed a similar interest in the metal and some selected alloys. We are not even talking about an iron-based SMA, if it was made by July 1947. Any such alloy would clearly not satisfy the U.S. Army's requirements for developing a tough alloy despite the apparent toughness and extremely lightweight nature of the Roswell foil. The alloy clearly had to be titanium-based for some reason. Unless, of course, the USAF never did manufacture an iron-based SMA (or did not want to tell the U.S. Army about it). But the fact that the USAF also wanted to look at titanium would also support the view that the Roswell foil probably had to be titanium-based. Therefore, it means the USAF would not have known about any iron-based SMA and, by implication, the U.S. Army as well. The Roswell foil, whatever its composition, had to be titanium-based.

At any rate, the U.S. Army had to have alloys made with titanium after 1947. Even if the U.S. Army had been unofficially interested in titanium in 1947, it would have required the USAF to tell the U.S. Army about titanium, since the first evidence of any interest in titanium by any U.S. military establishment showed up in Project Air Force RAND. In that case, the USAF must have been privy to important discussions about building a secret missile containing a significant amount of a titanium-based SMA. As such, the USAF would not only have known about the missile test, it would have given them the right to ask the U.S. Army about such an experiment.

Remarkably, we find that the USAF didn't. Unless the U.S. Army had been (and still is) more secretive about this missile test than the USAF, it seems unlikely the USAF would not have known. But even if the U.S. Army could have conducted the missile test in secret, it still had the problem of manufacturing this titanium-based alloy in 1947. More crucially, the U.S. Army decided to focus on titanium alloys after 1947 no matter what kind of composition was used in the Roswell foil or how tough and lightweight the alloy was at the time. Essentially, any problems in manufacturing this alloy for the U.S. Army would be the same as the USAF, which means it is highly unlikely the U.S. Army could have manufactured this missile using a powerful and tough SMA and tested it.

So, either the U.S. Army never did conduct a missile test, or the USAF and the U.S. Army are just not saying.

What about for other U.S.-based organizations?

Apart from the U.S. Bureau of Mines' interest in titanium, we have seen no evidence of any other U.S.-based organization expressing a deep and unremitting interest in titanium before, during, or immediately after Project Air Force RAND and then following it up with a secret missile test to confirm the usefulness of the new metal for aerospace applications. If any entity outside of the USAF had even vaguely known of the importance of titanium in building a missile, it is doing a far better job at keeping it secret than did any of the scientists at Battelle in 1948 after their embarrassing leak on the importance of highly pure titanium in the development of new hi-tech alloys. And if an entity had known something about titanium and its alloys, it would be because the USAF had told them.

Thus, the USAF may have stated in its Roswell Report that it had no knowledge of any missile tests during the required timeframe claiming "much of this testing done at nearby White Sands was secret at the time"[11]. However, given the fact that the USAF was the only entity that expressed interest in titanium through Project Air Force RAND, it is unlikely the USAF would not have been at least indirectly involved in a secret missile test by suggesting titanium as a suitable metal with which to construct a missile.

Furthermore, there would have had to have been further communication with the USAF about the location of the missile. As history tells us, it was only after a period of several days had passed before the unidentified object was eventually picked up, and only then did the USAF get told about it by a rancher. If the object was a missile, at the very least this tells us the missile had to have been lost during its flight. Now if it had been the U.S. Army that had performed the missile test, the USAF would have been told of the sudden disappearance of the missile and helped with the search from the air. However, we find that the USAF was not asked. As the Roswell Report stated:

"The USAF found no indicators or even hints that a missile was involved in this matter."[12]

It was either that, or both the U.S. Army and USAF kept quiet about a missile test; so secret was this missile test that the USAF and/or U.S. Army had to act totally unaware of the location of the missile after it crashed, only for the USAF to threaten the witnesses to silence once the debris was discovered. It must be a very special type of missile to treat witnesses this way.

Special in what sense? Did the missile contain an atomic bomb and the USAF did not want to get into trouble if the public found out? Only problem is, the USAF and the U.S. Army were not acting as if the missile had an atomic bomb inside. Rather, the military was acting like it was surprised by the discovery itself as if no atomic bomb or other explosive was ever present in the missile.

So why the surprise over a bit of metallic foil and other debris? Assuming the materials were part of a missile, surely someone in the military would have been aware of what had been sent up into the air at the time. Okay, one might forget a weather balloon or two, as there were many of these things released at the time, but a missile? Seriously, there can't be a whole heap of these things flying around with the USAF or Army choosing not to keep track of them or forgetting their whereabouts as they flew through the air. Too difficult to find a missile? The U.S. Army had no trouble tracking and finding the disc in the Plains of San Agustin nearly 200 kilometers away, assuming Barnett was correct in his claims. Plus, if it contained anything that would cause destruction in the case of a crash, this must surely be a good reason to track a missile after launch. However, in this case, on July 2, 1947, it appears that the object was not tracked, or if it was being tracked, it must have gone out of tracking range but, for some reason, finding it was not considered a priority for the U.S. Army (or USAF). Indeed, if it had been the latter case and the U.S. Army was responsible, the Army decided not to inform the USAF of the situation, either.

Are we to infer from this that the missile was not important enough to track? Well, it should have been important. For some reason, the USAF went to a lot of trouble to threaten witnesses to keep quiet, especially the rancher. Clearly something sensitive was found on the rancher's property, or the U.S. Army said something sensitive to the

USAF about the object in the Plains of San Agustin. Furthermore, a memo stating that there were victims is a *serious* issue. A discovery like this would certainly explain the USAF's behavior. But just the potential for a missile to cause injury should have been enough reason for the USAF or Army to pay close and undivided attention to the whereabouts of the missile, from the moment of launch to the moment it hit the ground, wherever it landed.

In fact, so much attention should have been given to the missile by the military that should it have strayed off-course and tracking was lost, the military would be expected to look for it with everything they had. Sure, if it the Army lost a missile and there was a risk to civilians, then bring in the USAF for assistance. It is the most reasonable thing to do considering that the USAF would have the aircraft and manpower to conduct an aerial search to find the missing missile. Yet, somehow, as it appears, the USAF was not asked, or it did not see a need for it and instead chose not to respond, looked elsewhere, expressed surprise when told about the debris, and upon learning about the location of the debris, threatened the witnesses into silence.

This seems to imply the missile was not that special after all. Or was it really more a question of not knowing about a missile because there were no missiles, or any other types of tests, conducted in the required "early July" timeframe?

If it was not so, there must be a reason to keep quiet over a secret missile test in New Mexico to this day. Is this because of the work on SMAs? Unlikely. We know a lot of this shape-memory foil was found lying around in the New Mexico desert, but attempting to keep this work secret to this day is pointless given how much the scientific community and the public now knows about SMAs.

Another possible reason for the continuing secrecy is that the USAF used a lot of this alloy, and it did not want the public to know just how much money had been spent in building what is presumed to be the world's most expensive missile. If we accept this argument, it raises several questions. For example, how did the USAF make so much of this titanium-based SMA by mid-1947? And why a SMA? The only conceivable purpose of a missile is to destroy itself and everything around it, so it kind of defeats the purpose of having a significant amount of SMA for keeping the original shape of the missile intact.

Maybe the SMA was used as part of the nose cone, just as Dr. Buehler had been looking for when working at NOL? Well, if you think this is plausible, think again. The amount of shape-memory foil found strewn over the rancher's property and in the second wreckage site where the "disc" was allegedly found would make this nose cone enormous and potentially make the missile itself look like it was the size of the Washington Monument, or larger. Such a large size would only have increased the likelihood of tracking and determining its location and, given its risk to the public, the motivation to go out there and search for it. The fact that the military did nothing of the sort would suggest that the object was not oversized in any way.

Furthermore, the USAF was happy to spend up big on titanium studies and to encourage businesses to ramp up production of the metal after 1947. So why stop spending money on titanium after conducting the missile test? Was this out of respect for the victims who had died in the crash?

As for the scientific knowledge we have on SMAs, especially NiTi, there is nothing to maintain a secret about. We know enough about these types of alloys to realize that there isn't anything worth keeping a secret. So why keep a secret on the missile experiment if the testing involved new materials?

Seriously, the alloy is not a dangerous explosive device. Nor would the USAF or the U.S. Army have been silly enough to put live explosives inside the missile for an experimental test of a new type of missile and then walk away and forget about it because they could not find it. That would be bizarre at the very least and make the military look like real dummies. Forget the dummies talked about in the USAF Roswell Report. This was supposedly a test of a SMA and, perhaps, a study of the aerodynamics of a new shape via a disc. If the missile somehow contained explosives in order to test something else, then the military would have tracked it and looked for it with all they had if they had lost it. In all probability, the missile could have landed on someone else's property.

And guess what? It did. On a rancher's property as the story goes, which would be more of a good reason to be out there looking for the inane missile; neither the USAF nor the U.S. Army would have walked away from the missile test knowing an unexploded missile could have landed on private property. Or if it did explode, the military would not have pretended it did not know how many civilians had been killed after the missile hit the ground and waited on the public to inform them of the whereabouts of the debris. No military establishment would be dumb enough to walk away from a missile test of this sort while the public was at considerable risk of an exploding flying missile contained inside a titanium-based SMA skin.

Therefore, the only way the USAF and the U.S. Army could have acted the way they did (i.e., not know the whereabouts of the missile and choose to ignore it for more than four days) is if the missile did not contain any explosives and the test was not important enough to search for it. In other words, there was nothing secret involved in the testing. It was just a shell for holding a few instruments and an engine to propel the object through the air.

Yet the secrecy continues.

Okay. So, perhaps the shape of the object might be the big secret. Fine. In that case, the military can just keep quiet about it, say thanks to the rancher for discovering the wreckage, and be grateful it was found.

Mind you, it isn't the first time civilians have seen a disc-shaped flying object from the USAF before. A case in point is the Avro car built in the 1950s. So, that kind of defeats the purpose of hiding the shape of the object from the public. Yet the secrecy continues to this day. And witnesses were threatened with their lives.

Why?

There is only one other explanation: some people were killed, either inside the missile and/or on the ground, by accident, and the military tried to hide it in order to avoid embarrassment. If we can rely on the word "aviators" in the second paragraph of Ramey's secret memo, according to Mr. Rudiak, then it is most likely that this refers to people inside the missile.

Unfortunately, we have yet another conundrum. You see, the identity of American pilots inside the missile would be known to the USAF and eventually to the public. There is no way the military could ignore or pretend not to know the whereabouts of the debris, or the identities of the people involved, because if this was a manned test missile, these would have been very brave military test pilots who gave their lives to serve their country to achieve some unusual outcome.

Perhaps they were trying to prove exactly how dangerous it is to fly a missile? Obviously the test was an absolute success.

Whatever the purpose of such an experiment, the natural reaction would be for military personnel to get off their backsides and intensely search for the missile and the men, even if the missile had flown off course. But as history has it, this is not the case. The USAF looked totally unprepared.

Maybe those who died in the incident were prisoners of war from Japan or Germany, or were highly trained monkeys? Even so, the USAF is not willing to reveal the names of these victims or mention whether there were animals onboard (not even an indication of the species). This is rather unusual, considering that the U.S. military is usually quite happy to mention the names of enemy combatants if it serves a particular purpose.

For example, we know there had been talk by some members of the public after the September 11, 2001 events in New York of strapping Osama Bin Laden onto a missile with a camera attached so the American people could watch him have a direct smack down with the ground, and explode. This is an understandable reaction, given how many people were hurt by the terrible attacks in New York. If, by 1947, the USAF had captured the worst of Adolf Hitler's men who had performed hideous experiments on the Jews in concentration camps, it probably would not have outraged the public in the slightest if the USAF had done something similar in New Mexico. Worse things could easily be committed on prisoners of war. Dying in one of these secret missile tests would, at least, be quick and painless.

Yet, we are not told about this possibility by the USAF.

In that case, these victims must be monkeys or some other animal, except the USAF still remains incredibly tight-lipped over the possible use of animals in a missile test to this day. Surely, it cannot be because of a concern by the USAF that it might create outrage among the animal rights activists. Using animals in secret experiments is not exactly new to anyone. Animals have been used in secret experiments many times in the past for many purposes, much of it probably ending up being horrible to the animals. Furthermore, no one is willing to say for sure whether such secret experiments are not occurring today. If we hear about it, the public will not like it and will demand changes take place; but if it had occurred in the past, say, 60 years ago, there is nothing the public could do or say about it. Maybe a slap across the USAF's wrist, and that's about it. Life would still go on for the military, and the Roswell incident would be forever solved.

There is something about the lingering nature of this secrecy to suggest that on the balance of probabilities, these victims were not animals.

Or are we dealing with American people on the ground who were killed by the missile? Even so, there is still a problem. If the USAF and the U.S. Army initially did not know about the victims, when the military did find them, they continued to keep it secret. This brings us to the same problem mentioned earlier. And for how long can a secret like this be kept? Not very long, since one would think that by now, the families of these unfortunate victims would have come forward to explain that their loved ones were missing. Even if these victims were a group of ordinary campers, the names of these unfortunate civilian victims would have been disclosed by now. They have not. Why? This is strange for a long-standing secret military experiment we still do not know about to this day.

Something does not add up.

All of this suggests that the missile itself, and any victims present at the crash site, had to be of foreign origin. Otherwise, nothing else would make sense. However, as we have seen earlier, the Russians did not have the knowledge to make a titanium-based (or iron-based) SMA for a missile by July 1947. And no one else in the world appears to have had the technology to achieve this, either. So, how could a titanium-based SMA missile from another country have been built at the time? And how could another country keep the names of the victims a secret to this day?

What is more unusual is how a news release and two secret memos described a missile as a disc, which is a rather odd shape for a missile. Again, we must return to the arguments presented in the previous case about aerodynamic stability.

Looking at the available evidence rationally, it is highly unlikely the object was a missile.

Could the object have been a high-speed airship?

In this final "man-made" explanation, there is a possibility that we could be dealing with a new type of airship. It would have had to have been symmetrical in shape to explain the disc in Ramey's secret memo, and one without wings; and it had to be fast-moving to explain the use of a significant amount of a super-tough, lightweight, high-temperature resistant, titanium-based SMA covering the entire skin section.

There are two issues we need to discuss here. Firstly, and logically, a fast-moving airship would require an engine. Secondly, by the process of elimination, the engine must have been located at the second wreckage site, since no complete machine was visible at the first crash site, however, the USAF claims a "disc" was found. With no indications how the shape could be determined from the first crash site, it seems reasonable to suggest that the engine and rest of the body must have been located at the second crash site.

Already we have two problems:
1. The USAF has not mentioned the location of the second crash site in its official reports on the incident in 1994 and 1997.
2. The USAF mentioned nothing about an engine, other than the object was a "disc".

It is looking like either this object had no engine, or it was not one that the USAF was familiar with (perhaps it was incorporated into the entire body of the object in a manner that made it difficult to determine how it moved). No engine? Then it cannot be a fast-moving airship. End of story. But if it did have an engine and the USAF could not mention it because it was hidden or designed to merge with the body of the object and worked on some unknown propulsion principle, then the object has to be foreign-made. Except no one has claimed responsibility for the object, and that would not look good for the USAF if it wanted to avoid anything embarrassing or highly sensitive. In that case, let us try to stick to a man-made explanation as best we can.

Well, for a start, the USAF could not even confirm whether this was a new type of fast-moving airship it had been testing. If it is man-made, the USAF is not willing to tell the public what type of object it was testing.

Again, not a good sign for the USAF that it hadn't found anything unusual or important. Perhaps we have to assume the object contained something secret?

Is this because SMAs remain a matter of utmost secrecy to the military? Unlikely. As mentioned previously, we have had unfettered access to information about this class of alloys in the scientific literature since the late 1950s.

Is the secrecy more to do with the shape of the object? No, because people have already seen a symmetrical airship in the past. For example, we know the USAF had built a symmetrical aircraft known as the Avro car in the 1950s and showed it to the public during a brief operational run. There is nothing secret about this shape for a flying vehicle that the public has not seen.

Therefore, the secret must be in either the engine or the victims.

Assuming that the secrecy relates to the engine, why the continuing secrecy over the technology? If the engine had worked, we would have known about it through its eventual application to other military aircraft. If not, what's the harm in talking about it? We are of the view that the engine did work. As we can gather, the object flew off the ground and traveled at high speeds over a good distance. The engines must have worked extremely well, no matter how unusual the design. A little tweaking here and there to prevent a lightning strike, and we should have had this technology incorporated in military jets as early as the late 1940s, but we haven't. And, if the engine was a failure, why has it been kept a secret to this day? The USAF could have reported that the engine idea was not feasible if they had wanted to, and so end the Roswell controversy. Apparently not.

Apart from these concerns, there remains the issue of how a disc-shaped airship can fly through the air at high speeds without losing aerodynamic stability. Was the air ionised and pushed away by strong electromagnetic fields? The fact that the airship flew with some reasonable stability over a fair distance would suggest that the USAF had already solved this technical challenge. Therefore, the technology should be available in the latest military jets, and we should be seeing man-made symmetrical flying airships roaming the skies by now; but we don't.

Leaving aside a discussion of how a disc-shaped airship could fly with reasonable stability through the air at high speeds, the amount of titanium needed, and the considerable purity required to build an SMA, would again bring us back to the arguments as outlined in the military jet scenario for the 1947 period. Forget an iron-based SMA; we have already eliminated that possibility once the USAF decided to look at titanium-based alloys after 1947 despite how great the Roswell foil was and its expected lower costs. So, it must be a titanium-based SMA. However, any kind of significant titanium in the alloy's composition to reveal a shape-memory effect would require phenomenal purity. It means that the USAF could not have made such an alloy (but, of course, the public would be extremely happy to await for the official report dated prior to July 1947 to be released to help the USAF explain that one to us).

What strikes one as more strange is the fact that all the work that went into this secret, high-speed military airship has ground to a halt and has never since seen the light of day simply because of a lightning strike. Why? Surely the world would be blessed with seeing a new range of wingless supersonic airships if this type of flying object was superior to any conventional winged aircraft. On the other hand, if the USAF is to claim that the airship design is flawed and it is not a viable replacement of winged aircraft, then why stop using the SMA itself? Clearly the alloy worked. And we know that it is not a question of cost (either the alloy was cheap and easy to make, or the USAF had no trouble raising the funds needed to build any kind of titanium-based alloy it wanted after seeing how much money was spent on titanium after 1947). Remember, the U.S. DoD was prepared to spend hundreds of millions of dollars to get titanium into the state in which they could use it to make new types of high-tech alloys. When it did, the latest American military fighter jets did not get the special dark-grey SMA treatment for the skin section. Given how expensive the 1947 experiment was and the technical knowhow allegedly available to build a SMA, something must have come out of all this work. Surely it must have been applied to some future military aeronautical and aerospace projects, and put into long-term practical use, and in reasonable quantities, after getting the titanium industry started for the military after 1950. And once we all learned about SMAs, there would be no need for secrecy in the material. Unfortunately, for some reason, it did not happen. Unless those secret U.S. spy planes with a dark-grey skin are examples of using a titanium-based SMA. But why the continuing secrecy if we already know about the alloy?

This leaves us with the only remaining reason considered vaguely rational for maintaining secrecy: the victims. But, once more, we have to question why the USAF is remaining quiet over what we presume were accidental deaths of a few military pilots or civilians on the ground? Surely, the names of these victims could not have remained a total secret to this day. Even more remarkable is the fact that the "victims" appear to have no family members coming forth to grieve the loss of their loved ones. Of course, we assume these were humans to maintain such great secrecy and not highly-trained monkeys or some other animal. Otherwise, we must consider the USAF to be extraordinarily concerned about not raising the ire of animal rights activists.

Or, could the victims and the airship have been foreign? But again, as discussed before, there is absolutely no evidence in the scientific literature to support another nation having the knowledge and technology to build a large quantity of highly pure titanium-based SMAs in order to construct a new type of fast-moving airship. And if anyone else had discovered a super tough iron-based SMA of dark-grey appearance, it would have been used in foreign aircraft. Again, we don't see this.

It looks like not even a new type of man-made, symmetrical, high-speed airship could have been responsible for the Roswell crash.

Where does science stand in relation to the Roswell case?

With all the likely explanations for a man-made flying object exhausted, and no other known super-tough dark-grey SMA in existence by July 1947 because of the subsequent decision by the USAF to study titanium-based alloys after 1947, where does this leave us in relation to the Roswell case? Indeed, what does the evidence in this book actually reveal about the object that crashed in the New Mexico desert in early July 1947?

What we do know from this research is that what crashed was definitely a foreign object, traveling at high speeds, that was constructed of extremely lightweight, but very tough, materials; we also know it was carrying more than one individual inside. It had to be foreign in nature because the USAF did not have the equipment to produce small samples of highly pure titanium-based alloys by early July 1947, and the military also required external help from the Battelle Memorial Institute to at first create the vacuum furnace needed to produce sufficiently pure titanium-based alloys and to later provide scientific help in analyzing the properties of these alloys. Even if this is untrue, and what was found was an iron-based SMA, the decision not to use it (and instead choose to look at titanium-based alloys) after July 1947, and to nor mention the names of the "victims", tell us that the object (and the "victims") have to be foreign. The victims could not be dummies because of the terminology used by the USAF in its secret memo: the term "victims" represents some type of biological entities who were once alive. Just to add to the foreign nature of this object, the USAF acted like they had no idea what had crashed. Not even the U.S. Army or U.S. Navy would claim responsibility for the object. Well, who else could it be? If this object and the people associated with it are not American-made, then who is responsible?

The scientific literature does not reveal the Russians as having the knowledge to construct this object with its extensive use of a dark-grey SMA. There is no evidence, anywhere, to suggest the manufacture of this object was accomplished by someone else. This is not surprising, considering that if the alloy was titanium-based, the only entity involved in the construction of the world's first vacuum furnace—the essential technology to allow anyone on Earth to manufacture highly pure titanium-based alloys—was the Battelle Memorial Institute. No one else in the world was even remotely in the running to create this valuable piece of equipment.

Since no one else is willing or able to show responsibility for the construction of this non-American-made object with its large quantities of a dark-grey shape-memory metallic skin, we are not left with very many other options to consider. In fact, there is only one option for explaining its origin: the object came from outside of the Earth. As Sherlock Holmes once said, when we have eliminated all the possible explanations, the impossible must be true. Certainly, the object we are dealing with is not natural. The difficulty in manufacturing the shape-memory metallic foil tells us that the object must be artificially made. This thought naturally creates several pertinent questions deserving of answers, What exactly was this object, and who could possibly have constructed something as sophisticated and expensive as this object? It seems like whoever built this thing had no concerns about money. What was more important to the people who created this object was to build the best by way of the lightest in weight and toughest flying object imaginable. Also, for what purpose was it constructed given the amount of effort and money necessarily spent in doing so? Clearly it must have been designed to travel reasonable distances in the fastest possible speed and the least amount of time. Otherwise, much cheaper materials would have been used. Either we are dealing with aliens from another world who visited the Earth in an object that had been constructed with a highly pure titanium-based SMA for its external skin as the witnesses are claiming, or we have an anonymous organization or government on Earth with highly advanced knowledge of titanium-based SMAs by mid-1947. Which is it?

Whatever the truth, only the USAF can answer this crucial question. The USAF must produce a report dated prior to July 1947 to support some form of secret SMA research it was doing and that was then utilised in a radical "disc-shaped" aircraft test, and show why the military thought such an interesting shape-memory effect would exist in a titanium-based SMA. Or, if it wasn't a titanium-based SMA, what similar kind of dark-grey SMA had existed back then? And while the USAF is at it, why not explain the following questions:

1. How is it possible for this dark-grey SMA to exist in sheet form in quantities sufficient to cover the exterior of a large flying object by mid-1947?

2. What is the precise composition of this SMA?

3. What type of high-speed flying object was it that needed so much of this extremely tough and lightweight SMA to cover at least the skin of the object?

4. Who were the pilots who flew the object, and why have their names been withheld for so long?

5. Why did the USAF ask for outside scientific assistance if the alloy had already been manufactured by the USAF or someone else?

6. And why did the USAF stop using the alloy after the crash, and then suddenly go back to analyzing its properties again after 1947 (or look at another similar, but much more expensive, dark-grey alloy through NiTi), but not use it to build military fighter jets (or at least not for a long time)?

However, if no such report exists to answer these questions, we would have no choice but to consider the witnesses' claims as the most probable explanation[13]. And then scientists may finally have something worthy of closer scrutiny in the Roswell case, as it may mean discovering something truly amazing for the scientific community and the rest of humanity.

CHAPTER 11

Conclusion

U NLESS THERE has been an extraordinary level of forgetfulness on the part of the USAF over their secret work on shape-memory alloys (SMAs), it is evident from this research how the secrecy over the Roswell incident endures to the present day. The official Roswell Reports of 1994 and 1997 make no mention of the work into SMAs. Only the First and Second Progress Reports make a point of mentioning no less than two, and probably four, SMAs in the titanium family considered of great interest to the USAF at the time. This is crucial considering it was the witnesses, both military and civilian, who explained the dark-grey Roswell metallic foil as exhibiting a shape-memory response. Actually, it was the military witnesses who made it clear the foil was definitely a metal of some sort. Not a polymer. There is really no way the USAF could not have known about SMAs by July 1947.

If the official history of nitinol (or NiTi) as revealed in this book is true and nitinol is meant to be just another common SMA, then surely the USAF can freely discuss their secret military experiment into SMAs without hesitation.

This does not appear to be the case.

Something is restricting the USAF from discussing its work; and similarly for the Battelle scientist who headed the team for the First and Second Progress Report of 1949.

In 2008, U.S. researcher Mr. Bragalia claimed to have spoken briefly to Dr. Craighead by telephone at his home in Columbus, Ohio, soon after the release of the official USAF Roswell Report in 1997. What he found was that the scientist was not able to freely provide information on the precise content of the Second Progress Report and the nature of his nitinol work, and refused to give comment on the relationship of nitinol with the Roswell materials.

Perhaps there is no relationship between the two? Maybe. Unfortunately, Dr. Craighead seems to have made it worse, assuming the conversation did take place, when Mr. Bragalia stated:

> "I asked if we were on the 'right path' about Roswell and Battelle, and he said 'that is a path you should not even be traveling.' !!! Very brief conversation—not much, but quite interesting..."[1]

How interesting?

Why is Dr. Craighead stopping all interest in this field by saying we should not go there? If there is not meant to be a relationship between Roswell and Battelle through the materials found and, ultimately, between nitinol and the Roswell metallic foil, just say so? Just don't add to the controversy by saying people should *not* go there. Otherwise, people will ask, "Why can't we go there?"

Is this because we cannot handle the truth? Well, let us find out. Or else, just tell us SMAs were used in the Roswell object in early July 1947? How hard could that be?

But the USAF cannot even do this. Why?

Let us put it this way. We are told it is a man-made event, is it not? Well, the USAF believes it is so. And SMAs have been known to other scientists and the public for over 50 years. Therefore, we must assume the secret work done in this field in the late 1940s by the USAF was part of a military experiment. So, what is so special about SMAs and the work carried out by Battelle for the USAF in this specific area that cannot be revealed?

Surely it can't be the flying object itself? What's the big deal about a shape-memory metallic flying object? It sounds like the most reasonable thing to do at the time in order to give the U.S. the technological advantage over other nations. However, this is today. The knowledge of SMAs and the possible use of such a material in a flying object is not exactly new or riveting information, nor will it create a state of hysteria among the public if the truth came out.

Even Dr. Wang and Dr. Buehler made a point of stating how some of their nitinol manufactured by their companies is going into components for making a modern U.S. fighter jet. Not exactly news to cause mass panic among the public, is it?

Yet the USAF is feeling a little restricted in not providing the information to the public.

The limited freedom to discuss the Second Progress Report, and Dr. Craighead's thoughts on what the USAF had been doing, appears to extend to Dr. Wang when interviewed by Mr. Bragalia, and when SUNRISE contacted him by email. But at least we can understand why: he is there to protect his own commercial interests in his nitinol companies. He has patents and he wants to make a profit. Fair enough. However, Dr. Craighead as well?

Dr. Craighead was just a scientist with no vested business interests. There is no good reason to keep quiet his scientific work on nitinol for the USAF after all this time.

We all know about nitinol's interesting property. We know a nitinol-like shape-memory metallic foil was discovered near Roswell. In fact, we can be fairly certain it is NiTi (or some NiTi-based variant), and certainly it must contain titanium given the extraordinary focus by the USAF and later the U.S. Army to look at titanium-based alloys. And all the while, the military continued to ignore what it already had and had presumably manufactured by July 1947. It does not make sense to ignore this foil even if it was composed of other elements. The only way to explain it is to say the Roswell foil was made with a significant amount of titanium in its composition. And if, as the witnesses claimed, the foil was dark-grey, the alloy that naturally comes in this color is NiTi (or NiTi-X, where X is another element added in small quantities). What? The foil was anodised and hence another type of alloy? Surely not, and it would not explain the decision to look at

413

titanium-based alloys. It has to be a titanium-based alloy just as General Exon said in order to explain the high toughness and how the USAF had a specific interest in titanium, and most likely it contained a significant amount of NiTi to explain the color. Scientific literature and the release of the Battelle/USAF reports after 1947 support this. And from history, we see it is the metal of choice for an early study ending in March 1947 according to Air Force Project RAND, and subsequently after 1947. This leaves us with only three main metals of interest in the military and civilian aerospace industry at the time: aluminium, stainless steel and titanium. Stainless steel and aluminium were already being used for aircraft and rocket components. But for some reason the USAF was not happy to continue using any kind of iron-based or aluminium-based alloy, even if it was an SMA. This leaves titanium as the only other metal to gain the interest of the USAF. Therefore we must be dealing with a titanium-based SMA for the Roswell flying object.

Fine.

Just say it was a secret military experiment involving the use of a titanium-based SMA in a top-secret flying object and be done with it. Dr. Craighead doesn't have to give more details. Not even the names of the alleged victims who presumably died in the crash and about which we remain totally ignorant. The public can understand the embarrassment. Just confirm whether we are dealing with a man-made event; and confirm the fact that the work involved SMAs. How hard could that be?

But he can't.

And more strangely, Dr. Craighead says we can't travel along that path. Now really? Why can't we? Because all he is doing is successfully raising the prospect of something more exotic in the nature of the mysterious flying object.

Are we to infer from his terse response that the object found near Roswell is rather *more* secret than we dare imagine because it is exotic? Or is he worried any hint of confirmation in the use of a SMA involving titanium could bring down another kind of secret?

Could this have to do with the manufacturing requirements to build a titanium-based SMA? And if the USAF could not build this alloy, who did?

Maybe we do have something truly scientifically significant after all, and the witnesses were telling the truth.

Not so? Okay, well what's so surreptitious about a military experiment going wrong in the middle of the desert that cannot be mentioned to the public today? Surely it can't be the shape of the flying object. A disc-shaped flying object is not exactly going to shock the public into a state of hysteria, nor will the presence of a significant amount of a titanium-based SMA.

Indeed, should the USAF claim in their next Roswell report that the secrecy really has to do with the SMA, it would be the worse kept secret in living memory considering what we know about SMAs, the work carried out by Battelle on NiTi and other notable SMAs on behalf of the USAF after 1947, and the enthusiasm expressed by U.S. scientists in the same period over titanium and its alloys in highly pure form.

Furthermore, the public is aware of the possibility of a SMA being somehow related in the Roswell case and is starting to question the link between nitinol and Roswell. Take, for instance, this online quote posted on March 2, 2001:

> "Shape-memory alloys - a Roswell consequence? Well sitting in my Manufacturing materials lecture today, our lecturer starts by giving us a brief history into shape-memory alloys.
>
> For those of you who don't know, these are materials that can react to external stimuli, pressure, heat, magnetism etc.
>
> Now apparently these materials were first discovered in 1953 while trying to find a suitable material that could be used in the nose cones of missiles, when the scientist who had just taken the material out of the forge noticed that the material gave off a different ring to the cooled material, showing a change in crystal structure within the material.
>
> During this lecture I couldn't get the idea out of my head that it sounded just like the material they 'gathered' from the supposed crash in roswell, so I did a bit of research and found this site:
> http://www.ufos-aliens.co.uk/morphingmetals.html

Take it all with a pinch of salt but its definitely got me thinking some more about the ramifications of these materials and their supposed origins.

Any thoughts?"[2]

David Rudiak added his thoughts on the issue when he said:

> "Another thing I consider 'miraculous' is that so many people would describe such a property when the existence of the class of 'memory metals' would have been generally unknown. Now we commonly have eyeglass frames made of Nitinol, but that wasn't the case 20 years ago or even 10 years ago. And an extreme memory-metal like Nitinol wasn't around in 1947 (extreme meaning it has a very pronounced shape-memory effect, unlike earlier memory alloys in which the effect was very weak)."[3]

Why is the Roswell metallic foil still held in such high regard and considered a material of great national secrecy? Well, it has to be, since the Roswell Reports of 1994 and 1997 make no mention of nitinol or SMAs in general.

Or is it that the secret isn't on SMAs and NiTi? Could the secret be with something else? If so, what kind of secret are we dealing with here?

The only logical explanation would be to say the USAF had accidentally killed military personnel on the ground, or inside the flying object, at the time of the crash late one night and they wanted to cover it up because they chose to forget the whereabouts of the object and not realize there had been casualties. Strange that they would forget about their own military personnel being lost in the desert. Sounds more like the victims have to be civilian. Perhaps human civilians, possibly midgets if claims of small bodies can be given any credence, were having a camping trip in the middle of the night? Surely the death of experimental monkeys flown up in a secret flying object and crashing won't cut it given the level of secrecy the Roswell case is still showing. It would have to be humans. Either that, or the original witnesses are telling the truth, and rumors of something exotic are definitely on the cards.

Only problem is, the secret memo held by General Ramey suggests that the humans involved were "aviators". Must be something very special about these pilots. The fact that we still do not know the identities of these pilots certainly makes people wonder what exactly is the USAF hiding from the public? And the fact that the USAF was unprepared for this event until it was told about the wreckage tells us the pilots and the object itself must be foreign. In other words, the pilots cannot be Americans.

If we accept even the *slightest* possibility that the firsthand testimony of the original military and civilian witnesses regarding a super-tough shape-memory foil is reliable, and just a fraction of the scientific evidence presented in this book regarding the focus for the military on titanium-based alloys at the time is true, then the type of object that crashed near Roswell cannot be a U.S. military or scientific experiment of any kind, secret or otherwise, from any nation on Earth.

Supporting this claim is how the technology to make pure titanium and its alloys for displaying a possible shape-memory effect and in adequate quantities wasn't available at any time during 1947. The words of Dr. Craighead made it very clear in his own article the true situation for titanium and the technology for making alloys with it in early 1947. This is reinforced by the report's conclusion from Air Force RAND when it stated nothing about the need to manufacture higher purity titanium to reveal exotic properties such as a shape-memory effect. Even if the work had somehow been done in secret because someone had realized the importance of higher purity titanium for alloy purposes, no scientist was showing excitement over titanium in 1947. Not even the USAF was expressing excitement or presenting a report in 1947 claiming the titanium samples were of poor quality and had to be immediately improved. In fact, the USAF did not have the equipment to create their own titanium alloys because of how Dr. Nielsen had to come out of the USAF to work at NYU in the "summer of 1947" to study NiTi and some other titanium-based alloys. Very strange. Yet lo and behold, we find an expression of interest from the USAF to do just that, that is to look at highly pure titanium-based alloys, followed by a request to Battelle to secretly study titanium from May 18, 1948. This was soon followed within the same year by evidence of scientific excitement leaking out over the importance of higher purity titanium for alloy work as the *Scientific American* article on the metal published in April 1949 has revealed. And this resulted in a sudden change in the status of the metal after 1947 like no other, as mentioned in the *Encyclopedia Britannica* of 1969 and several more editions, before it was eventually removed and the metal kept to its uninspiring level of development and discovery we see today.

And among all this activity were two titanium-based SMAs that caught the interest of the USAF in not one progress report, but several reports over a lengthy period of time. At the same time, the need to establish the dual crystalline structures at various temperatures and composition was important. Dual crystalline structures are exactly what you need to find if one wanted to establish any form of shape-memory effect in alloys.

How odd?

Clearly it can't be the scientists, whether secretly or otherwise, who had been responsible for the object, or else the leak would have occurred in 1947 and probably before the Roswell crash had even occurred. Then we would know about SMAs since 1947. This is not the way history tells us. SMAs began after 1958 (more specifically in 1962-63). We can, therefore, cross the scientists off the list. So it must be the U.S. military. Except the only ones who could have done this kind of work must be the USAF. And, unfortunately, we see the USAF had to ask for help from Battelle to study titanium and selected titanium-based alloys, including some notable shape-memory types of interest to the USAF after 1947. This is doubly odd. Because, surely, the USAF would have figured this one out on its own if they had built the flying object in July 1947, especially considering the amount of a titanium-based SMA that was used. So why ask U.S. scientists to help the military after the event?

Furthermore, the object had to be artificial in nature, given the level of technology required to produce a super-tough "nitinol-like" SMA in such high quantities, covering what is believed to be the skin of a 9 meter disc-shaped object. You certainly don't find this kind of material lying around even in trace quantities inside a meteorite. Someone clearly had to have manufactured the object. And, from the way the USAF were acting towards the object when they discovered it, it was an artificial object the USAF wasn't expecting to find—as if the military

418

was definitely not the inventor, nor the creator of this mysterious flying object. Either that, or they were so embarrassed or worried by something they saw that they had to keep it quiet. But no amount of embarrassment can hide the fact that the USAF had to ask for scientific assistance after 1947. Whatever it is the USAF saw, the embarrassment can't be in what they did. It must be in what someone else did when the object was manufactured, flown and eventually crashed in the New Mexico desert. This leaves open the possibility it could have been another country responsible for this object. But, again, we find not one country outside of the U.S. willing to claim responsibility for the object. And not surprising too. There is nothing in scientific literature to suggest any country was studying SMAs in the same way we see the USAF was doing in 1948.

Not even the Russians were involved.

With all the countries off the radar on this kind of work, there aren't too many options left to consider for the origin of this object. Once we cross off all nations, all there is left is outer space.

This brings us back to the exotic explanation as the most plausible scientific explanation for the Roswell case. And this happens to be the consistent explanation many witnesses keep coming back to when explaining what they saw.

Are we really dealing with something exotic?

Otherwise it has to be the responsibility of the USAF to show the public with complete honesty (if they are deliberately hiding something) or openness (if they had forgotten about it) what it had been doing, or had found, in early July 1947.

Seriously, how hard is it to provide the public with a report dated prior to July 1947 explaining what the USAF was doing with SMAs, and what kind of testing they performed? Or, at least, explain who in the world could have made the object? Just pick up the report, give a tick of approval by the head Defense Chiefs for immediate release to the public, and send it through to DTIC who, in turn, can publish it online for all to see. Then all this talk of *aliens crashing to Earth* in 1947 can be given the appropriate ceremonial burial in the back of people's minds.

No?

Well, if the USAF refuses to release the all-important report to quell such concerns among the public, don't be surprised if people continue to think something exotic has crashed all those many years ago. How else are people going to think? By continuing to maintain the USAF façade that they are not involved in any SMA work? Or pretend they don't know anything about it? Such a move would only heightened the extraterrestrial belief among the public for the Roswell case.

And, with a book like this one, would also make their original Roswell Reports look decidedly short of the required scientific standard of quality when any scientist can walk in, look at the Roswell testimony of the key firsthand witnesses, do a check of the scientific abstracts and articles, and then point out the obvious fallacy in the USAF "aluminium foil" argument after extracting the history of nitinol and titanium from historical records and sources.

Indeed, this research has reached a point where it will now be imperative for the USAF to release a report dated prior to the Roswell event to explain how they managed to produce a highly pure titanium-based shape memory alloy. Or else major implications for society are expected to be realized from this research if the USAF does not release the crucial report.

Should the USAF refuse to release the report and instead want to uphold this silly secrecy stance over what we are led to believe is a secret military experiment involving the use of a titanium-based SMA (possibly with human victims involved), then the only way to be sure about this is for another investigation to take place. This time it should be done impartially and with total independence from the USAF, the U.S. Government, and anyone else involved with this case. There is no point in allowing the USAF to conduct another investigation of their own affairs knowing how easily they can botch the whole process by selecting another couple of witnesses of their choice to support their preferred position. It needs to be wide-ranging and comprehensive enough to gather all the evidence and make the proper analysis. It has to be able to gain access to the most secret USAF locations in the country, and to allow people working in these locations to speak freely if there is anything the world needs to know of an extraterrestrial nature.

Nothing should be left unturned in this case.

The whole thing needs to come out once and for all. No pretending like the USAF is trying to do a serious job of investigating its own affairs but really doesn't care about the outcome, or wishes it would go away. Otherwise, the American people will continue to believe something alien has crashed in New Mexico and, all the while, the USAF chooses to throw to the public, at taxpayers expense, biased and expensive reports claiming they are comprehensive and complete.

Far from it.

These reports, at least the ones we have received, are not comprehensive. They are not even balanced scientific reports. These are technical reports to support a preferred position for the USAF. If the USAF wishes to take on this view, it is time they explain the secret military experiment on SMAs and tell us exactly what they were doing. Or else explain the discovery that was made in New Mexico all those many years ago.

Nothing less would do.

Appendices

A   *FBI Teletype, July 8, 1947*
Official teletype summarizing the USAF position to the media regarding the type of object that crashed on a rancher's property near Roswell. Also mentions the Roswell object as a "hexagonal-shaped" weather balloon. Words censored refer to Major Curtan.

B   *Brigadier-General Ramey's Memo, July 8, 1947*
Best available interpretation of the most readable words in Ramey's secret memo.

C   *Memo from General Nathan Twining, July 15, 1947*
Memo written for General Vandenberg under a Presidential Directive discussing the internal structure and unusual fabrication techniques of a disc allegedly recovered near Socorro, New Mexico. Also mentions a possible second disc near Victorio Peak, probably not far from the rancher's property.

D   *Letter from General Nathan Twining, July 17, 1947*
Letter from General Nathan Twining to Mr. Julius Earl Schaefer.

E   *Memo from Research and Development Board, January 28, 1949*
Discusses the importance of U.S. scientific study into super high temperature materials including titanium-based alloys first instigated by the USAF in its contract work with the Battelle Memorial Institute since May 18, 1948.

F   *Battelle/USAF Reports on Titanium Alloys*
First page details of selected and formerly classified Battelle/USAF reports on the study of titanium and its alloys after 1947.

G   *Shape-Memory Alloys (SMAs)*
Table of known one-way and two-way SMAs.

H   *NiTi Crystalline structure*
Austenite and martensite

I   *NiTi Unit Cell Lattice Dimensions*
Austenite and martensite

J   *Physical Properties of Nickel and Titanium*
Melting point, atomic radius etc.

K   *Converting between atomic % and weight %*
With examples

L   *Alloys studied by Prof. John P. Nielsen, 1947-58*
Metallurgy work carried out at NYU by Prof. John P. Nielsen.

APPENDIX A

FBI Teletype
July 8, 1947

TELETYPE

FBI DALLAS   7-8-47

6-17 PM

DIRECTOR AND SAC, CINCINNATI

URGENT

FLYING DISC, INFORMATION CONCERNING. [CENSORED], HEADQUARTERS EIGHTH AIR FORCE, TELEPHONICALLY ADVISED THIS OFFICE THAT AN OBJECT PURPORTING TO BE A FLYINF DISC WAS RE COVERED NEAR ROSWELL, NEW MEXICO, THIS DATE. THE DISC IS HEXAGONAL IN SHAPE AND WAS SUSPENDED FROM A BALLOON BY CABLE, WHICH BALLOON WAS APPROXIMATELY TWENTY FEET IN DIAMETER. [CENSORED] FURTHER ADVISED THAT THE OBJECT FOUND RESEMBLES A HIGH ALTITUDE WEATHER BALLOON WITH A RADAR REFLECTOR, BUT THAT TELEPHONIC CONVERSATION BETWEEN THEIR OFFICE AND WRIGHT FIELD HAD NOT BORNE OUT THIS BELIEF. DISC AND BALLOON BEING TRANSPORTED TO WRIGHT FIELD BY SPECIAL PLANE FOR EXAMINATION. INFORMATION PROVIDED THIS OFFICE BECAUSE OF NATIONAL INTEREST IN CASE AND FACT THAT NATIONAL BROADCASTING COMPANY, ASSOCIATED PRESS AND OTHERS ATTEMPTING TO BREAK STORY OF LOCATION OF DISC TODAY. [CENSORED] ADVISED WOULD REQUEST WRIGHT FIELD TO ADVISE CINCINNATI OFFICE RESULTS OF EXAMINATION. NO FURTHER INVESTIGATION BEING CONDUCTED.

END

CXXX ACK IN ORDER

UA   92 FBI CI MJW

BPI   H8

8-38 PM

6-22 PM OK FBI WASH DC

APPENDIX B

Brigadier-General Ramey's
Memo, July 8, 1947

OPERATION AT THE

AND THE VICTIMS OF THE WRECK YOU FORWARDED TO THE

AT FORT WORTH, TEX.

IN THE "DISC" THEY WILL SHIP FOR A1-8TH ARMY

B29 ST OR C47, WRIGHT AF ASSESS AIRFOIL AT ROSWELL. ASSURE

SAID THIS MISSTATE MEANING OF STORY AND THINK

NEXT SENT OUT DE MORAWIN CREWS.

PR OF WEATHER BALLOONS SENT ON THE

AND LAND D

TEMPLE

APPENDIX C

Memo from
General Nathan Twining
July 15, 1947

HEADQUARTERS ARMY AIR FORCE

Air Accident Report on Flying Disc Aircraft that
Crashed near White Sands Proving Ground, New Mexico
D333.5   ID   15 Jul 47
H1., Air Material Command, Wright Field, Ohio,
15 July 1947

To: Commanding General, Army Air Forces,
        Washington 25, D.C.
        HQ. AIR DEFENSE COMMAND
        ATTN: AC/15-2

        Forwarded for your information

        FOR THE COMMANDING GENERAL

                            N. F. TWINING
                            Lieutenant General,
U.S.A.

                            Commanding

1. As ordered by Presidential Directive, dated 9 July 1947, a preliminary investigation of a recovered "Flying Disc" and remains of a possible second disc, was conducted by the senior staff of this command. The data furnished in this report was provided by the engineer staff personnel of T-2 and Aircraft Laboratory, Engineering Division T-3. Additional data was supplied by the scientific personnel of the Jet Propulsion Laboratory, CIT and the Army Air Forces Scientific Advisory Group, headed by Dr Theodore von Karman. Further analysis was conducted by personnel from Research and Development.

2. It is the collective view of this investigation body, that the aircraft recovered by the Army and Air Force units near Victorio Peak and Socorro, New Mexico, are not of U.S. manufacture for the following reasons:

a. The circular, disc-shaped "planform" design does not resemble any design currently under development by this command nor of any Navy project.

b. The lack of any material propulsion system, power plant, intake, exhaust either for propeller or jet propulsion, warrants this view.

c. The inability of the German scientists from Fort Bliss and White Sands Proving Ground to make a positive identification of a secret German V weapon out of these discs. Though the possibility that the Russian have managed to develop such aircraft, remains. The lack of any markings, ID numbers or instructions in Cyrillic, has placed serious doubt in the minds of many, that the objects recovered are not of Russian manufacture either.

d. Upon examination of the interior of the craft, a compartment exhibiting a possible atomic engine was discovered. At least this is the opinion of Dr Oppenheimer and Dr von Karman. A possibility exists that part of the craft itself comprises the propulsion system, thus allowing the reactor to function as a heat exchanger and permitting the storage of energy into a substance for later use. This may allow the converting of mass into energy, unlike the release of energy of our atomic bombs. The description of the power room is as follows:

(1) A doughnut shaped tube approximately thirty-five feet in diameter, made of what appears to be a plastic material, surrounding a central core (see sketch in TAB1). This tube was translucent, approximately _____ inch thick. This tube appeared to be filled with a clear substance, possibly a heavy water. A large rod centered inside the tube, was wrapped in a coil of what appears to be of copper material, run through the circumference of the tube. This may be the reactor control mechanism or a storage battery. There were no moving parts discernible within the power room nor in _____.

(2) This motivation of a electrical potential is believed to be the primary power to the reactor, though it is only a theory at present. Just how a heavy water reactor functions in this environment is unknown.

(3) Underneath the power plant, was discovered a ball-turret, approximately ten feet in diameter. This turret was encompassed

by a series of gears that has a unusual ratio not known by any of our engineers. On the underside of the turret were four circular cavities, coated with some smooth material not identified. These cavities are symmetrical but seem to be movable. Just how is not known. The movement of the turret coincides with the dome-shaped copula compartment above the power room. It is believed that the main propulsion system is a bladeless turbine, similar to current development now underway at AMC and the Mogul Project. A possible theory was devised by Dr August Steinhoff (a Paperclip scientist), Dr Wertner von Braun and Dr Theodore van Karman: as the craft moves through the air, it somehow draws the oxygen from the atmosphere and by a induction process, generates a atomic fusion reaction (see TAB 2). The air outside the craft would be ionized, thus propelling the craft forward. Coupled with the circular air foil for lift, the craft would presumably have an unlimited range and air speed. This may account for the reported absence of any noise and the apparent blue flame often associated with rapid acceleration.

(4) On the deck of the power room there are what resembles typewriter keys, possibly reactor/powerplant controls. There were no conventional electronics nor wiring to be seen connecting these controls to the propulsion turret.

e. There is a flight deck located inside the copula section. It is round and domed at the top. The absence of canopy, observation windows/blisters, or any optical projection, lends support to the opinion that this craft is either guided by remote viewing or is remotely controlled.

(1) A semi-circular photo-tube array (possibly television).

(2) Crew compartments were hermetically sealed via a solidification process.

(3) No weld marks, rivets or soldered joints.

(4) Craft components appear to be molded and pressed into a perfect fit.

APPENDIX D

*Letter from
General Nathan Twining
July 17, 1947*

BOEING AIRPLANE COMPANY

WICHITA DIVISION
WICHITA, KANSAS

JULY 17, 1947

Dear Earl:

    I have received your letter in which you asked us to drop by at Wichita for a brief visit. With deepest regrets we had to cancel our trip to the Boeing factory due to a very important and sudden matter that developed here. All of us were considerably disappointed as Mr Allen had planned a very fine trip for us; however, we hope to go out at a later time. Will remember your invitation and get out to see you just as soon as we can, as I am very anxious to see the XL-15.

    I have been busy quite a bit the last couple of weeks so have not had a chance to submit any information to you that you asked for in your round robin letter. I will get on this very shortly.

Best regards,

N. F. TWINING
Lieutenant General,
U.S.A.

P.S. Unification looks like a sure thing now.

Mr J. E. Schaefer
Boeing Airplane Col,
Michita, Kansas

436

APPENDIX E

Memo from Research and Development Board

Research and Development Board
Committee on Basic Physical Sciences

MEMORANDUM   28 January 1949

To: The Chairman, Committee on Basic Physical
             Sciences
Subject: Statement on Program Planning, Panel on
             Metallurgy

1.   The organization of sub-panels supporting
this Panel was begun in November. Of the groups
now organized, none has not more than two times,
and six of the groups projected have not yet had
their first meetings. Accordingly, this statement
cannot undertake to deal with two metallurgical
programs in detail. The Panel has considered only
the most urgent problems and the programs directed
to their solution.

2.   The following subpanels have submitted
reports of initial surveys of work underway in
their respective fields:

Super High Temperature Materials
Ferrous-Base High Temperature Materials
Ceramic Coatings for Metals
Ceramics
Aluminium
Magnesium
Titanium and Zirconium
Carbon and Alloy Steel
Welding and Joining
Corrosion and Surface Treatment
Magnetic Materials

Those reports, some of which are quite detailed,
are being circulated among all interested
committees of the Board, as well as among the
Technical Services of the Departments.

3.   The performance of weapons and carriers under development depends largely on the materials used in their construction, and better materials are essential to meet the new requirements. The properties of the materials are usually the limiting factors. Besides, most of the unique designs are based on the availability of materials with new and improved characteristics. For example, improvements must be sought in the following categories: (a) high temperature materials of all kinds and the techniques of welding and joining them, (b) metals and alloys of increased strength-weight ratios and toughness and (c) magnetic materials.

4.   Accordingly, the Panel recommends that emphasis be so placed as to bring to completion, within a few years at most, all projects relating to the development of the following:

a. Super high temperature materials for use in the temperature range, 1800-3000°F, including methods of production and development of molybdenum, chromium, tungsten and other unusual metals and their alloys in forms suitable for use.
b. Nickel-base and cobalt-base alloys for use at temperatures up to 1800°F.
c. Ceramics and ceramic coatings for metals for unusually high temperatures up to 5000°F.
d. Iron, steel, aluminium and magnesium alloys for higher strength and toughness at both high and low temperatures.
e. Methods of producing titanium on a commercial scale and the development of high strength and good corrosion properties at both room and elevated temperatures.

5.   The Panel also recommends intensification, wherever possible, of long-range research directed toward new and better materials, as well as an extension of knowledge of the properties of known metals and alloys to facilitate better use of available metals in improved designs of military material of all kinds.

6.   In connection with all those researches and developments, grave consideration should be given to the conservation of critical materials by such measure as the substitution of more plentiful materials, the use of smaller amounts of scarce materials and more efficient methods of extraction and recovery.

/s/ Clyde E. Williams

**Chairman, Panel on Metallurgy,
Committee on Basic Physical
Sciences**

CONFIDENTIAL

APPENDIX F

*Battelle/USAF Reports
on Titanium Alloys*

WADC TR-52-334

SUMMARY REPORT

Covering the Period May 19, 1952, to December 7, 1952

on

DEVELOPMENT OF TITANIUM-BASE ALLOYS

Contract No. AF 33(038)-3736

to

WRIGHT-PATTERSON AIR FORCE BASE
OHIO

from

BATTELLE MEMORIAL INSTITUTE

December 31, 1952

INTRODUCTION

This is the Final Report on Contract No. AF 33(038)-3736 covering the work done during the period from May 19, 1952, to December 7, 1952. The termination date of this contract was December 31, 1952.

This alloy-development program was started under Contract No. W 33-038 ac-21229, May 18, 1948. The present contract became effective May 18, 1949. In addition to this report, four previous summary reports on alloy development have been issued under these contracts:

Summary Report-Part III, July 30, 1949
AF Technical Report No. 6218-Part II, June 30, 1950
AF Technical Report No. 6623, June 18, 1951
WADC Technical Report No. 52-249, June 18, 1952

This work is being continued under Contract No. AF 33(616)-384 which started December 5, 1952.

First page of the final Summary Report for the Second-Stage Contract Number AF 33(038)-3736 in the development of titanium-based alloys, December 31, 1952

DEVELOPMENT OF TITANIUM-BASE ALLOYS

Battelle Memorial Institute
Columbus 1, Ohio

June 18, 1952

Contract No. AF 33(038)-3736
RDO No. 615-11

WRIGHT AIR DEVELOPMENT CENTER
AIR RESEARCH AND DEVELOPMENT COMMAND
UNITED STATES AIR FORCE
WRIGHT-PATTERSON AIR FORCE BASE, OHIO

First page of Battelle/USAF Report in the Development of Titanium-based Alloys for Second-Stage Contract Number AF 33(038)-3736, June 18, 1952

Research and Development on Titanium Alloys

BATTELLE MEMORIAL INST COLUMBUS OH

31 OCT 1949

The Second Progress Report in the Development of Titanium-based Alloys for Second-Stage Contract Number AF 33(038)-3736, October 31, 1949

APPENDIX G

Shape-Memory Alloys (SMAs)

Table 1: Nitinol (NiTi)

Alloy	Atomic Weight	M_s (°C)	A_s (°C)	Hysteresis (approx.)
Ti-Ni$_{49\text{-}51\ at.\%}$	106.56	-50	110	20 to 50
Ti$_{50}$-Ni$_{50}$	106.56	-19	-4	40

Table 2: NiTi-X SMAs

Alloy	Atomic Weight	M_s (°C)	A_s (°C)	Hysteresis (approx.) °C
Ti$_{49}$-Ni$_{44}$-Ag$_7$	214.48	63	23	14
Ti$_{49}$-Ni$_{42}$-Ag$_9$	214.48	61	20	14
Ti$_{50}$-Ni$_{48}$-Au$_2$	214.48	75	-	-
Ti$_{50}$-Ni$_{50\text{-}x}$-Cu$_x$ (x=0-25)	170.11			5 to 25
Ti$_{42.5}$-Ni$_{42.5}$-Cu$_{15}$	170.11	-34	-30	10
Ti$_{50}$-Ni$_{30}$-Cu$_{20}$	170.11	58	60	6
Ti$_{50}$-Ni$_{25}$-Cu$_{25}$	170.11	61	62	5
Ti$_{50}$-Ni$_{46}$-Cu$_3$-Mo$_1$	266.05	-30	-	-
Ti$_{49.5}$-Ni$_{39.5}$-Cu$_{10}$-Cr$_1$	222.10	25	30	14
Ti$_{50}$-Ni$_{48.5}$-Co$_{1.5}$	165.49	-238	160	-
Ti$_{50}$-Ni$_{43}$-Co$_7$	165.49	-100	-	-
Ti$_{50}$-Ni$_{48.5}$-Cr$_{1.5}$	158.61	-95	-	-
Ti$_{50}$-Ni$_{48}$-Mn$_2$	161.50	-95	-	-
Ti-Ni-Cr-Al	185.54			
Ti-Ni-Cu-Al-Mn	252.05			
Ti$_{50}$-Ni$_{47}$-Fe$_3$	162.46	-98	-	-
Ti$_{50}$-Ni$_{47}$-Fe$_6$	162.46	-100	-	-
Ti$_{51}$-Ni$_{48.3}$-Mo$_{0.7}$	202.50	-13	20	-
Ti$_{50}$-Ni$_{49}$-Mo$_1$	202.50	10	20	36
Ti$_{50}$-Ni$_{47.5}$-Fe$_2$-Mo$_{0.5}$	258.34	-90	5	52
Ti$_{40}$-Ni$_{50}$-Hf$_{10}$	285.10	120	165	65
Ti$_{50}$-Ni$_{40}$-Hf$_{10}$	285.10	140	185	65
Ti$_{49}$-Ni$_{36}$-Hf$_{15}$	285.10	230	-	
Ti$_{50}$-Ni$_{35}$-Hf$_{15}$	285.10	235	256	55

Ti$_{30}$-Ni$_{50}$-Hf$_{20}$	285.10	290	317	60
Ti$_{49}$-Ni$_{31}$-Hf$_{20}$	285.10	350	-	-
Ti$_{40}$-Ni$_{44}$-Hf$_{10}$-Cu$_6$	199.46	72	90	40
Ti$_{35.5}$-Ni$_{49.5}$-Hf$_{7.5}$-Zr$_{7.5}$	199.46	195	230	42
Ti$_{30.5}$-Ni$_{49.5}$-Hf$_{10}$-Zr$_{10}$	199.46	290	315	50
Ti$_{46.4}$-Ni$_{47.6}$-Nb$_6$	199.46	-50	-29	-
Ti$_{50}$-Ni$_{45}$-Pd$_5$	213.01	20	-	-
Ti$_{51}$-Ni$_{29}$-Pd$_{20}$	213.01	85	-	20
Ti$_{50}$-Ni$_{30}$-Pd$_{20}$	213.01	90		
Ti$_{51}$-Ni$_{26.5}$-Pd$_{22.5}$	213.01	95	-	
Ti$_{52.5}$-Ni$_{22.5}$-Pd$_{25}$	213.01	170	-	15
Ti$_{50}$-Ni$_{15}$-Pd$_{35}$	213.01	225	-	15
Ti$_{50}$-Ni$_{40}$-Pt$_{10}$	301.70	90	100	20
Ti$_{50}$-Ni$_{30}$-Pt$_{20}$	301.70	266	271	31
Ti$_{50}$-Ni$_{25}$-Pt$_{25}$	301.70	430	450	40
Ti$_{50}$-Ni$_{20}$-Pt$_{30}$	301.70	560	594	55
Ti$_{50}$-Ni$_{15}$-Pt$_{35}$	301.70	680	750	-
Ti$_{50}$-Ni$_{10}$-Pt$_{40}$	301.70	810	850	-
Ti$_{50}$-Ni$_5$-Pt$_{45}$	301.70	900	950	-
Ti$_{46.74}$-Ni$_{52}$-Re$_{1.26}$	292.77	37	87	55
Ti$_{45}$-Ni$_{50}$-Ta$_5$	287.51	44	76	53
Ti$_{47}$-Ni$_{50}$-Ta$_3$	287.51	58	90	48
Ti$_{50}$-Ni$_{46}$-V$_4$	157.50	-50	-	-
Ti$_{50}$-Ni$_{45}$-V$_5$	157.50	-80	-	-
Ti$_{49}$-Ni$_{50}$-W$_1$	290.40	-28	-12	36
Ti$_{48}$-Ni$_{50}$-W$_2$	290.40	-37	-26	39
Ti$_{50}$-Ni$_{48}$-W$_2$	290.40	55	70	29
Ti$_{45.5}$-Ni$_{49.5}$-Zr$_5$	197.83	70	110	66
Ti$_{40.5}$-Ni$_{49.5}$-Zr$_{10}$	197.83	90	170	68
Ti$_{35.5}$-Ni$_{49.5}$-Zr$_{15}$	197.83	190	235	41

Table 3: Other SMAs.

Alloy	Atomic Weight	Transformation Temperature		Hysteresis (approx.) °C
		M_s (°C)	A_s (°C)	
Ag-Cd$_{44\text{-}49 \text{ at.\%}}$	220.27	-190	-50	15

Au-Cd$_{46.5-50\,at.\%}$	309.37	30	100	
Cu-Al	90.50			
Cu$_{70.38}$-Al$_{26.56}$-Mn$_{3.06}$	145.44	82.51	115.78	
Cu$_{67.45}$-Al$_{27.96}$-Mn$_{4.60}$	145.44	115.95	136.17	
Cu$_{72.50}$-Al$_{24.70}$-Mn$_{2.81}$	145.44	152.67	165.61	
Cu$_{74.15}$-Al$_{20.28}$-Mn$_{5.57}$	145.44	498.58	511.59	
Cu-Al-Nb	183.41			
Cu$_{67.4}$-Al$_{26.8}$-Nb$_{1.0}$-Mn$_{4.8}$	238.34	-32	68	100
Cu-Al$_{14-14.5\,wt\%}$-Ni$_{3-4.5\,wt\%}$	149.21	-140	100	35
Cu$_{86}$-Al$_4$-Zn$_{10}$	155.88	-100		
Cu$_{60}$-Al$_{10}$-Zn$_{30}$	155.88	300		
Cu-Au-Zn	325.88			
Cu$_{5-35\,at.\%}$-Mn	118.48	-250	180	25
Cu-Sn$_{approx.\,15\,at.\%}$	182.23	-120	30	
Cu-Zn$_{38.5-41.5\,wt\%}$	128.91	-180	-10	10
Cu-Zn-Al$_{a\,few\,wt\%}$	155.87	-180	200	10
Cu-Zn-Ga	198.63			
Cu-Zn-Si$_{a\,few\,wt\%}$	157.00	-180	200	10
Cu-Zn-Sn$_{a\,few\,wt\%}$	247.60	-180	200	10
Fe-Mn$_{32\,wt\%}$-Si$_{6\,wt\%}$	138.87	-200	150	100
Fe-Mn-Si-Cr	190.87			
Fe-Ni	114.56			400
Fe-Ni-Co-Al	200.47			
Fe-Pt$_{approx.\,25\,at.\%}$	250.94	-130		4
In-Cd	227.22			
In-Tl	319.19			
Ni-Al$_{36-38\,at.\%}$	85.67	-180	100	10
Ni-Al-Re	271.87			
Ni-Co-Al	144.61			
Ni-Co-Ga	187.36			
Ni-Fe	114.56			
Ni-Fe-Ga	184.28			
Pt alloys	221.09			
Ti$_{18-23\,at.\%}$-In	162.72	60	100	4
Ti-Mo-Al	170.80			
Ti-Mo-Ga	213.56			
Ti-Nb	140.81			
Ti-Nb-Al	167.77			

Ti-Nb-O	156.81			
Ti-Nb-Pt	335.90			
Ti-Nb-Ta	321.75			
Ti-Nb-Zr	232.03			
$Ti_{50}-Au_{50}$	244.83	597	615	38
$Ti_{50}-Pd_{50}$	154.30	510	520	70
$Ti_{50}-Pt_{50}$	242.95	1070	1040	65
Ti-Zr	139.12			

All composition ratios are in atomic percent (at.%) unless otherwise indicated.

Table 4: Known two-way SMAs.

Alloy	*Atomic Weight*
Cu-Al	90.502
Cu-Al-Ni	149.21
Cu-Al-Zn	155.872
In-Tl	319.19
Ti-Ni	106.61
Ti-Ni-Co	165.49
Ti-Ni-Cu-Al-Mn	252.05

Al: aluminium, Ag: silver, Au: gold, Cd: cadmium, Co: cobalt, Cr: chromium, Cu: copper, Fe: iron, Ga: gallium, Hf: hafnium, In: indium, Mn: manganese, Mo: molybdenum, Ni: nickel, Nb: niobium, O: oxygen, Pd: palladium, Pt: platinum, Re: rhenium, Si: silicon, Sn: tin, Ta: tantalum, Ti: titanium, Tl: thallium, Zn: zinc, Zr: zirconium.

Data obtained from Encyclopedia of Chemical Technology Volume 20, p.728, http://en.wikipedia.org/wiki/Shape_memory_alloy and Google.com results and other sources.

Transformation temperature range and hysteresis temperature data from Hodgson et al. "Shape Memory Alloys" 1990, ASM Handbook, Volume 2, pp.897-902.

APPENDIX H

NiTi Crystalline Structure

AUSTENITE PHASE

Position of nickel (green) and
titanium (grey) atoms

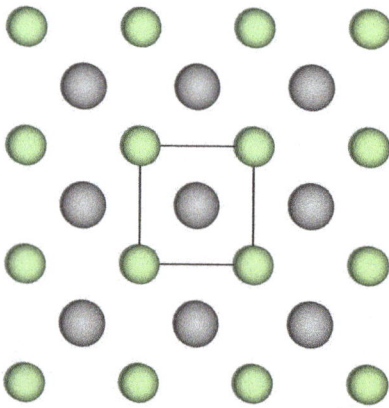

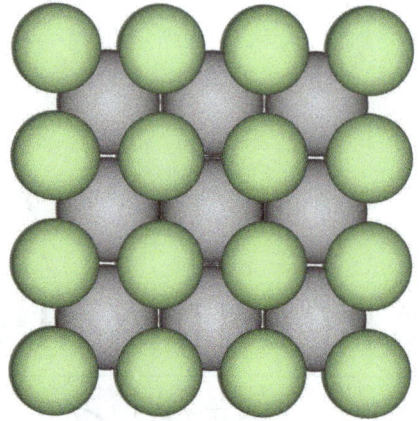

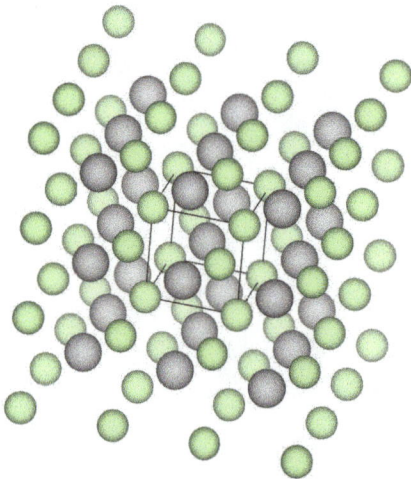

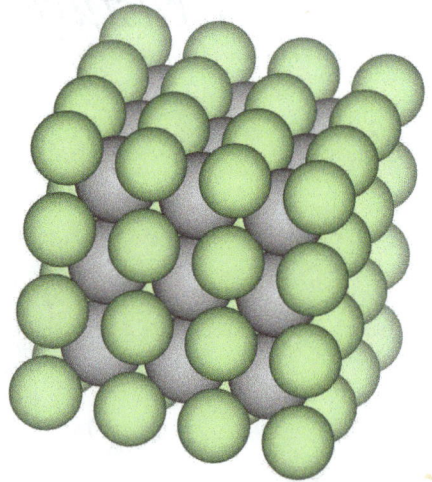

Nickel radii: 125pm
Titanium radii: 147pm

AUSTENITE PHASE

Cubic B2

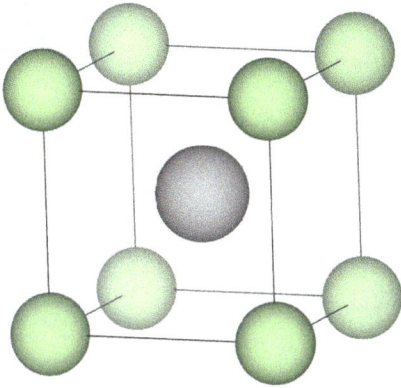

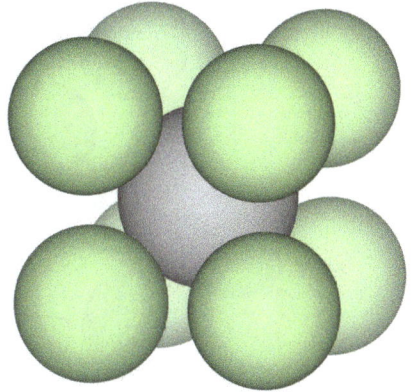

Nickel-based B2 (b.c.c.) structure

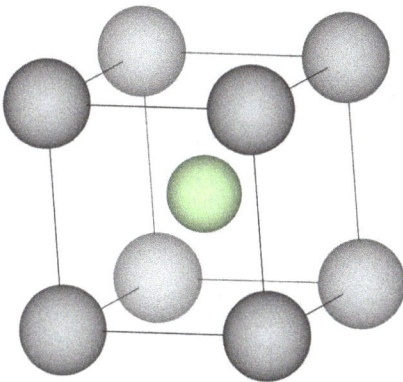

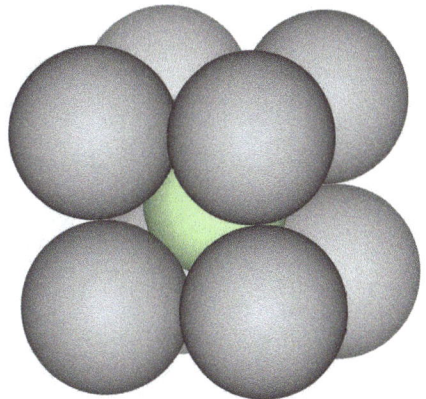

Titanium-based B2 (b.c.c.) structure

b.c.c. : body centred cubic

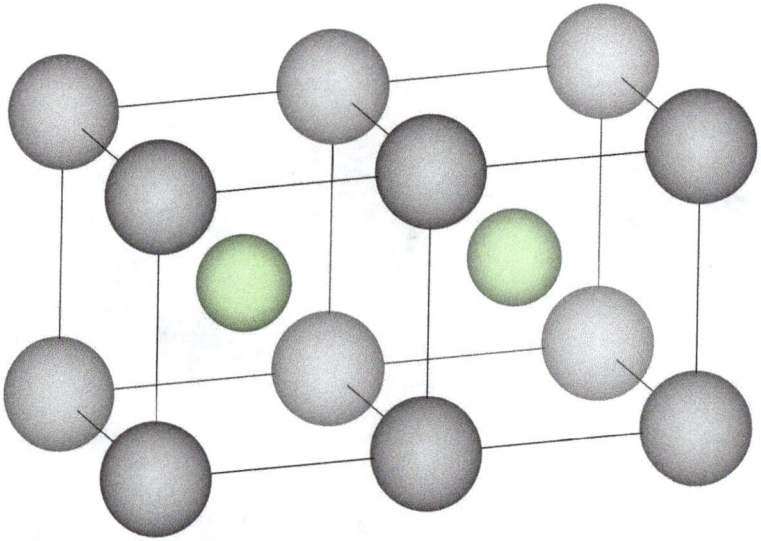

Two unit cells

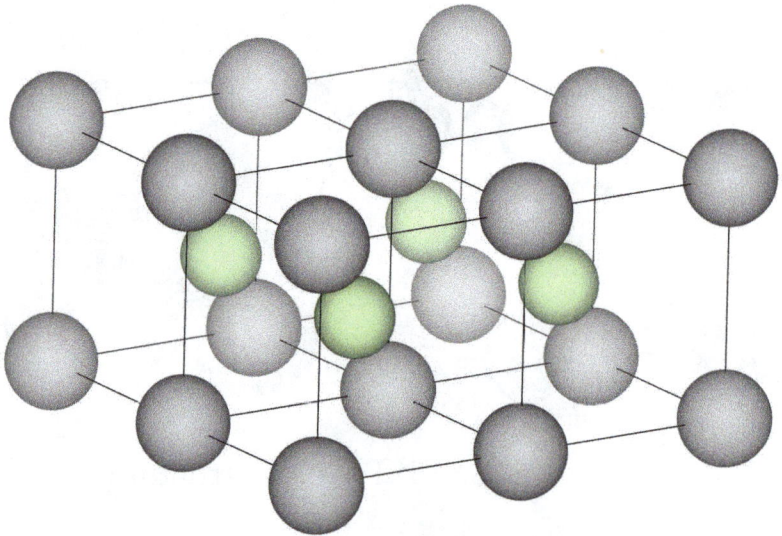

Four unit cells rotated 90°

AUSTENITE PHASE

Cubic B2 (or Tetragonal)

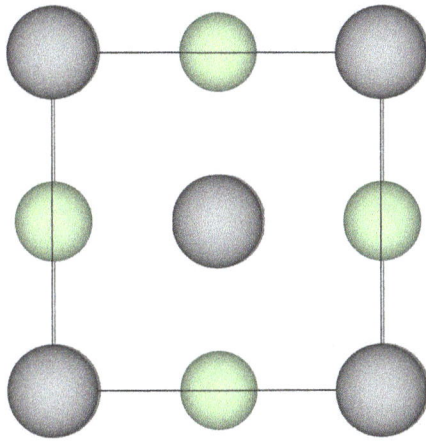

Top view

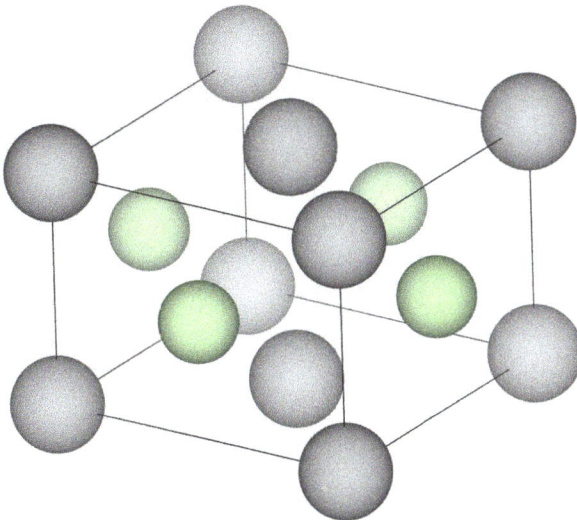

3D view

MARTENSITE PHASE

Orthorhombic B19

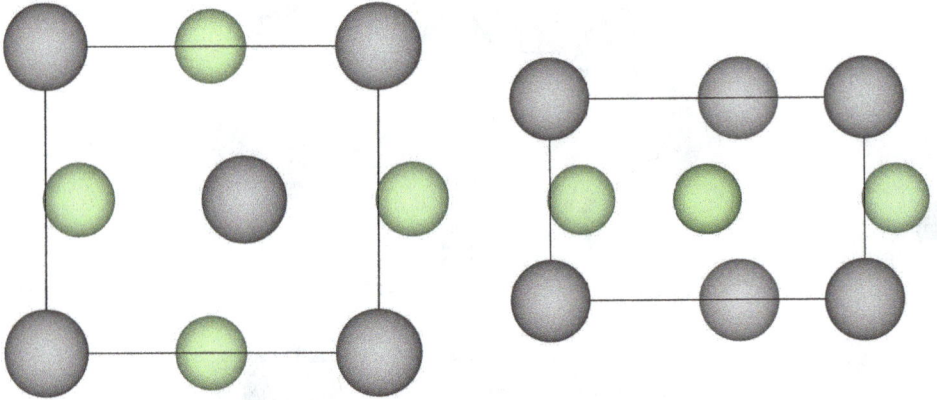

Top view on left, side view on right

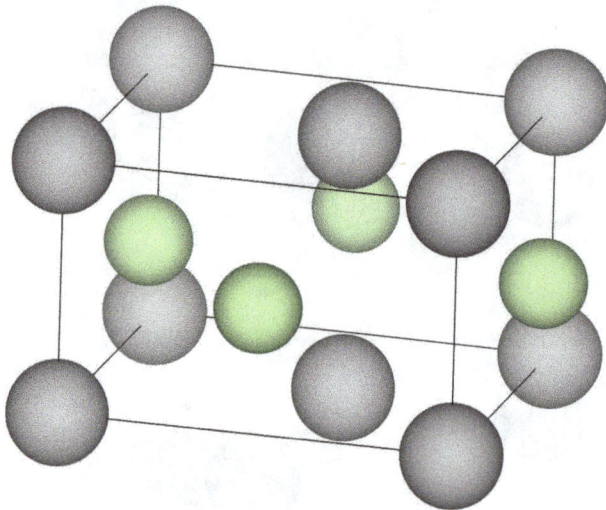

3D view

MARTENSITE PHASE

Monoclinic B19'

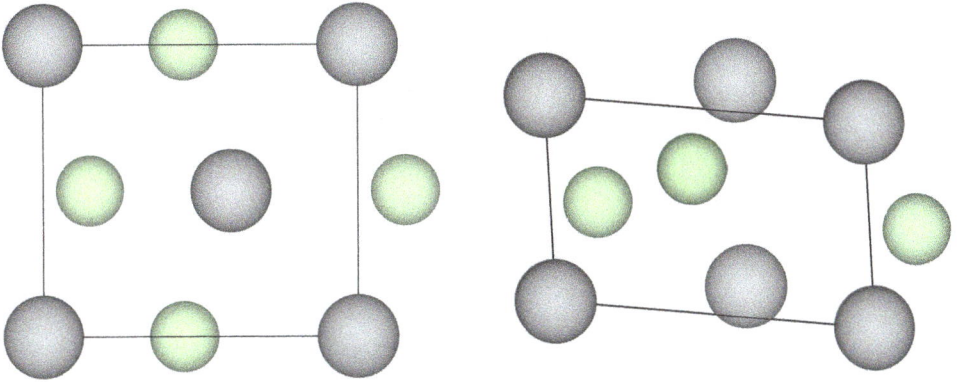

Top view on left, side view on right

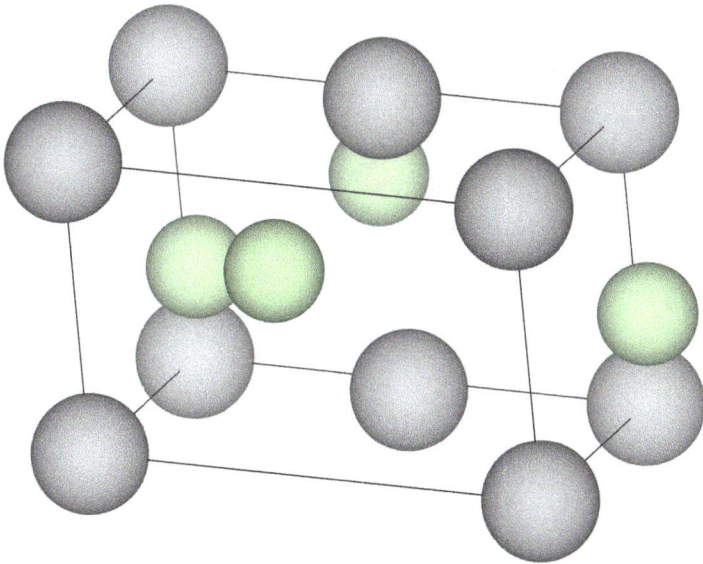

3D view

SIMPLIFIED UNIT CELL

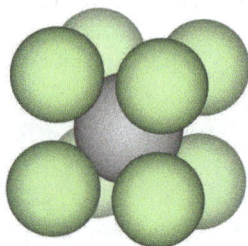

Austenite Phase

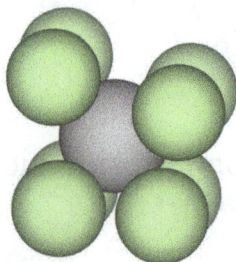

Martensite Phase

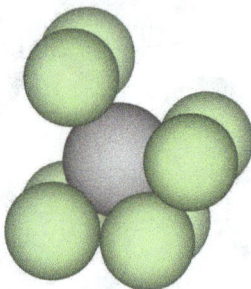

Martensite Phase (Stress-Induced)

These diagrams show four nickel atoms and one titanium atom from inside one of the titanium unit cell "cages" and another four nickel atoms from the next "cage". The two nickel atoms that move forward in the previous pages are facing the top left corner.

Supporting Evidence 1

for Cubic B2 (or Tetragonal) in the

Austenite Phase

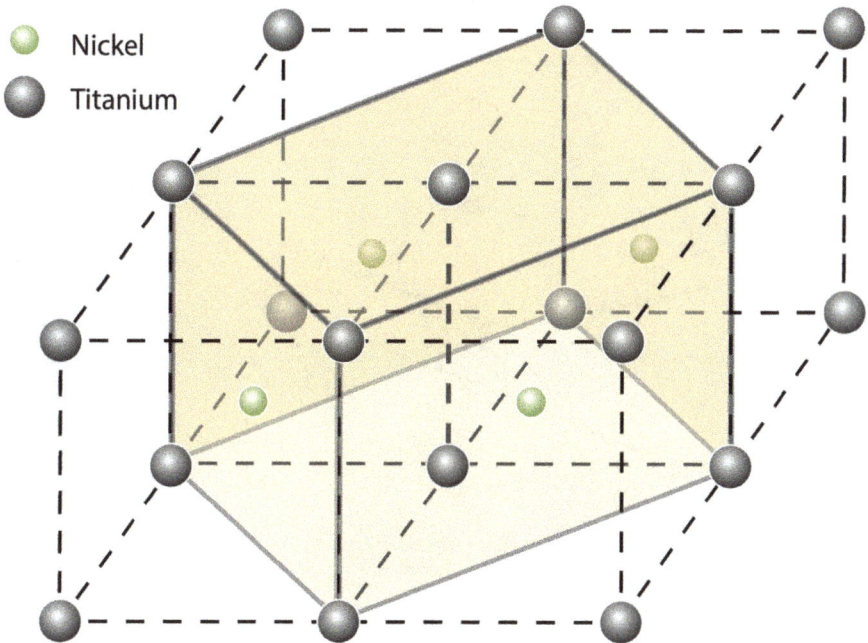

This Austenite unit cell structure supported by Otsuka et al. (1999). This combines the cubic and tetragonal unit cells. The same structure is also used by Xiangyang (2003) in "Crystal Structures and Shape-Memory Behavior in NiTi".

Supporting Evidence 1

for Orthorhombic B19 in the

Martensite Phase

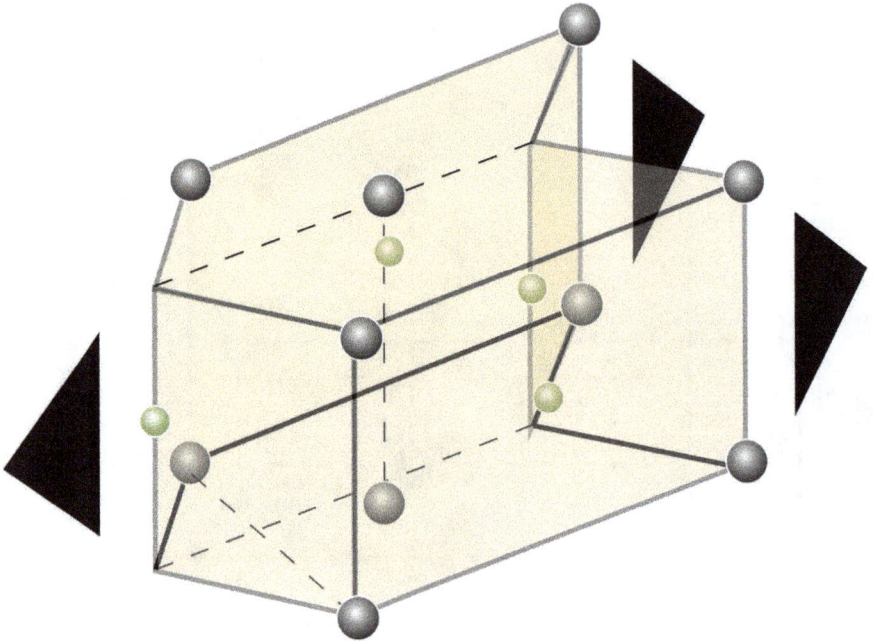

The Martensite orthorhombic B19 unit cell structure supported by Otsuka (1999). According to these researchers, two nickel atoms move forward and eight titanium atoms move backwards (hence the use of the arrows by the researchers).

Supporting Evidence 1

for Monoclinic B19' in the

Martensite Phase

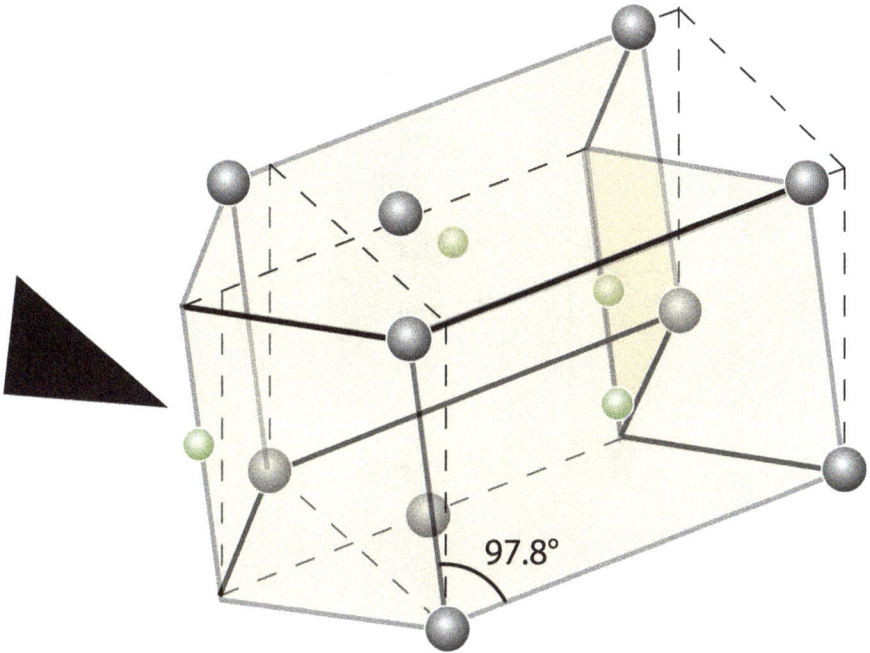

The Martensite monoclinic B19' unit cell structure supported by Otsuka et al. (1999). The angle observed by the researchers at the time of the study was 97.8°.

Supporting Evidence 2

for Cubic B2 (or Tetragonal) in the

Austenite Phase

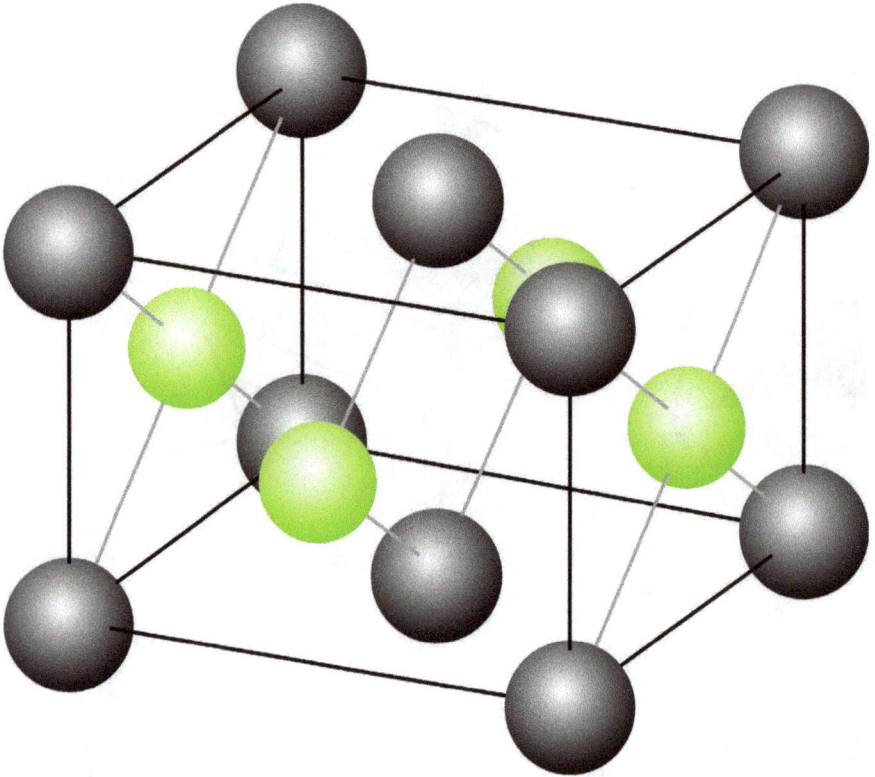

The Austenite unit cell structure supported by Huang et al. (2003) from the article, "Crystal Structures and Shape-Memory Behavior of NiTi".

Nickel atoms: Green
Titanium atoms: Grey

Supporting Evidence 2

for Orthorhombic B19 in the

Martensite Phase

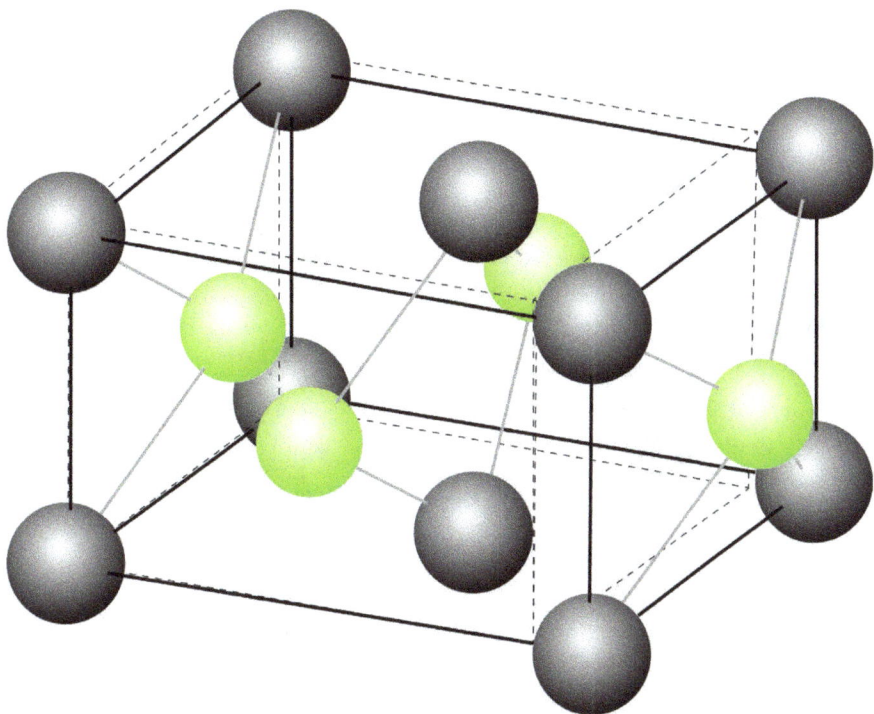

The Martensite orthorhombic B19 unit cell structure supported by Huang et al. (2003).

Supporting Evidence 2

for Monoclinic B19' in the

Martensite Phase

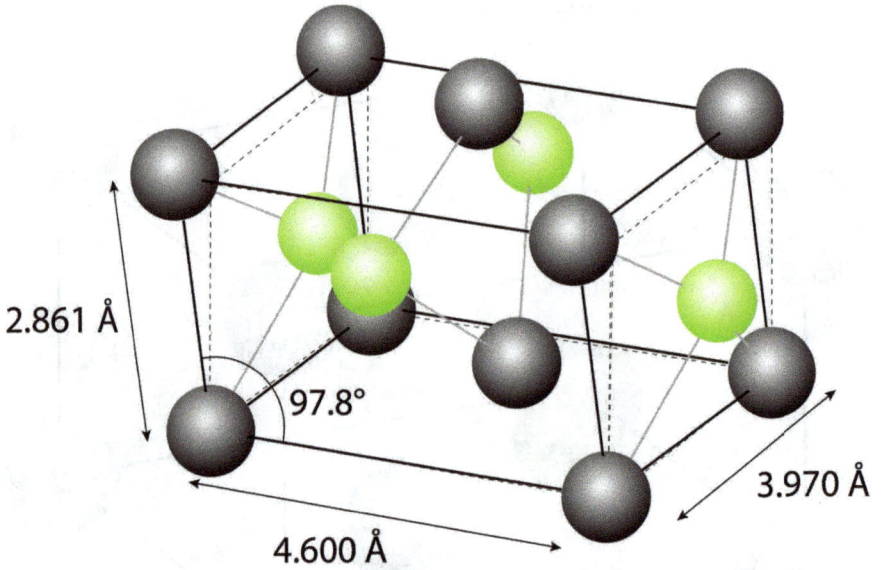

The Martensite monoclinic B19' unit cell structure supported by Huang et al. (2003).

Supporting Evidence 2

for Monoclinic B19' Stress-induced in the

Martensite Phase

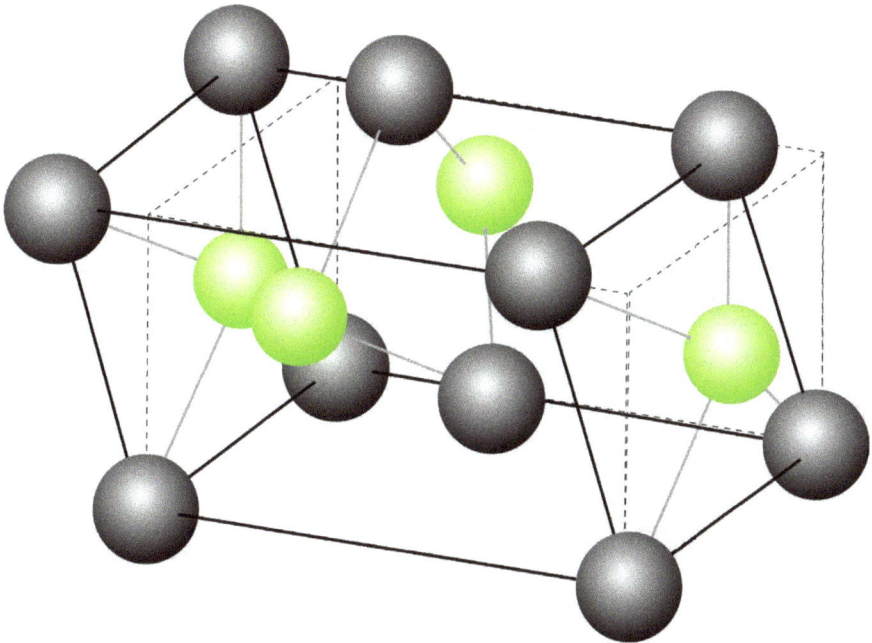

The stress-induced Martensite monoclinic B19' unit cell structure as provided by Huang et al. In the most recent studies, this is very similar to the B19" and BCO structures reported by Vishnu et al. except the displacements of the atoms are more significant and the angle of slanting is greater.

The angle of slanting for this unit cell is assumed to be part of the bending of the NiTi alloy in a direction that enhances the monoclinic structure (i.e., increases the angle and creates further displacement of the atoms).

The Rhombohedrally-distorted (R-phase)

in the Austenite unit cell

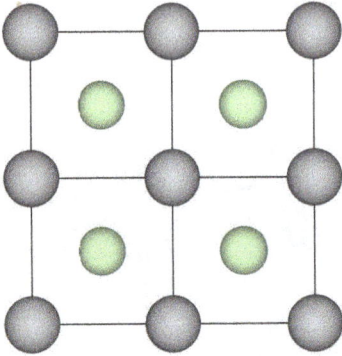

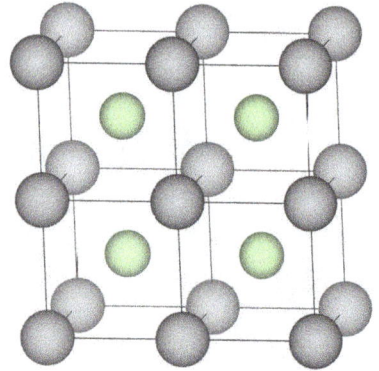

Step 1: Cubic B2 in the Austenite phase
(Flat view on left, 3D view on right)

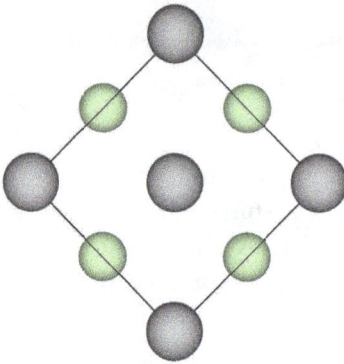

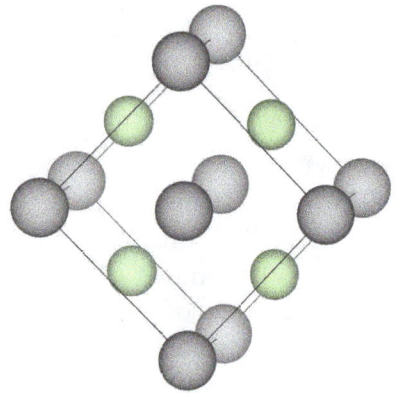

Step 2: Tetragonal in the Austenite phase

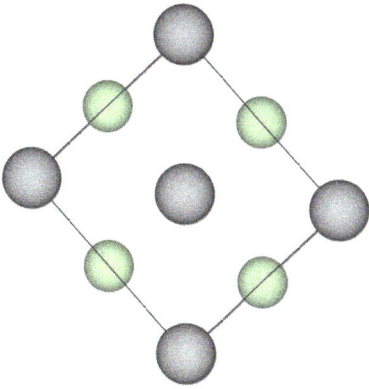

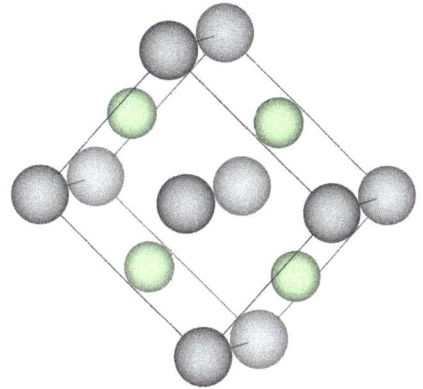

Step 3: R-phase

Supporting Evidence

for

R-Phase

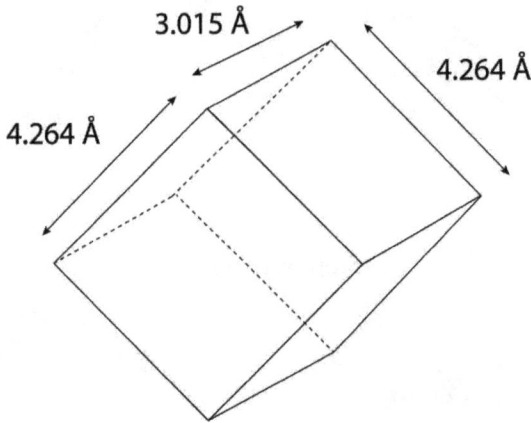

Tetragonal

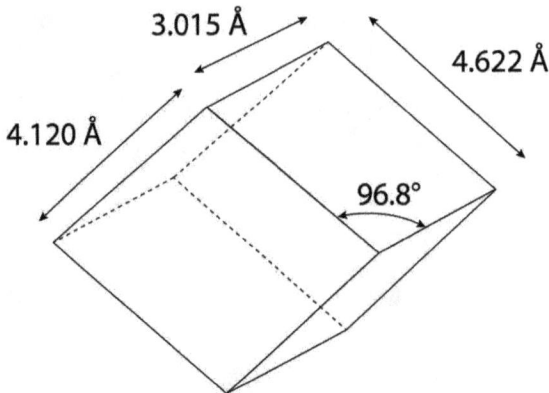

R-Phase

The diagrams on this page were used by Duerig & Bhattacharya in 2015 showing the tetragonal unit cell for the Austenite phase and the rhombohydrally distorted austenite unit cell (known as the R-phase).

APPENDIX I

*NiTi Unit Cell
Lattice Dimensions*

Lattice parameters for cubic B2 (b.c.c.) unit cell austenite crystalline structure from scientific literature

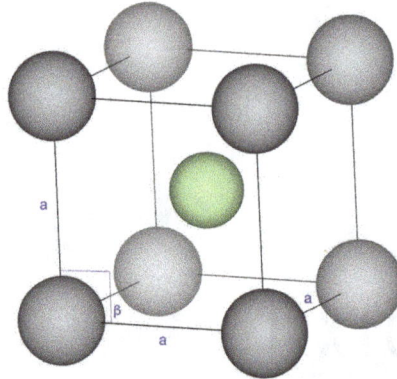

Source	Year	a (Å)	β (°)
Ye et al.	1997	2.977	90
Huang et al.	2001	2.960	90
Sengupta et al.	2009	3.015	90
Hatcher et al	2009	3.019	90
Moitra et al.	2011	3.016	90
Holec et al	2014	3.007	90
Shi et al.	2014	3.010	90
Min		2.960	90
Max		3.019	90
Average		3.000	90

Length is measured in Angstroms (Å). Divide by 10 for nanometer (nm)

Lattice parameters for tetragonal (f.c.c.) unit cell austenite crystalline structure from scientific literature

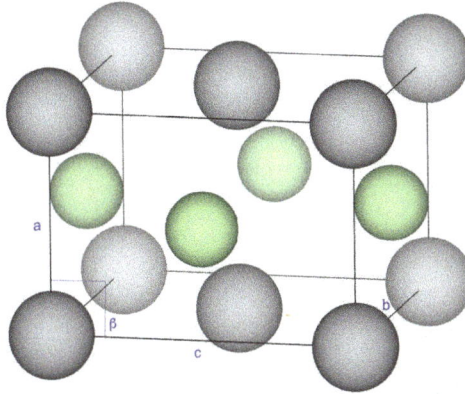

Source	Year	a (Å)	b (Å)*	c (Å)*	β (°)
Ye et al.	1997	2.977	4.210	4.210	90
Huang et al.	2001	2.960	4.186	4.186	90
Sengupta et al.	2009	3.015	4.263	4.263	90
Hatcher et al	2009	3.019	4.269	4.269	90
Moitra et al.	2011	3.016	4.265	4.265	90
Holec et al	2014	3.007	4.252	4.252	90
Vishnu et al. (Experi.)	2010	3.014	4.262	4.262	90
Vishnu et al. (DFT-GGA-3p0)	2010	3.009	4.255	4.255	90
Shi et al.	2014	3.010	4.257	4.257	90
	Min	2.960	4.186	4.186	90
	Max	3.019	4.269	4.269	90
	Average	3.003	4.246	4.257	90

* Lattice dimensions for b and c are mathematically calculated based on a perfect tetragonal unit cell using the B2 cubic unit cell length provided in a.

NOTE: B2 cubic unit cell is part of the tetragonal unit cell.

Lattice parameters for monoclinic B19' unit cell martensite crystalline structure from scientific literature

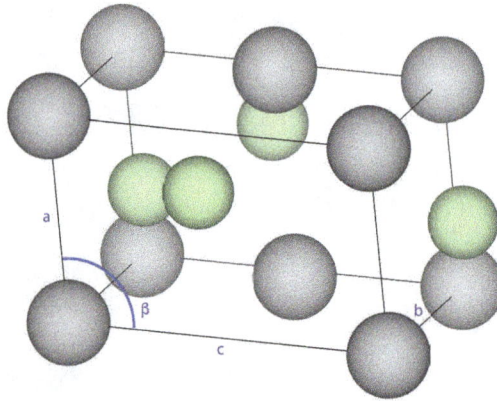

Source	Year	a (Å)	b (Å)	c (Å)	β (°)
Otsuka et al.	1971	2.889	4.120	4.622	96.8
Hehemann & Sandrock	1971	2.883	4.117	4.623	96.8
Michal & Sinclair	1981	2.885	4.120	4.622	96.8
Buhrer et al.	1983	2.884	4.110	4.665	98.10
Kudoh et al.	1985	2.898	4.108	4.646	97.78
Lai and Liu	2000	2.956	4.189	4.455	93.26
Huang et al.	2003	2.861	3.970	4.600	97.8
USPP-GGA	2003	2.929	4.048	4.686	97.8
YCH-LDA	2003	2.892	4.049	4.598	97.80
Vishnu et al. (Exper.)	2010	2.898	4.108	4.646	97.80
Vishnu et al. (DFT-GGA-3p0)	2010	2.933	4.108	4.678	98.26
Min		2.861	3.970	4.455	93.26
Max		2.956	4.189	4.686	98.26
Average		2.900	4.095	4.678	97.18

Length measured in Angstroms (Å). Divide by 10 for nanometer (nm)

472

Lattice parameters for orthorhombic B19" martensite unit cell crystalline structure from Vishnu et al.

Source	Year	a (Å)	b (Å)	c (Å)	β (°)
DFT-GGA-3p6	2010	2.923	4.042	4.801	102.44
DFT-GGA-3p0	2010	2.926	4.034	4.819	103.20
Min		2.923	4.034	4.801	102.44
Max		2.926	4.042	4.819	103.2
Average		2.925	4.038	4.810	102.82

Lattice parameters for BCO martensite unit cell crystalline structure from Vishnu et al.

Source	Year	a (Å)	b (Å)	c (Å)	β (°)
DFT-GGA-3p6	2010	2.928	4.017	4.923	106.64
DFT-GGA-3p0	2010	2.926	4.012	4.925	106.50
DFT-GGA	2010	2.940	3.997	4.936	107.0
Min		2.926	3.997	4.923	106.50
Max		2.940	4.017	4.936	107.0
Average		2.931	4.008	4.928	106.71

APPENDIX J

*Physical properties of
nickel and titanium*

Table 1: Melting Point, Density and Ionization Energy

Element	Symbol	Melting Point* °C	Density g cm^{-3}	Ionization Energy** eV
Titanium	Ti	1670	4.50	6.8281
Nickel	Ni	1453	8.91	7.6398
Zirconium	Zr	1855	6.50	6.6339
Hafnium	Hf	2231	13.20	6.8251
Platinum	Pt	1769	21.30	8.9587
Palladium	Pd	1555	12.02	8.3369
Copper	Cu	1085	8.93	7.6398
Iron	Fe	1538	7.86	7.9024

* Melting points are provided by ASM Alloy Phase Diagram Database.

** This is the *energy* required to remove the most loosely held electron from one mole of gaseous atoms to produce 1 mole of gaseous ions each with a charge of 1+. Data provided by Lenntech (http://www.lenntech.com/periodic-chart-elements/ionization-energy.htm).

Table 2: Atomic radius*

Element	Symbol	Empirical	Calculated
Titanium	Ti	140	176
Nickel	Ni	135	149
Zirconium	Zr	155	206
Hafnium	Hf	155	208
Platinum	Pt	135	177
Palladium	Pd	140	169
Copper	Cu	135	145
Iron	Fe	140	156

* Also known as the metallic radius. For a non-alloy material, this is ½ the distance between the nuclei of two atoms just touching each other in a crystalline structure.

Data provided by WebElements (https://www.webelements.com/iron/atom_sizes.html)

The unit of measurement is in picometer (pm).

Table 3: Covalent radius[*]

Element	Symbol	Empirical	2008 values	Molecular single bond[**]	Molecular double bond[*]	Molecular triple bond[*]
Titanium	Ti	136	160	136	117	108
Nickel	Ni	121	124	110	101	101
Zirconium	Zr	148	175	154	127	121
Hafnium	Hf	150	175	152	128	122
Platinum	Pt	128	136	123	112	110
Palladium	Pd	131	139	120	117	112
Copper	Cu	138	132	112	115	120
Iron	Fe	125	132[***]	116	109	102

[*] Covalent radius is the distance between two nuclei.
[**] Also called the ionic radii, which in this case is for the cations (i.e., the atom has lost electrons). Ionic radii is the distance between two nuclei of the ionised atoms.
[***] This value is for low spin; 152pm for high spin.

Data provided by WebElements (https://www.webelements.com/iron/atom_sizes.html).

The unit of measurement is in picometer (pm).

1 pm = 1×10^{-12} meter (m)

100 pm = 1 Ångstrom (Å)

1000 pm = 1 nanometer (nm)

APPENDIX K

Converting Between atomic % and weight %

From weight % to atomic %

Weight is measured in grams, and atomic is measured by the number of atoms. The other thing to remember when converting is to use a constant called Avogadro's number, which is $6.0221408577474 \times 10^{23}$ mol^{-1} (where the unit is in moles).

To convert weight % to atomic %, consider a 100 grams sample of the alloy Ti-Ni$_{56.06 \text{ wt%}}$. This means that for every 100 grams of the alloy we have in our hands, we know it will have 56.06 grams of Ni and 43.94 grams of titanium.

To calculate atomic %:

1. Obtain from a reference book the atomic weight of Ni (= 58.6934) and Ti (= 47.867).
2. Number of atoms in 56.06 grams of Ni = 56.06 x Avogadro's Constant / atomic weight of Ni = 56.06 x 6.0221408577474 x 10^{23} / 58.6934 = 5.75194513 x 10^{23}.
3. Similarly, number of atoms in 43.94 grams of Ti = 43.94 x Avogadro's Constant / atomic weight of Ti = 43.94 x 6.0221408577474 x 10^{23} / 47.867 = 5.52808551 x 10^{23}.
4. Total number of atoms in steps 2 and 3 = 5.75194513 x 10^{23} + 5.52808551 x 10^{23} = 1.12800306 x 10^{24}.

Therefore, atomic % of Ni = Number of Ni atoms / Total number of atoms) x 100% = 5.75194513 x 10^{23} / 1.12800306 x 10^{24}) x 100% = 50.99%. The alloy composition can now be written as Ti-Ni$_{51 \text{ at.%}}$

From atomic % to weight %

To convert atomic % to weight %, consider a sample containing 100 atoms. As an example, let us use Ti-Ni$_{51\ at.\%}$. In other words, for every 100 atoms, this alloy would contain 51 nickel atoms and 49 titanium atoms.

To calculate the weight %:

1. Weight of 51 Ni atoms = (51 x atomic weight of Ni) / Avogadro's Constant = (51 x 58.6934) / $6.0221408577474 \times 10^{23}$ = 2993.3634 / $6.0221408577474 \times 10^{23}$ = $4.97059679 \times 10^{-21}$ atomic mass units.

2. Weight of 49 Ti atoms = (49 x atomic weight of Ti) / Avogadro's Constant = (49 x 47.867) / $6.0221408577474 \times 10^{23}$ = 2345.483 / $6.0221408577474 \times 10^{23}$ = $3.89476609 \times 10^{-21}$ atomic mass units.

3. Total weight of Ni and Ti for 100 atoms = $3.89476609 \times 10^{-21}$ + $4.97059679 \times 10^{-21}$ = $8.86536288 \times 10^{-21}$

Therefore, weight% of Ni = ($4.97059679 \times 10^{-21}$ / $8.86536288 \times 10^{-21}$) x 100% = 56.06%. The alloy composition can now be written as Ti-Ni$_{56\ wt\%}$

APPENDIX L

Alloys Studied by
Prof. John P. Nielsen
1947-58

Name(s)	Air Force Involvement	Date	Document	Aim	Contains NiTi?	Useful to NiTi?
Nielsen, J.P. Work, H.K.	Contract Number W 300 69 ORD 4477	Aug 25, 1949	"Titanium–Carbon and Titanium–Nitrogen Alloys" Interim Technical Report No.2 http://www.stormingmedia.us/25/2 561/A256159.html	Phase diagrams of TiC and TiN systems	No	Maybe
Stover, Edward Roy (MIT) Read and signed by Nielsen, J.P.	Financial support from Wright–Patterson AFB	1950	"The Binder Phase in Titanium Carbide-Nickel Cermets" Thesis submitted to the Department of Metallurgy, NYU in 1950	Presumed to be related to the TiNiC system	Yes	Yes
Nielsen, J.P. Hibbard Jr, W.R.		Sep 1950	An X-ray Study of Thermally Induced Stresses in Microconstituents of Aluminium-Silicon Alloys	X-ray diffraction study of heat-induced aluminium–silicon (AlSi) system	No	No
Nielsen, J. P. Margolin, H	Research Division of NYU for AF Technical Report 6596 Contract No. AF 33(038)–8725	Dec 1, 1950 (covering work from Sep 30, 1949 to Dec 1, 1950)	"Titanium-Nickel Phase Diagram" (Part 1)	Phase diagram, crystalline structures and a description of melting techniques for the titanium-nickel (NiTi) system up to 90 percent nickel in Part 1.	Yes	Yes
Nielsen, J. P. Margolin, H Ence, E.	Research Division of NYU for AF Technical Report 6597 Contract No. AF 33(038)–8725	Oct 1951 (covering work from Dec 1, 1950 to Sep 30, 1951)	"Titanium-Nickel Phase Diagram" (Part 2)	A complete phase diagram from 0 to 100 percent nickel.	Yes	Yes

Nielsen, J.P.	U.S. Army	Mar 1952	"Equilibrium Diagram of Titanium Alloy Systems" Information Bulletin No.14, published by Watertown Arsenal, Watertown, Mass.	Phase diagrams and properties of various titanium-based alloys	No	Yes
Stover, Edward Roy (MIT) Read and signed by Nielsen, J.P.	Financial support from Wright–Patterson AFB	(May) 1952	"The Binder Phase in Titanium Carbide – Nickel Cermets" Thesis submitted to the Department of Metallurgy, NYU on May 16, 1952.	Focuses on nickel-rich portion of the nickel-titanium-carbon (NiTiC) above 1,200°C. Also looks at TiC, TiN, TiO and NiC systems.	Yes	Yes
Nielsen, J.P. Margolin, H. Ence, E.		Feb 1953	"Titanium–Nickel Phase Diagram" Transactions of the Metallurgical Society of AIME (Journal of Metals), Volume 197, pp.243-247.	Phase diagram and crystalline structures of the titanium-nickel (NiTi) system. Also found no evidence to support Rostoker's claim in 1951 for a decomposition of NiTi into Ti₂Ni and TiNi₃.	Yes	Yes
Nielsen, J.P. Margolin, H. Ence, E. Kirk, W.	U.S. Army	1953	"Final Report to Watertown Arsenal on Contract No. DA-30-069-ORD-208." New York University, Engineering Research Division.	Relating to titanium and titanium-based alloys	No	Maybe

Author	Notes	Date	Citation	Description		
Nielsen, J.P. Cadoff, L.	Definitely of interest to Wright–Patterson AFB in April 1966 through AF Technical Report No. AFML–TR–65–2, Part II, Volume XIII when looking at Ti-B-V, Zt-B-C and Hf-B-C systems.	Feb 1953	"Titanium–Carbon Phase Diagram" Journal of Metals, Volume 5, pp.248-252.	Phase diagram and crystalline structures of the titanium–carbon (TiC) system	No	Maybe
Nielsen, J.P. Cadoff, L.	Definitely of interest to Wright–Patterson AFB in April 1966 through AF Technical Report No. AFML–TR–65–2, Part II, Volume XIII when looking at Ti-B-V, Zt-B-C and Hf-B-C systems.	1953	"Titanium–Carbon Phase Diagram" Trans. Am. Institute Mining Metallurgy Engineering, Volume 197, pp.248-252.	Phase diagram and crystalline structures of the titanium–carbon (TiC) system	No	Maybe
Margolin, H. Stone, L.		1953	"Titanium-Rich Regions of the Ti-C-N, Ti-C-O and Ti-N-O Phase Diagrams" Journal of Metals, Volume 5, pp.1498-1502.	Phase diagram and crystalline structures of the titanium–carbon (TiC) system	No	Maybe

Author	Notes	Date	Title	Description	No	Yes/Maybe
Nielsen, J.P. Margolin, H. Paltry, A. E.	Definitely of interest to Wright–Patterson AFB in April 1966 through AF Technical Report No. AFML–TR-65-2, Part II, Volume XIII when looking at Ti-B-V, Zt-B-C and Hf-B-C systems.	1954	"Titanium-Nitrogen, Titanium–Boron System" Transaction American Society of Metals, Volume 46, pp.312-329.	Phase diagram, crystalline structures and size of atoms in close-packed metals of the titanium-boron (TiB) and titanium-nitrogen (TiN) systems.	No	Yes
Nielsen, J.P.		Feb 1954	"The Geometric Coalescence Mechanism in Grain Boundary Migration" Rudarsko-Metallurskega Zomika (Yugoslavia)		No	Maybe
Nielsen, J.P.		Sep 1954	"A Mechanism for the Origin of Recrystallization Nuclei" Journal of Metals		No	Maybe
Nielsen J. P. Cadoff, I. Miller, E.		Jun 1955	"Properties of Arc-Melted vs Powder Metallurgy Titanium Carbide" Plansee Proceedings, page 10 and presented at the Second International Plansee Seminar, Reutte, Austria.	Phase diagram and crystalline structures of the titanium-carbon (TiC) system	No	Maybe

| Stover, Edward Roy (MIT) Read and signed by Nielsen, J.P. | Financial support from Wright-Patterson AFB | (Feb) 1956 | "The Nickel-Titanium-Carbon System" Thesis submitted to the Department of Metallurgy, NYU on January 9, 1956 | Discusses nickel-rich portion of the nickel-titanium-carbon (NiTiC) above 1,200°C. Also looks at TiC, TiN, TiO and NiC systems. | Yes | Yes |

Bibliography

Adenstedt, Henry K. & Freeman Jr., William R. "The Tentative Titanium-Silver Binary System": *WADC Technical Report 53-109 Part 1*. April 1953.

Barnes, Clive. July 1999, *Shape Memory and Superelastic Alloys*. Copper Development Association Inc. Available from http://www.copper.org/publications/newsletters/innovations/1999/07/shape.html.

Benedicks, Carl. "On the Elasticity of Solid Solutions, in particular those of AuCd": *Arkiv för Matematik, Astronomi och Fysik (Archive for Mathematics, Astronomy and Physics)*. May 1940. Volume 27A, Number 8, pp.1-11.

Berlitz, Charles & Moore, William. 1980, *The Roswell Incident*. London: Granada Publishing Limited.

NOTE: There are slight variations in the wording of some of the witnesses' quotes in the 1988 Berkley edition and the 1980 Granada edition of this book. Both were used to create this research document.

Blundell, Nigel & Boar, Roger. 1984, *The World's Greatest UFO Mysteries*. London: Octopus Books Ltd, pp.102-105.

Boehm, George A. W. "Titanium": *Scientific American*. April 1949, Volume 180, Number 4, pp.48-51.

Brachet, Jean-Christophe, Olier, P., Brun, G., and Dubuisson, P. "Superelasticity and Impact Properties of Two Shape Memory Alloys: Ti50Ni50 and Ti50Ni48Fe2": *Journal de Physique IV (Proceedings) 07(C5)*. November 1997, DOI: 10.1051/jp4:1997589. Available from https://www.researchgate.net/profile/P_Olier/publication/450672

84_Superelasticity_and_Impact_Properties_of_Two_Shape_Memor y_Alloys_Ti50Ni50_and_Ti50Ni48Fe2/links/00b49525299673c539 000000.pdf?origin=publication_detail

Brookesmith, Peter (editor). 1980-84, *The UFO Casebook —Appearances and Disappearances: The Unexplained Series*. London: Orbis Publishing Limited.

Brookesmith, Peter (editor). 1984, *The Age of the UFO*. London: Orbis Publishing Limited, pp.44-54.

Buehler, William J. & Rozner, Alexander G. U.S. Patent Number 3,351,463 (November 7, 1967): "High Strength Nickel-Base Alloys".

Buehler, William J. & Wang, Frederick E. U.S. Patent Number 3,558,369 (January 26, 1971): "Method of Treating Variable Transition Temperature Alloys".

Buehler, William J. & Wiley, Raymond C. U.S. Patent Number 3,174,851: "Nickel-Titanium Alloy".

Buhrer, W., Gotthardt, R., Kulik, Andrzej J., and Staub, F. "Powder Neutron-Diffraction Study of Nickel Titanium Martensite": *Journal of Physics F: Metal Physics*. May 1983, Volume 15 Number 5, pp..

Burkart, M. W. & T. A. Read, T. A. "Diffusionless phase change in the indium-thallium alloys": *Transactions of the Metallurgical Society of AIME (Journal of Metals)*. 1953. Volume 197, pp.1516–1524.

Burleson, Dr. Donald. "Looking Up": *Roswell Daily Record* (Vision Magazine). January 7, 2000.

Bystrom, Anders & Almin, Karl Erik. "X-ray Investigation of Gold-Cadmium Alloys Rich in Gold": *Acta Chemica Scandinavica*. 1947. Volume 1, pp.76-89.

Cardarelli, François. 2008, *Materials Handbook – A Concise Desktop Reference*. Second Edition. London, U.K.: Springer-Verlag London Limited.

Carey, Thomas J. & Schmitt, Donald R. 2009, *Witness to Roswell*. Franklin Lakes, New Jersey: The Career Press, Inc.

Chang, L. C. & Read, T. A. "Plastic Deformation and Diffusionless Phase Changes in Metals—The AuCd β-phase": *Transactions of the Metallurgical Society of AIME*. January 1951. Volume 189, pp.47-52.

Chemical Abstracts, 1932 (Volume 26) 5467, "The crystal structure of AuCd"; 1941 (Volume 35) 3581(3-5), "Modulus of elasticity and internal friction of the intermediary phases in the system gold-cadmium"; 1949 (Volume 43) 3766(c-d), "A tentative titanium-nickel diagram".

Chluba, Christoph, Ge, Wenwei, Lima de Miranda, Rodrio, Strobel, Julian, Kienie, Lorenz, Quandt, Eckhard, and Wuttig, Manfred. "Ultralow-fatigue shape-memory alloy films": *Science*. May 29, 2015, Volume 348, Issue 6238, pp.1004-1007.

Claiborne, William. "GAO turns to Alien Turf in Probe": *The Washington Post*. January 14, 1994, p.A22.

Craighead, C. M., Simmons, O.W. and Eastwood, L.W. "Titanium Binary Alloys": *Transactions of the Metallurgical Society of AIME (Journal of Metals)*. Volume 188, March 1950, pp.485-513.

Crain, E.J. (editor). "Titanium in Aircraft": *Chemical Abstracts*. Volume 45 (columns 7381-10680). September 10 - November 25, 1951. The Ohio State University, Columbus, Ohio, USA: American Chemical Society, 9441d.

Crain, E.J. (editor). "Titanium alloys": *Chemical Abstracts*. Volume 45 (columns 1-3656). January 10 - April 25, 1951. The Ohio State University, Columbus, Ohio, USA: American Chemical Society, 107f.

Crain, E.J. (editor). "Microradiographic study of the distribution of alloying components in a nickel solid solution": *Chemical Abstracts*. Volume 45 (columns 1-3328). January 10 - April 10, 1952. The Ohio State University, Columbus, Ohio, USA: American Chemical Society, 1416c.

Duerig, T. W., and Bhattacharya, K. "The Influence of the R-Phase on the Superelastic Behavior of NiTi": *Springer*. May 16, 2015, DOI 10.1007/s40830-015-0013-4.

Duerig, T. W., and Pelton, A. R. "Ti-Ni Shape Memory Alloys": *Materials Properties Handbook Titanium Alloys*. 1994, pp.1035-1048.

Friedman, Stanton T. & Berliner, Don. 1992, *Crash at Corona*. New York: Paragon House.

Encyclopedia Britannica. Chicago, USA: Encyclopedia Britannica Inc., 1969, 1980, 1986 and 1988 editions.

Evans, Hilary & Spencer, John (eds.). 1988, *Phenomenon: Forty Years of Flying Saucers*. New York: Avon Books Printing.

Fernandes, Daniel J., Peres, Rafael V., Mendes, Alvaro M. & Elias, Carlos, N. "Understanding the Shape-Memory Alloys used in Orthodontics": *ISRN Dentistry*. October 3, 2011, Volume 2011, Article ID.132408, pp.1-6. Available from https://www.ncbi.nlm.nih.gov/pmc/articles/ PMC3185255/#B38.

Gong, F. F., Shen, H. M., and Wang, Y. N. "Crystallization of Amorphous Sputtered NiTi Shape-Memory Alloy Films": *Materials Research Society Symposium Proceedings*. 1996, Volume 398, pp.405-410.

Good, Timothy. 1997, *Beyond Top Secret*. London: Pan Books, pp. 317, 455-483.

Good, Timothy. 1991, *Alien Liaison*. London: Arrow Books Ltd, pp.83-94.

Good, Timothy (editor). 1990, *The UFO Report 1991*. London: Sidgwick & Jackson, pp.132-191.

Gou, Liangliang, Liu, Yong and Ng, Teng Yong . "Effect of Cu content on Atomic Positions of $Ti_{50}Ni50_{-x}Cu_x$ Shape Memory Alloys Based on Density Functional Theory Calculations": *Metals*. November 26, 2015, Volume 5, Issue 4, pp.2222-2235.

Grant, Julius (editor). 1969, *Hackh's Chemical Dictionary*. Fourth Edition. New York, USA: McGraw-Hill, Inc.

Grayson, Martin (ed.). 1983, *Encyclopedia of Chemical Technology*, Third Edition. Volume 23. New York: John Wiley & Sons, Inc., pp.98-130 (Section on "Titanium and Titanium Alloys").

Greninger, Alden B. & Mooradian, Victor G., "Strain Transformation in Metastable Beta Copper-Zinc and Beta Copper-Tin Alloys": *Transactions of the Metallurgical Society of AIME*. Volume 128, 1938, pp. 337–368.

Guthikonda, Venkata Suresh & Elliott, Ryan S. "Toward an effective Interaction potential model for the shape-memory alloy AuCd": *AEM Report Number 2008-1*. April 2, 2008. pp.1-29.

Hehemann, R. F. and Sandrock, G. D. "Relations between the Premartensitic Instability and the Martensite Structure in TiNi": *Scripta Metallurgica.* September 1971, Volume 5, Number 9, pp.801-805.

Hesemann, Michael & Mantle, Philip. 1997, *Beyond Roswell.* New York: Marlowe & Company.

Hatcher, Nicolas B., Kontsevoi, Oleg Yu, and Freeman, Arthur J. "Role of elastic and shear stabilities in the martensitic transformation path of NiTi": *Physical Review B.* October 2009, Volume 80, Issue 14, 144203.

Holec, David, Friák, Martin, Dlouhý, Antonin, and Neugebauer, Jörg. "Ab initio study of point defects in NiTi-based alloys": *Physical Review B.* March 20, 2014 (Revised version), Volume 89, Issue 1, id.014110 (pp.1-7).

Horzewski, Michael and Giba, Jeffrey. U.S. Patent No.5730741 A: "Guided spiral catheter". Publication date: March 24, 1996. Also available from https://www.google.com/patents/US5730741.

Hough, Peter & Randles, Jenny. 1991, *Looking for the Aliens: A Psychological, Scientific and Imaginative Investigation.* London: Blandford Press.

Hough, Peter & Randles, Jenny. 1994, *The Complete Book of UFOs.* London: Judy Piatkus (Publishers) Ltd, pp.67-76.

Huang, Xiangyang, Bungaro, Claudia. Godlevsky, Vitaliy. and Rabe, Karin M. "Lattice Instabilities of Cubic NiTi from first Principles": *Physical Review.* Section B. May 14, 2001, Volume 65, Number 1. pp.1-11.

Huang, Xiangyang, Ackland, Graeme J. & Rabe, Karin M. "Crystal Structures and Shape-Memory Behavior of NiTi": *Nature Materials.* April 20, 2003, Volume 2, pp.307-311.

KettelKamp, Larry. 1996, *ETs and UFOs: Are They Real?* New York: William Morrow and Company, Inc, pp.30-49.

King, R. Bruce (editor-in-chief). 1994, *Encyclopedia of Inorganic Chemistry.* Volume 1. New York, USA: John Wiley & Sons, pp.98-99 & 102.

Klaproth, Martin Heinrich. 1801, *Analytical Essays Towards Promoting the Chemical Knowledge of Mineral Substances.* English Translation. London: G. Woodfall (Printer).

Kleinherenbrink, P. H., and Beyer, J. "Control of the transformation temperatures of TiNi shape memory alloys by ternary additions": *University of Twente (Netherlands) from the Faculty of Mechanical Engineering.* 1989, pp.187-190.
Article available at http://dx.doi.org/10.1051/esomat/198904001.

Knittel, Donald, and Wu, James B. C. 1998, Chapter 6 on "Titanium and its Alloys": *Mechanical Engineers' Handbook* (edited by Myer Kutz). Second Edition. John Wiley & Sons, Inc., pp.91-108.

Korff, Karl K. 1994, *The Roswell UFO Crash.* New York: Prometheus Books.

Kudoh, Y., Tokonami, M., Miyazaki, S., Otsuka, K. "Crystal structure of the Martensite in Ti-49.2 at.%Ni Alloy Analyzed by the Single Crystal X-ray Diffraction Method": *Acta Metallurgica*, 1985. Volume 33, Issue 11, pp.2049-2056.

Laves, F., and Wallbaum, H. J. "Zur Kristallchemie von Titan-Legierungen": *Die Naturwissenschaften.* October 1939, Volume 27, Issue 40, pp.674-675.

Ling, H. C. and Kaplow, Roy. "Stress-Induced Shape Changes and Shape Memory in the R and Martensite Transformations in Equiatomic NiTi": *Metallurgical Transactions A.* December 1981, Volume 12, Issue 12, pp.2101-2111.

Machado, L. G., & Savi, M. A. "Medical applications of shape memory alloys": *Brazilian Journal of Medical and Biological Research.* June 2003. Volume 36, Number 6, pp.683-691.

Margolin, Harold, and Nielsen, John P. *Titanium Metallurgy* (In Chapter titled "Modern Materials, Advances on Development and Application"). Edited by H. H. Hausner and published in 5 volumes by Academic Press: New York. Publication Date: March 1960. Chapter in Volume 2, pp.225-325.

Melton, K. N., Proft, J. L., and Duerig, T. W. "Wide Hysteresis Shape-Memory Alloys based on the Ni-Ti-Nb System": *Materials Research Society International Symposium Proceedings*. 1989. pp.165-170. Available from http://www.intrinsicdevices.com/27_Wide_Hysteresis_NiTiNb.pdf.

McGraw-Hill Encyclopedia of Science and Technology, 1987. Volumes 1 and 16 (Shape-memory alloys).

Margolin, Harold, and Farrar, Paul. "The Physical Metallurgy of Titanium Alloys": *Ocean Engineering* (published by Pergamon Press). Volume 1, pp.329-345.

Matsumoto, O., Miyazaki, S., Otsuka, K., and Tamura, H., "Crystallography of martensitic transformation in Ti-Ni single crystals": *Acta Metallurgica*. August 1987, Volume 35, Issue 8, pp.1929-2175.

Michal, G. M. and Sinclair, R. "The Structure of TiNi Martensite": *Acta Crystallographica Section B*. 1981, Volume 37, Issue 10, p.1803-1807.

Missler, Chuck & Eastman, Mark. 1997, *Alien Encounters The Secret Behind the UFO Phenomenon*. USA: Koinonia House.

Mohri, Maryam, Chakravadhanula, Venkata Sai Kiran, and Nili-Ahmadabadi, Mahmoud. "Crystallization study of amorphous sputtered NiTi bi-layer thin film": *Materials Characterization*. May 2015, Volume 103, pp.75-80.

Moitra, Amitava, Solanki, Kiran N., Horstemeyer, Mark F. "The location of atomic hydrogen in NiTi alloy: A first principles study": *Computational Materials Science*. January 2011, Volume 50, Number 3, pp.820-823.

Mysteries of the Unknown: The UFO Phenomenon, 1987. Richmond, Virginia, USA: Time-Life Books Inc, pp.39-40.

Nespoli, Adelaide, Biffi, Carlo Alberto, Casati, Riccardo, Passaretti, Francesca, Tuissi, Ausonio & Villa, Elena. "New Developments in Mini/Micro Shape Memory Actuators":*InTech*. 2012, pp.35-52.

Nicolson, Iain & Moore, Patrick. 1985, *The Universe*. London: William Collins Sons & Co Ltd.

Nielsen, John P., Margolin, Harold, and Ence, E. "Titanium-Nickel Phase Diagram": *Transactions of the Metallurgical Society of AIME (Journal of Metals)*. February 1953, Volume 197, p.243-247.

Ogden, H. R., and Jaffee, R. L. *Report on the Effects of Carbon, Oxygen, and Nitrogen on the Mechanical Properties of Titanium and Titanium Alloys*. TML Report No.20, October 19, 1955. Prepared at the Titanium Metallurgical Laboratory at the Battelle Memorial Institute on behalf of the Office of Assistant Secretary of Defense for Research and Development.

Ölander, Arne. "An Electrochemical Investigation of solid Cadmium-Gold Alloy": *Journal of the American Chemical Society (published by ACS Publications)*. 1932. Volume 54, pp.3819-3833.

Ölander, Arne. "The Crystal Structure of AuCd": *Zeitschrift Fur Kristallographie (Journal for Crystallography)*. 1932. Volume 83, pp.145-148.

Otsuka, Kazuhiro, and Ren, Xiaobing. "Physical metallurgy of TiNi-based shape memory alloys": *Progress in Materials Science*. 2005. Volume 50, Number 5, pp.511-678.

Otsuka, Kazuhiro, and Ren, Xiaobing. "Recent Developments in the Research of Shape Memory Alloys": *Intermetallics 7*. May 6, 1998, Volume 7, Issue 5, pp.511-528.

Otsuka, Kazuhiro, Sawamura, T. and Shimizu, K. "Crystal Structure and Internal Defects of Equiatomic TiNi Martensite": Physica Status Solidi. May 16, 1971, Volume 5, Issue 2, pp.457-470.

Otsuka, Kazuhiro, Nakai, K., and Shimizu, K. "Structure Dependence of Superelasticity in Cu-Al-Ni Alloy": *Scripta Metallurgica*. 1974, Volume 8, pp.913-918.

Parker, Steve. 1992, *Collins Eyewitness Science: Electricity*. Pymble, NSW: Harper Collins Publishers.

Pearson, W.B. 1958 & 1967, *A Handbook of Lattice Spacings and Structures of Metals and Alloys*. New York: Pergamon Press.

Perkins, Jeff. "Residual Stresses and the Origin of Reversible (Two-Way) Shape Memory Effect": *Scripta Metallurgica et Materialia*. 1974. Volume 8, pp.1469-1476.

Potter, Marion E. et al. (eds.). 1946, 1947, 1948, *The Industrial Arts Index*. New York: The H.W. Wilson Company.

Rachinger, William A. "A super-elastic single crystal calibration bar": *British Journal of Applied Physics*. June 1958, Volume 9, Issue 6, pp.250-252.

Randles, Jenny. 1987, *The UFO Conspiracy: The First Forty Years*. London: Blandford Press, pp.18-22.

Randle, Kevin D. & Schmitt, Donald R. 1991, *UFO Crash at Roswell*. New York: Avon Books.

Randle, Kevin D. & Schmitt, Donald R. 1994, *The Truth about the UFO Crash at Roswell*. New York: M. Evans and Company, Inc.

Ramaiah, K. V., Saikrishna, C. N., Gouthama, and Shaumik, S. K. "$Ni_{24.7}Ti_{50.3}Pd_{25.0}$ High Temperature Shape-Memory Alloy with Narrow Thermal Hysteresis and High Thermal Stability": *Materials & Design*. April 2014, Volume 56, pp.78-83.

Reece, Peter L. (editor). 2007, *Progress in Smart Materials and Structures*. New York: Nova Science Publishers, Inc.

Reynolds, J. E., and Bever, Michael B. "On the Reversal of the Strain induced Martensitic Transformation in the Copper-Zinc System": *AIME Transactions: Journal of Metals*. 1952, Volume 194, pp.1065–1066.

Sachs, Margaret. 1980, *The UFO Encyclopedia*. London: Transworld Publishers Limited.

Saitoh, Ken-ichi, Sato, Tomohiro, and Shinke, Noboru. "Atomic Dynamics and Energetics of Martensitic Transformation in Nickel-Titanium Shape Memory Alloy": *Materials Transactions*. 2006, Volume 47, Number 3, pp.742-749.

Saler, Benson, Ziegler, Charles A. and Moore, Charles B. 1997, *UFO Crash at Roswell: The Genesis of a Modern Myth*. Washington, USA: Smithsonian Institution Press.

Sanders, Kevin. October 1982, "Miracle Metal – It Generates Unlimited Energy": *Omega Science Digest*. U.S. Edition. New York: The Hearst Corporation, pp.93-96.

Saunders, Keith (sic). March/April 1984, "Miracle Metal – There is Absolutely Nothing Like It": *Omega Science Digest*. Australian Edition. New York: The Hearst Corporation, pp.11-13 & 119.

Scheil, Erich. "Über die Umwandlung des Austenits in Martensit in gehärtetem Stahl" ("On the transformation of austenite to martensite in hardened steel"): *Zeitschrift für anorganische Chemie (Journal of Inorganic Chemistry)*. 1929, Volume 183 (1), pp.98-120.

Scheil, Erich. "Über die Umwandlung des Austenits in Martensit in Eisen-Nickellegierungen unter Belastung" ("On the transformation of austenite to martensite in iron-nickel alloys under load"): *Zeitschrift für anorganische Chemie (Journal of Inorganic Chemistry)*. July 22, 1932. Volume 207 (1), pp.21-40.

Sehitoglu, H., Hamilton, R., Maier, H. J., and Chumlyakov, Y. "Hysteresis in NiTi Alloys": *Journal de Physique IV*. June 2004, Volume 115, pp.3-10.

Sengupta, Arkaprabha and Papadopoulos, Panayiotis. "Constitutive Modeling and Finite Element Approximation of B2-R-B19' Phase Transformations in Nitinol Polycrystals": *Computer Methods in Applied Mechanics and Engineering*. September 1, 2009, Volume 198, Issues 41-44, pp.3214-3227.

Shi, H., Frenzel, J., Martinez, G. T., Rompaey, S. Van, Bakulin, A., Kulkova, S., Van Aert, S., Schryvers, D. "Sit Occupation of Nb Atoms in Ternary Ni-Ti-Nb Shape Memory Alloys": *Acta Materialia*. August 1, 2014, Volume 74, pp.85-95.

Spencer, John. 1991, *UFOs: The Definitive Casebook*. London: Hamlyn Publishing Group Limited.

Spencer, John (editor). 1991, *The UFO Encyclopedia*. London: Headline Book Publishing PLC, pp.66-67, 257-258 & 330-331.

Smithells, G. J. *Metal Reference Book*, 1950s editions. Formerly available from Dr. Andrzej Calka's office at the Material Science Engineering section of the Research School of Physical Sciences at the ANU (1995).

Stemman, Roy. 1991, *Mysteries of the Universe*. London: Bloombury Books.

Story, Ronald D. (ed.). 1980, *The Encyclopedia of UFOs*. Garden City, New York: Dolphin Books, Doubleday & Company, Inc.

Swords, Michael D. "A Different View of 'Roswell — Anatomy of a Myth'": *Journal of Scientific Exploration*. 1998, Volume 12, Number 1, pp.103-125.

Tambling, Richard. 1978, *Flying Saucers: Where do they come from?*. Hong Kong: Horwitz Publications. The New Colombia Encyclopedia. Colombia University Press, 1967.

Teeple, H. O. "Nickel and High-Nickel Alloys": *Industrial and Engineering Chemistry*. 1950, Volume 42, Number 10, pp.1990-2001.

Teramoto, Dr. Alberto. 2010, *SENTALLOY: The Story of Superelasticity (A White Paper Report)*. Bohemia, NY: Dentsply GAC.

The World's Greatest UFO & Alien Encounters: Conspiracies, Abductions, Little green men—Encounters beyond belief. London: Octopus Publishing Group Ltd. / Bounty Books / Chancellor Press 2002.

Tottle, Prof. C. R. 1984, *An Encyclopedia of Metallurgy and Materials*. Estove, Plymouth, England: Macdonald and Evans / The Metals Society.

Thomas, Dave. "The Roswell Incident and Project Mogul": *The Skeptical Inquirer*. July/August 1995 (available from http://www.csicop.org/si/9507/roswell.html).

Urbina, C., De la Flor, S., Ferrando, F.. "R-phase influence on different two-way shape memory training methods in NiTi shape memory alloys": *Journal of Alloys and Compounds*. February 4, 2010, Volume 490, Issues 1-2, pp.499-507.

Wang, Frederick E. *On the TiNi (Nitinol) Martensitic Transition Part 1*. January 1, 1972. U.S. Naval Ordnance Laboratory, White Oak, Maryland, USA.

Weaver, R. L. & McAndrews, J. September 8, 1994, *Executive Summary: Report of Air Force Research Regarding the 'Roswell Incident'*. Washington, D.C.: Government Printing Office.

Weaver, R. L. & McAndrews, J. July 1995, *The Roswell Report: Fact or Fiction in the New Mexico Desert*. Washington, D.C.: Government Printing Office.

Weaver, R. L. & McAndrews, J. July 1997, *The Roswell Report: Case Closed.* Washington, D.C.: Government Printing Office.

Ye, Y. Y., Chan, C. T., Ho, K. M. "Structural and Electronic Properties of the Martensitic Alloys TiNi, TiPd, and TiPt": *Physical Review.* August 15, 1997, Volume 56. Issue 7, pp.3678-3689.

Yenne, Bill. 1997, *UFO: Evaluating the Evidence.* Rowayton, CT (USA): Saraband Inc.

Yurko, G. A., Barton, J. W., and Parr, J. Gordan. "The Crystal Structure of Ti2Ni": *Acta Crystallographica.* March 23, 1959, Volume 12, pp.909-911.

Zarinejad, Mehrdad. & Liu, Yong. "Dependence of Transformation Temperatures of NiTi-based Shape-Memory Alloys on the Number and Concentration of Valence Electrons": *Advanced Functional Materials.* September 23, 2008, Volume 18, Issue 18, pp.2789-2794.

NOTES

Chapter 1

1   *The Advertiser* , a conservative daily newspaper published in the city of Adelaide, Australia, gathered the latest information about the reward being offered in the U.S. The newspaper article reported on July 9, 1947 how Mr. J. Culligan, president of an Illinois company put forward a reward of US$1,000 "for the capture of a flying disc if it is tangible, or a true explanation of the phenomena". According to *The Milwaukee Journal*, dated July 8, 1947, a further $2,000 was also offered by "the Spokane Athletic Round Table, a group of gagsters, and the Los Angeles world inventors' exposition".

2   From an interview conducted by investigators with Colonel Thomas J. DuBose.

3   It is the nature of people in the military to keep quiet when ordered to do so as part of their job— at least in the case of Marcel, Cavitt and others at the Roswell AAF. However, oddly enough, we learned the rancher was also forced to keep quiet after he was tracked down and interrogated by the military over some issue about whether he discovered or saw something else on his property. This is unusual unless the USAF had discovered something extraordinary and highly sensitive. Or did the military realize the wreckage was part of a highly secret military experiment with the potential of casualties being present? Or was it merely the fact that the object contained weapons of mass destruction? Or was it a new type of object that crossed the borders of the United States from another nation without detection that was of greatest concern for the USAF at the time?

4   Some of those events were kept quiet because of pressure from higher military headquarters to maintain the status quo or risk punishment, as well as a loss in the radio license and employee's position.

5   Quote from the GAO web site at http://www.gao.gov/about/. GAO is also called the General Accounting Office.

6   It is almost as if the USAF was confirming the Roswell object was a disc. So why use Mogul balloons as the official explanation for the Roswell incident? Clearly these balloons are not disc-shaped. If anyone would know the shape of the balloon and reflector in a Mogul balloon, it would have to be Dr. Charles B. Moore, the scientist who formerly worked on Project Mogul. In fact, in an official picture of him holding up the reflector of a Mogul balloon as his best recollection of what he observed, we can see it is not a disc. Or is there another more secret Mogul balloon experiment with an unclassified reflector component shaped like a "disc"? If so, it must be one type of balloon that Dr. Moore does not know about. Or is the USAF trying to say Project Mogul is a cover for yet another secret project (perhaps the object was really an experimental aircraft or missile of some type) conducted near Roswell that they don't want to talk about? If so, there is something else associated with this object that the USAF seems intent in keeping quiet at all costs (i.e., the "victims").

7   Personal statement made at his web site at http://www.v-j-enterprises.com/jbond.html.

8   As revealed by the blowtorch, cutting and sledgehammer blow tests performed by Roswell AAF personnel and other witnesses. Quotes supporting this will be revealed in Chapters 2 and 3.

9   USAF *Executive Summary of the Report of Air Force Research Regarding the "Roswell Incident"*, July 1994, p.4.

10   As we shall see in Chapters 2 and 3, the USAF have a reputation of threatening civilian and military witnesses to keep quiet, which would explain why this observation did not come out earlier. As the years go by, and with many retired and nearing the end of their lives, more and more witnesses have decided to reveal details to investigators, family members, or close friends. The only complicating factor in this revealing of their experiences is evidence of some false witnesses emerging in the 1980s and 90s for fame, money, or the possibility they could be helping the USAF to maintain the secrecy. We will look at a couple of examples in Chapter 2, and possibly another in Chapter 8.

Chapter 2

1   *Roswell Daily Record*: "RAAF Captures Flying Saucer on Ranch in Roswell Region", July 8, 1947, p.1.

2   Son of William Ware Brazel Sr. (1873 – 1954) and Anne Jane Wiggins (1875 – 1959).

3   A check of weather reports have confirmed there were thunderstorms in late June and early July, specifically on 2 and 4 July, but not during the first three weeks of June. Together with the news release of July 8 from the Roswell AAF public information officer claiming the "flying disc" was picked up "sometime last week", the night of July 2 for the crash when Brazel heard an odd explosion has to be considered most reasonable.

4   Berlitz & Moore 1988, pp.85-86.

5   Formerly named Maggie Mae Wilson (1902-1975).

6   Within an hour according to witnesses' statements.

7   The words "burst" and "dirty stainless steel" were the words allegedly used by Barnett, according to his close friend L. W. "Vern" Maltais (taken from Randle & Schmitt 1994, p.149 and Randle & Schmitt 1991, p.90).

8   This may indicate the possibility that the victims may have had no male reproductive organs or other means of indicating gender. Either Barnett was looking at an all female crew with no breasts, all males with no male reproductive organs, or a combination of the two. However, nothing would indicate to Barnett that he saw plastic dummies. The length of time spent by all the witnesses in observing the bodies would suggest these victim were real biological entities. We will discuss in Chapter 10 a possible explanation for this observation assuming all these bodies are real individuals.

9   Beginning in 1991, UFO investigators and authors Kevin Randle and Donald Schmitt uncovered new witnesses and gathered further information suggesting the second crash site may be much closer to the initial debris field than originally thought. The distance to this second crash site is allegedly 40 miles northwest of Roswell. Since then, an Air Force veteran with former top-secret crypto-clearance, Thomas J. Carey, has teamed up with Schmitt to present a more convincing case for the new second crash site. Among the reasons stated for supporting the new location, the suggestion that the Plains of San Agustin is too far away to be of relevance and Barnett's testimony is at best second-hand information from his closest friends and his boss. Furthermore, authors Berlitz and Moore later told they were not sure on the precise date for when Barnett saw the object other than stating to his friends that it occurred sometime in the "summer of 1947". Berlitz and Moore assumed it was related to the initial Roswell debris site. For the purposes of this book, we shall stick to the original second location as published by Berlitz and Moore since Barnett managed to observe a dark-grey or "dirty stainless steel" metallic object for the outer skin which is consistent with the dark-grey metallic foil pieces found on the local rancher's property. Indeed, the new witnesses who have appeared after 1991 are not yet claiming they have seen anything dark-grey in the metal of the symmetrical disc allegedly found at the new second location (now described as bright silvery-colored like aluminium foil). Does this mean there is another object that crashed? Or are some bogus witnesses pushing the case for another crash site?

10   Hasemann & Mantle 1997, p.242.

11   Berlitz & Moore 1988, pp.92-93.

12   Berlitz & Moore 1988, p.93.

13   Quote from Karl Pflock, *Roswell in Perspective*, 1994 and http://roswellproof.homestead.com/Proctor.html.

14   Hasemann & Mantle 1997, p.16. The Proctors understood a reward had been posted by a newspaper for the recovery of a flying saucer or parts of one as the public began to talk about the phenomenon sweeping the nation at this time.

15   Could this foil with its slightly darker on one side than the other be a composite of two shape-memory alloys to help retain its shape at different temperatures (i.e. very cold and very hot)?

16   The account of what happened on Sunday morning appears to be changing in more recent times. Now it is thought that Brazel had spotted birds of prey circling off in the distance just after 6:00 AM. As he proceeded to investigate, he noticed a strange odor in the air. Nearly 2.5 miles east of the Hines Draw debris field, he dismounted and ascended the bluff. It is alleged at this point that Brazel had found three small humanoid bodies. They were all dead. However, original investigations by Berlitz and Moore do not confirm this account. Could more recent witnesses be confused, or providing false accounts of what happened (possibly as a ploy to keep the curious away from the main crash sites in the Plains of San Agustin and on the Foster's ranch)? Or were there additional bodies found later by the USAF (possibly during the recovery operation)?

17   Berlitz & Moore 1988, p.69

18   *The World's Greatest UFO & Alien Encounters: Conspiracies, Abductions, Little green men...Encounters beyond belief* 2002, p.142.

19   *The World's Greatest UFO & Alien Encounters: Conspiracies, Abductions, Little green men...Encounters beyond belief* 2002, p.142.

20   *The World's Greatest UFO & Alien Encounters: Conspiracies, Abductions, Little green men…Encounters beyond belief* 2002, p.142.

21   A copy of Marcel Jr. affidavit is available from the SUNRISE web site.

22   *The World's Greatest UFO & Alien Encounters: Conspiracies, Abductions, Little green men…Encounters beyond belief* 2002, p.142.

23   http://www.theroswelllegacy.com/ (Marcel, Jr.'s quote as of October 2008).

24   Randle & Schmitt 1991, pp.53-54.

25   Carey & Schmitt 2009, p.207.

26   Hesemann & Mantle 1997, p.23.

27   Highway #285 right up beyond the Corona road was also cordoned off by the USAF.

28   Carey & Schmitt 2009, p.42. Among the witnesses named were Leonard "Pete" Porter, Bill Jenkins and L.D. Sparks, all local ranchers in the area.

29   *Roswell Daily Record*: "Harassed Rancher who located 'Saucer' Story sorry he told about it", July 9, 1947.

30   It has been claimed in later investigations after 1991 that three alien bodies had been allegedly recovered on the Foster Ranch not far from the initial debris field. As of 2009, this theory has been expanded to include an egg-shaped bright-silvery aluminium metallic object hidden under a canvas tarp being transported on a truck by the USAF to Roswell AAF (and passed through the main street in the city of Roswell). While there exists a photograph of an egg-shaped bright-silvery object covered with a tarp and sitting on a trailer in the middle of Roswell, it is believed any such recovery of bodies and the object by the USAF would have taken place at the same time as the initial debris was being cleared. It has been rumored one of the victims was found alive and had allegedly survived long enough to be escorted into a field ambulance and brought to Hangar P-3 (or Building 84 as it is known today). Such a claim has been the subject of some fake photographs depicting an alien standing between two U.S. military officials as if someone wanted to let the public know there may be some truth to the claim. As for the other occupants, one was found dead and badly damaged, while the third was dead but showed no obvious signs of damage. The location for these bodies were thought to be only a few miles from the initial debris field.

31   *Roswell Daily Record*: "Harassed Rancher who located 'Saucer' Story sorry he told about it", July 9, 1947.

32   *Roswell Daily Record*: "Harassed Rancher who located 'Saucer' Story sorry he told about it", July 9, 1947.

33   Are there more missing pieces waiting to be discovered in the New Mexico desert? Or is the military worried just the tiniest fragment of this stuff could get found and analyzed by scientists and so make it harder for the military to cover up their tracks?

34   Quote from Lydia Sleppy as she heard it from Johnny McBoyle from Berlitz and Moore "Berkley Edition" November 1988, p.15-16.

35   The alleged statement from McMullen to concoct the story of a weather balloon comes from the affidavit of Brigadier General Thomas J. DuBose (1903-1992).

36   Quote from his personal web site at http://www.v-j-enterprises.com/jbond.html obtained in 2008.

37   A transcript of the taped interview was published in *International UFO Reporter*, November/December 1990.

38   Photographs marked with a UTA number are courtesy of the Fort Worth Star-Telegram Photograph Collection, Special Collections at the University of Texas Arlington Library. Photographs not in the UTA collection but taken by Mr. Johnson are owned by UTA. However, since the original plates are missing, they must be considered in the public domain. Photographs with a UTA number showing Major Jesse A. Marcel with the "substitute" wreckage were actually taken by a U.S. government employee by the name of Major Charles A. Cashon, and, therefore, should be considered in the public domain.

39   Quote from his personal web site at http://www.v-j-enterprises.com/jbond.html.

40   Quote from http://en.wikipedia.org/wiki/Witness_accounts_of_the_Roswell_UFO_incident.

41   Quote from his personal web site at http://www.v-j-enterprises.com/jbond.html. Karl Pflock names Charles A. Cashon as the USAF photographer who took the picture of Irving Newton.

42   Quote from his personal web site at http://www.v-j-enterprises.com/jbond.html.

43   Quote from his personal web site at http://www.v-j-enterprises.com/jbond.html.

44   Major George A. Filer has also published a book titled *Filer's Files: Worldwide Reports of UFO Sightings*, co-authored by David E. Twichell

45   Just before his death, Mr. Johnson was using the email address of jbonjo@aol.com (originating IP address 205.188.144.207) obtained from his personal web site. He lived at Long Beach, Los Angeles in California.

46	The World's Greatest UFO & Alien Encounters: Conspiracies, Adbuctions, Little green men… Encounters beyond belief 2002, p.144.
47	Carey & Scmitt 2009, p.81.
48	Major Jesse Marcel was released in the evening of July 9, 1947, returning to Roswell in a B-29. Captain Frederick Ewing was the pilot in-charge and could confirm Marcel was a passenger on his flight.
49	George Filer's originating IP address at the time was 64.12.138.207.
50	*The World's Greatest UFO & Alien Encounters: Conspiracies, Adbuctions, Little green men…Encounters beyond belief* 2002, p.144.
51	Further details about the inconsistency in the size of the weather balloon when comparing the area of the debris had allegedly covered and what was shown in Ramey's office can be found at http://www.theblackvault.com/wiki/index.php/Roswell_UFO_incident.
52	See Appendix D; Hesemann & Mantle 1997, p.58.
53	*USAF Executive Summary: Report of Air Force Research Regarding the 'Roswell Incident',* September 8, 1994, p.14. *The Roswell Report* 1995, p.21.
54	See Appendix C.
55	*USAF Executive Summary: Report of Air Force Research Regarding the 'Roswell Incident',* September 8, 1994, pp.21-22. *The Roswell Report* 1995, pp.29-30.
56	Burleson 2000, p.22.
57	Photos courtesy of University of Texas Archives (UTA).
58	See Appendix B.
59	Teletype saved by radio announcer Frank Joyce of Roswell KGFL. Text is reproduced at http://www.roswellproof.com/Alamogordo_July9a.html.
60	See Appendix A.
61	http://www.phils.com.au/rosletters.htm.
62	See Appendix A; Hough & Randles 1994, p.71. All references to Major Curton in this FBI teletype refer to Major Edwin Kirton.
63	Good 1997, p.464.
64	Good 1997, p.464.
65	Missler & Eastman 1997, p.58.
66	Randle, Kevin. 1995, *Roswell UFO Crash Update.* Transcript of interview, June 18, 1990.
67	*Roswell UFO Crash Update*; Kevin Randle, 1995; transcript of interview, June 18, 1990.
68	Tambling 1978, p.24.
69	http://www.presidentialufo.com/harrys.htm (now available at http://www.presidentialufo.com/harry-s-truman/64-president-harry-s-truman).
70	http://www.presidentialufo.com/harrys.htm (now available at http://www.presidentialufo.com/harry-s-truman/64-president-harry-s-truman).

Chapter 3

1	Details of the encounter with Cavitt is revealed in Carey & Schmitt 2009, p.205. Also included are statements from Cavitt's wife, Mary, who claimed to have overheard conversations between his husband and Marcel (several days after returning to Roswell from Ramey's office). She claims the men were talking about applying as much heat as possible to something. The two were meant to be playing their weekly game of Bridge, but somehow ended up talking about their work. The men were sitting on the patio when this happened. Mary recalled how the situation ended when she said, "Cav [her husband] reminded Jess [Marcel] that the material was classified "top-secret" and he had better get rid of it". The men came back into the house and never mentioned the material Marcel had or what they were doing to anyone (p.82).
2	Affidavit, May 5, 1991.
3	Interview July 1990; quote from Stanton Friedman and Don Berliner, *Crash at Corona*, 1991.
4	Quote appeared in Kevin Randle and Don Schmitt, *UFO Crash at Roswell*, 1991.
5	Quote from Stanton Friedman and Don Berliner, *Crash at Corona*, 1991.
6	Quote from Stanton Friedman and Don Berliner, *Crash at Corona*, 1991.
7	Interview December 1979; quote from Charles Berlitz and William Moore, *The Roswell Incident,* 1980,
8	Interviews February, May, December 1979 with William Moore and Stanton Friedman; quote from Charles Berlitz and William Moore, *The Roswell Incident,* 1980.
9	Quote from Stanton Friedman and Don Berliner, *Crash at Corona*, 1991.
10	Quote from Stanton Friedman and Don Berliner, *Crash at Corona*, 1991.
11	Affidavit, May 6, 1991.

12	Quote from the interview televised in the documentary *The Roswell Events*, ed. Fred Whiting, sponsored by Fund for UFO Research, 1991; quoted in the *1994 USAF Report on Roswell*. Also quoted in Michael Hesseman and Philip Mantle, *Beyond Roswell*, 1997.
13	Quote from Berlitz & William 1980 edition of *The Roswell Incident*.
14	Berlitz & William 1980 edition of *The Roswell Incident*.
15	Affidavit, May 6, 1991.
16	Karl Pflock, *Roswell in Perspective*, 1994 or *Roswell: Inconvenient Facts and the Will to Believe*, 2001.
17	Berlitz & William 1980 edition of *The Roswell Incident*.
18	Interviewed June 1979; Quote from Berlitz & William 1980 edition of *The Roswell Incident*.
19	Berlitz & Moore 1988, p.72 & 73.
20	Hesemann & Mantle 1997, p.19.
21	Korff 1997, p.33 ('after a good rain').
22	Berlitz & Moore 1988, p.88.
23	Bill's statement originally obtained from Christopher Schmidt of the Northeastern University in Boston, Massachusetts, USA (cschmidt@lynx.dac.northeastern.edu). The quote also appears in Randle & Schmitt 1994, p.138 and 1991, p.132 but worded slightly differently. Remaining part of quote from Bill Brazel obtained from Kevin Randle and Don Schmitt, *UFO Crash at Roswell*, 1991.
24	Berlitz & Moore *The Roswell Incident* 1980.
25	Hasemann & Mantle 1997, pp.30-31.
26	Christopher Schmidt of the Northeastern University in Boston, Massachusetts, USA (cschmidt@lynx.dac.northeastern.edu), Hasemman & Mantle 1997, p.31, and Friedman & Berliner 1992.
27	Quote from the video documentary *UFOs, A Need to Know*, 1991.
28	Quote from the video documentary *UFOs, A Need to Know*, 1991.
29	Quote from Kevin Randle and Don Schmitt, *UFO Crash at Roswell*, 1991.
30	Quote from Robert Shirkey, *Roswell 1947: "I Was There"*, 1999, pp.94-95.
31	Randle & Schmitt 1991, p.85.
32	This statement obtained from Randle & Schmitt 1991, p.84.
33	This statement obtained from Christopher Schmidt of the Northeastern University in Boston, Massachusetts, USA (cschmidt@lynx.dac.northeastern.edu). The statements from Brigadier General Arthur E. Exon can also be found in Hesemann & Mantle 1997, pp.59-60.
34	Quotes from Brigadier General Arthur E. Exon obtained from *UFO Crash at Roswell*, 1991 & *The Truth About the UFO Crash at Roswell*, 1994, by Kevin Randle & Donald Schmitt (Based on phone and personal interviews from July 1989 - July 1990). Also in *Roswell UFO Crash Update*; Kevin Randle, 1995; transcript of interview, June 18, 1990.
35	*Roswell UFO Crash Update*; Kevin Randle, 1995; transcript of interview, June 18, 1990.
36	Brigadier General Arthur E. Exon's quote on bodies recovered from another location is obtained from *UFO Crash at Roswell*, 1991 & *The Truth About the UFO Crash at Roswell*, 1994, by Kevin Randle & Donald Schmitt.
37	Quote from Stanton Friedman and Don Berliner, *Crash at Corona*, 1991.
38	Combined from Stanton & Berliner's book and http://specialenquirer.informe.com/forum/viewtopic.php?f=21&t=14&start=0&st=0&sk=t&sd=a&view=print.
39	Quote from Stanton Friedman and Don Berliner, *Crash at Corona*, 1991.
40	Good 1997, p.474.
41	Friedman & Berliner 1992, p.85.
42	Good 1997, p.456.
43	Good 1997, p.463.
44	Carey & Schmitt 2009, p.207.

Chapter 4

1	Claiborne 1994, p.A22.
2	*Executive Summary of the Roswell Report*, September 8, 1994, p.2.
3	A copy of the memo from the Inspector General is available in Attachment 3 of Colonel Weaver's full Roswell Report.
4	A copy of this letter is included in the Laurance Rockefeller UFO Disclosure Initiative.
5	Quote from http://www.nicap.org/roswell/big_names_at_roswell.htm as of October 2016.

6   The report was downloadable at http://www.roswell.org/gao.html#report. Alternative link as of September 2016 is https://www.fas.org/sgp/othergov/roswell.html, or check the SUNRISE web site.

7   Good 1997, p.480.

8   Good 1997, p.480.

9   The title of the report changed to *The Roswell Report: Case Closed*. The title is interesting as it suggests the USAF is not expecting new evidence to become available in the future to question any aspect of this report. It is almost as if the military does not want anyone to look at the Roswell case anymore.

10   Saler et al. 1997, p.78.

11   *Executive Summary of the Roswell Report*, September 8, 1994, p.6. *The Roswell Report* 1995, p.14.

Chapter 6

1   In medicine, a composition range of 50.6 to 51.0 atomic % for nickel is particularly useful since this reveals an important superelastic phase in nitinol. By getting nitinol to be highly flexible, the alloy can be made into very thin wires and packaged into a tiny space where it can pass through blood vessels and later deployed in the right areas of the human body before the shape response can be automatically activated to open up blocked vessels, or to achieve other objectives as required.

2   Moitra et al. 2011, p.821.

3   NiTi gets physically harder over time with continuous cold-working of the alloy, and hardness is an indicator of the closeness of the atoms due to very short bond lengths. Therefore, those shorter dimensions measured for the cubic B2 structure are likely to be due to the scientists playing around with the alloy by bending it regularly and testing the shape-memory effect prior to using x-ray diffraction to measure the length between the atoms. In such cases, the maximum length should be used. For a conservative estimate, we use the figure of 3.015Å from Sengupta et al. 2009.

4   X-ray diffraction is a technique of casting shadows on a 2-dimensional film to help scientists determine the position of atoms in 3-dimensions and thus determine the crystalline shape of the solid.

5   These figures are the maximum values from a list of published values in scientific literature. Appendix I shows these published values and the maximum. Variability in these figures from other scientists probably depends on the accuracy of x-ray diffraction equipment, composition ratio of the elements (for example, Matsumoto et al. 1987, ,p.2137 used a sample of Ni-Ti(49.8 at. % in their study), and whether the alloy has been cold worked and/or gone through its thermal cycling for an unspecified number of times just prior to determining the crystalline structure and dimensions.

      As Saitoh et al. stated in their article, "Atomic Dynamics and Energetics of Martensitic Transformation in Nickel-Titanium Shape Memory Alloy", pages 742-743:

> "By extensive experimental study on identification of crystalline structure of B19′ structure, its detailed lattice parameters (unit cell length and monoclinic angle) have been tried to be determined. The exact parameters, however, still have relatively broad ranges."

6   Or with minor additions of certain ":compatible" elements that don't destroy the shape-memory effect (such as copper, as suggested by Liangliang et al. 2015, p.2222) to form an alloy called NiTi-X, there can be a straight transition from the high-temperature B2 cubic structure to the low-temperature B19 orthorhombic structure (i.e., no monoclinic structure or R-phase). For example, Liangliang et al. wrote that, in the alloy NiTi-Cu:

> "TiNiCu alloys with Cu content of less than 7.5 at. % are known to show the B2→B19′ transformation. For the alloys with Cu content between 7.5 at. % and 15 at. %, the B2→B19→B19′ transformation takes place, and for Cu ≥ 15 at. %, there is a direct B2→B19 transformation."

      How the addition of copper (or other elements) affect the martensite structure is yet to be fully elucidated. As Lianliang et al. put it:

"…the mechanism of Cu addition in modifying the martensite crystal structure is not clearly understood."

Part of the latest scientific research into shape-memory alloys today is aimed at understanding these changes to the martensite crystalline structure by different elements, so that one day the original crystalline structures of NiTi at different temperatures can be fully explained and, with it, the secret of the shape-memory effect.

7   Named after German metallographer Adolf Karl Gottfried Martens (1850–1914)

8   The alloy has to be freshly made to reveal this relatively soft and pliable physical characteristic in the cool state. As soon as the alloy is bent and twisted a few dozen times (in a process called *cold-working*), the alloy hardens, making it increasingly more difficult to bend or put a dent in it. Increased hardness is also achieved by activating the shape-memory response as this will create a highly compact crystalline structure. Other ways to harden NiTi include the process of *carburizing*, which is a method of adding small amounts of carbon to the surface of the alloy while heat treating it.

9   This is the moment when the alloy transforms enough of the monoclinic B19' crystalline unit cell structures to the slightly more compact and rigid orthorhombic B19 crystalline unit cell structures, and eventually returning to an even more compact cubic B2 crystalline structure of the austenite phase.

10   The martensite phase for NiTi also allows the alloy to be cut and pierced with a sharp instrument while in this softer, more pliable phase.

11   Named after English metallurgist William Chandler Austen (1843–1902).

12   *Wikipedia*, September 2016. https://en.wikipedia.org/wiki/R-Phase.

13   Holec et al. 2014, p.1.

14   Ling & Kaplow 1981, p.2101.

15   Brachet et al. 1997, p.1.

16   Duerig & Bhattacharya 2015, p.3.

17   Brachet et al. 1997, p.1

18   Urbina et al. 2010. Available from https://www.researchgate.net/publication/223047645_R-phase_influence_on_different_two-way_shape_memory_training_methods_in_NiTi_shape_memory_alloys.

19   The R-phase is seen more as an annoyance for engineers than anything else. Generally, the more the R-phase appears, the larger the austenite-martensite hysteresis, and this can reduce the precision of the transformation temperature M_s as needed to build highly efficient and precise thermal actuators using NiTi. With further experimentation, scientists have learned that additions of copper (Cu), platinum (Pt) and palladium (Pd) will suppress the R-phase. Additions of iron, cobalt and chromium will make the R-phase more pronounced (but will increase the superelastic temperature range for NiTi). Otherwise, continual cold-working and aging of NiTi will bring out more of the R-phase.

20   William J. Buehler and G. Rozner "High-Strength Nickel-Base Alloy", U.S. Patent No. 3,351,463. Publication Date: November 7, 1967.

21   It also means that as soon as you remove the source of the heat for getting it into the austenite crystalline structure, the alloy can feel unusually cool to the touch despite just a few seconds ago applying heat to do its transformation.

22   T. W. Duerig and K. Bhattacharya. "The Influence of the R-Phase on the Superelastic Behavior of NiTi": ASM International. May 16, 2015. Available from https://link.springer.com/article/10.1007/s40830-015-0013-4.

23   Originally obtained from http://www.billhammack.com/comm/4265.htm. Now available at http://www.engineerguy.com/comm/4265.htm as of September 2016.

24   The alloy will feel unusually cool to the touch when it springs back rapidly into its original shape. Applying a lot of heat from a blow torch and removing the heat may surprise some people in how quickly the alloy feels cool to the touch (usually within a few seconds). This is an observation considered unique to SMAs.

25   The strain limits do change over time as the alloy hardens. If the same strain or stress is applied to nitinol without adjusting for the changing and diminishing range of the strain limits, there is an increased risk the alloy will break. Reducing the imperfections and impurities in the crystalline structure can minimise this problem. However, keep within the strain limits, and there is absolutely no reason why the alloy cannot continue its shape-memory response for as long as it is required.

26   This is the temperature at which the Roswell foil was known to have exhibited its shape-memory effect. A graph supporting nitinol's room temperature activation is provided in this chapter. Other observations also suggest the possibility that the foil could have exhibited a superelastic response below body temperature.

27   Nespoli et al. 2012, p.38.

28   If one reduces the impurities in nitinol even further, the voltage needed to activate the shape-memory effect will be considerably less. You won't need more than three volts and with minimal electrical current for this to happen. The energy efficiency and dramatic shape-memory response in nitinol simply improves the purer it is (and the closer it is to perfection in its crystalline structure).

29   http://www.patentstorm.us/patents/4564395/fulltext.html.

30   A classic test of just how hard this material can get is as follows: if you dropped an amorphous alloy shaped as a ball onto another equally hard flat surface, the ball would continually bounce up and down as if it is a perfectly elastic material. The ball does not absorb the energy from the flat surface and dissipate this energy to the surroundings at a lower frequency (primarily as heat and light) to help diminish the height of the ball with each bounce. Rather, the energy is returned to the flat surface virtually exactly as it was received. In a sense, you can call the ball a shape-memory material because of its apparent perfect elasticity (even though the ball does not actually bend or change shape as a result of the collision with the flat surface). Scientists have created a term called the *coefficient of restitution* calculated using the formula $(H_1/H_2)^{1/2}$, where H_1 is the initial height of the ball when it was dropped, and H_2 is the maximum height reached on the first bounce. Using this formula, we can say an amorphous alloy ball has a very high coefficient of restitution compared to say a stainless steel or titanium ball. The formula is intended to help us make comparisons with other materials.

31   Gong et al. p.405. Also reaffirmed in Mohri et al. 2015, p.75.

32   Zarinejad & Liu 2008, p.2792.

33   The reason why electrons don't just fall in and crash into the positively charged nuclei is because the nuclei itself is spinning at a high rate. According to the laws of electromagnetism, a spinning positively charged object emits synchrotron radiation. In quantum mechanics, we call this a probability wave (which is a mathematical way of saying that the electrons are more likely to be found within certain regions). It is this wave of pure electromagnetic energy and the gravitational property it possesses (see Einstein's Unified Field Theory) that effectively controls the position of the electrons in specific "orbits" (which may also be described as shells, or orbitals since the region is not infinitely sharp). In addition to this, the waves do naturally redshift as it moves away from the nuclei. So with each successive cycle of the wave, the amplitude is reduced. Where the amplitude of the wave is at its maximum, a certain number of electrons are held in those regions. If the amplitude of the radiation is diminished, the electrons can still be held in position, but more weakly. In other words, less energy is required to remove the electrons from this region. This is why the outermost electrons are the easiest to remove compared to those located closer to the nuclei.

34   Zarinejad & Liu 2008, p.2791.

35   Zarinejad & Liu 2008, p.2792.

36   How well it does this probably depends on the level of impurities and microdefects in the crystalline structure.

37   If the area is large enough and maintained at the right temperature range, we have what is called a superelastic phase where two nickel atoms can slip in and out of the titanium "cage" with relative ease.

38   Zarinejad & Liu 2008, p.2789.

39   Zarinejad & Liu 2008, p.2789.

40   This current scientific situation with SMAs would have important implications for the USAF in its own study of SMAs. As we shall discover in the chapter on the history of SMAs from the U.S. military perspective, it is clear the USAF at Wright-Patterson AFB had been looking at NiTi and at least a couple of NiTi-X alloys at the same time just after 1947. Given the fact that the Roswell witnesses were certain that they had seen a shape-memory material of metallic appearance and with enough military witnesses confirming it was a metal (or more specifically an alloy) of some sort in July 1947, along with the way this material ended up at Wright-Patterson AFB for analysis, it is highly unlikely the USAF did not know anything about SMAs. It must have known something about SMAs. However, if scientists today are struggling to figure out how SMAs work, it is highly unlikely the scientists at Battelle could have figured it out in 1947. And, by implication, the USAF would have no hope of figuring it out either since it was the military that asked Battelle for assistance after 1947 to study titanium-based alloys, of which NiTi, TiZr and a couple of NiTi-X alloys were definitely on the radar for the USAF for some reason. In fact, the presence of two NiTi-X alloys would strongly suggest that the USAF was also trying to understand how SMAs work. But if scientists studying NiTi-X today still haven't found an explanation for the shape-memory effect, and with no evidence the USAF is using NiTi or NiTi-X in the construction of fighter jets in vast quantities to show that it had stumbled on the shape-memory effect by accident and could manufacture it with relative ease, we can reasonably surmise from this that the USAF has still not

figured out how SMAs work. However, we have a problem. If the USAF did conduct a secret military experiment involving exotic new materials in early July 1947, how did the Roswell foil get manufactured to show its shape-memory effect? We know it has to be made of titanium given the incredible toughness of the foil and ability to resist the high temperatures of a blow torch. Therefore, if the USAF did have a brief moment of creativity in its long history to randomly combine elements with titanium to help it discover a shape-memory effect in a super tough alloy, then the technology available to manufacture it with sufficient purity and quantity was just not available in 1947. It means the USAF could not have combined elements randomly or even as a planned event if a theory of SMAs had been developed by the USAF by early 1947. If this is true, it will have important implications for the type of object that crashed in 1947 and who actually manufactured it.

41   Holec et al. 2014, p.1.
42   Fernandes et al. 2011. Available from https://www.ncbi.nlm.nih.gov/pmc/articles/PMC3185255/#B38.
43   Please note that at this "room temperature" composition ratio, there is no need to apply more heat to the alloy to activate its "shape-memory" response. There is enough heat in the environment to activate it.
44   Sengupta & Papadopoulos 2009, p.3214.
45   Horzewski & Giba 1998, U.S. Patent No. 5730741 A: "Guided spiral catheter".
46   Duerig & Pelton 1994, p.1043.
47   Duerig & Pelton 1994, p.1036.
48   Duerig & Pelton 1994, p.1042.
49   Duerig & Pelton 1994, p.1036.
50   Robert M. Abrams and Sepehr Fariabi (Abbott Cardiovascular Systems Inc.). U.S. Patent No. 7244319: "Superelastic guiding member". Publication date: July 17, 2007.
51   Duerig & Pelton 1994, p.1037. Indeed, much of the work into SMAs nowadays is focused on finding NiTi-X alloys with a small hysteresis. It is driven by the commercial demand for precise control of the transformation temperature and seeing the shape-memory effect occur rapidly within a very small temperature range. Those alloys having a "narrow hysteresis" are mainly used in devices called *actuators*.
52   Liangliang et al. 2015, p.2223. At 20 at.% of copper in NiTi-Cu, not only is the hysteresis very narrow, but also the monoclinic B19' crystalline structure changes to the orthorhombic B19 crystalline structure. This means that the transformation from austenite to martensite is a straight B2 to B19 with no monoclinic structure appearing anywhere. Also, copper atoms have similar atomic radii to nickel, which may explain why the shape-memory effect of NiTi is not destroyed by the small addition of copper.
53   Otsuka & Ren 1998, p.515.
54   Ramaiah et al. 2014, p.78. Article is also available online at http://www.sciencedirect.com/science/article/pii/S0261306913010224.
55   Melton et al. 1989, p.165.
56   Sehitoglu et al. 2004, p.3.
57   Sehitoglu et al. 2005, p.5.
58   Brachet et al. 1997, p.1.
59   https://en.wikipedia.org/wiki/R-Phase
60   Fernandes et al. 2011, p.4. The concept of atomic size and its effect on thermal hysteresis is mentioned in Zarinedad & Liu 2008, p.2794.
61   Similar sizes of the copper and nickel atoms are mentioned by Liangliang et al. 2015, p.2233.
62   Ronald Dean Noebe, Susan L. Draper, Michael V. Nathal, and Edwin A. Crombie (NASA). U.S. Patent No. 7749341: "Precipitation hardenable high temperature shape memory alloy". Publication date: July 6, 2010.
63   Duerig & Pelton 1994, p.1037.
64   Dmitrii Victorovich Golberg from the Furukawa Electric Co., Ltd. U.S. Patent No. 5641364: "Method of manufacturing high-temperature shape memory alloys". Publication date: June 24, 1997.
65   Frederick E. Wang and William J, Buehler, U.S. Patent No. 3,558,369 "Method of Treating Variable Transition Temperature Alloys".
66   Duerig & Pelton 1994, p.1037.
67   Nespoli et al. 2012, p.35. Available from http://cdn.intechopen.com/pdfs-wm/39975.pdf.
68   Zarinejad & Liu 2008, p.2789.
69   Zarinejad & Liu 2008, pp.2789-2790.
70   Kleinherenbrink & Beyer 1989, p.187.
71   Kleinherenbrink & Beyer 1989, p.187.
72   Support for this claim is available from Chluba 2015, pp.1004-1007. Also available online from http://www.sciencemag.org/content/348/6238/1004.short.

73   Not even Ti-Al with aluminium (Al) being a lighter metal than nickel (Ni) is described as a shape-memory alloy

74   The NiTi article in the 1984 Australian edition of *Omega Science Digest* confirms that nitinol is the most powerful shape-memory alloy.

75   *Encyclopedia of Chemical Technology* Volume 20, p.728 and *Wikipedia* at http://en.wikipedia.org/wiki/Shape_memory_alloy.

Chapter 7

1   *An Encyclopedia of Metallurgy and Materials* 1984, p.291.
2   With cadmium composition at 47.5 atomic percent.
3   http://en.wikipedia.org/wiki/Pseudoelasticity as of July 2011.
4   http://www.chemeurope.com/en/encyclopedia/Shape_memory_alloy.html as of August 2016.
5   http://www.scielo.br/scielo.php?script=sci_arttext&pid=S0100-879X2003000600001 as of May 2015.
6   http://www.explainthatstuff.com/how-shape-memory-works.html as of May 2015.
7   http://smart.tamu.edu/overview/smaintro/simple/definition.html as of August 2016.
8   http://www.oxforddictionaries.com/definition/english/thermo-elastic as of May 2015.
9   http://www.copper.org/publications/newsletters/innovations/1999/07/shape.html as of May 2015.
10   http://cml.postech.ac.kr/z/3750-014.pdf as of May 2015.
11   http://cml.postech.ac.kr/z/3750-014.pdf as of May 2015.
12   http://www.explainthatstuff.com/how-shape-memory-works.html as of May 2015.
13   http://heim.ifi.uio.no/~mes/inf1400/COOL/Robot%20Prosjekt/Flexinol/Shape%20Memory%20Alloys.htm as of May 2015.
14   Ölander did report the elastic behavior at a meeting of the Swedish Metallurgical Society on May 27, 1932, but he never pursued it further.
15   Defined as a measure of the resisting power of a specified material to the flow of an electric current.
16   http://onlinelibrary.wiley.com/doi/10.1002/zaac.19322070103/abstract as of May 2015.
17   Steel is an alloy of iron and carbon. However, it should be noted that the presence of martensite and austenite phases in this alloy does not necessarily prove it is a shape-memory alloy. The only way to know for sure is to change the temperature to permit the martensite phase to exist, deform the alloy, and reverse the temperature to allow the austenite phase to exist. If there is any movement in the alloy during this temperature change, it is a shape-memory alloy. In the case of steel, this is definitely not a shape-memory alloy.
18   Greninger & Mooradian. *Transactions of the Metallurgical Society of AIME (Journal of Metals).* Volume 128 (1938), p.337.
19   Perkins 1974.
20   Chang & Read, *Transactions of the Metallurgical Society of AIME (Journal of Metals).* Volume 189 (1951), pp.47-52.
21   Reece 2007, p.30. The term "twinning" was not specifically mentioned by Chang and Read in their 1951 article.
22   Otsuka & Wayman. 1998, *Shape Memory Alloy.* UK: Cambridge University Press, p.2.
23   Indium-thallium (InTl) alloys in the range 15–38 atomic percent Tl display a shape-memory response.
24   Burkart & Read, "Diffusionless phase changes in the indium-thallium system": *Transactions of the Metallurgical Society of AIME (Journal of Metals).* Volume 197 (1953), pp.1516-1524.
25   Hornbogen & Wassermann. *Z. Metallkunde.* Volume 47 (1956), pp.427-433.
26   Chen. *Transactions of the Metallurgical Society of AIME (Journal of Metals).* Volume 9 (1957), pp.1202-1203.
27   Laves & Wallbaum 1939, p.674.
28   Yurko et al. 1959, p.909.
29   Klaproth 1801, p.210.
30   The company was looking to find a method to manufacture titanium in large quantities to a reasonable purity after discovering the metal's high melting point would make it a useful material for manufacturing filaments inside electric light bulbs.
31   Grayson 1983, p.98.
32   U.S. Patent Number 2,205,854: *Method for Manufacturing Titanium, Alloys Thereof Titanium.*
33   U.S. Patent Number 2,522,679.
34   Crain 1951, Column 107f and U.S. Patent No. 2,522,679.

35	Grayson 1983, p.98.
36	Grayson 1983, p.98.
37	The USAF asked Battelle after 1947 to test this and find ways to improve the purity of titanium and its alloys to a much higher level . In the recently declassified USAF/Battelle First Progress Report dated August 31, 1949, Battelle wanted to test the stability of the titanium tetrachloride structure.
38	*Encyclopedia Britannica* 1988, Volume 11, p.801.
39	The U.S. took control of the *Encyclopedia Britannica* after the 9th Edition (around the 1920s). It was previously a UK reference work.
40	*The Industrial Arts Index*. Annual Cumulation. White Plains, N.Y.: H.W. Wilson, 1913-1957 (period 1946-48).
41	http://metals-history.blogspot.com/2010/04/titanium-new-metal-for-aerospace-age.html as of June 22, 2010.
42	Grayson 1983, p.98.
43	Knittel & Wu, Chapter 6, p.91.
44	Adenstedt & Freeman April 1953, p.1 (PDF p.8).
45	Grayson 1983, p.98.
46	Document was available from http://www.scefiling.org/filingdocs/4910/19561/44779e_Tabx03.pdf.
47	http://www.titaniumeden.com/about_ti.php (3 tons of titanium in 1948).
48	http://www.titaniumeden.com/about_ti.php (3 tons in 1948 and 20,000 tons in the 1980s); http://www.cascadiadesignstudio.com/titanium-history.htm (Over 900 tons or 2,000,000 pounds in 1953).
49	U.S. Patent No. 2,819,194 (January 7, 1958).
50	Crain 1952, Column 1416c.
51	After 1998, Dr. Andrzej Calka moved to the Department of Materials Science and Engineering at the University of Wollongong (e-mail: aoc2107@uow.edu.au or acalka@uow.edu.au).
52	Y. C. Liu also extends his gratitude for "their initiative and criticism during the course of this investigation" as provided by Prof. John P. Nielsen and Harold Margolin in the Acknowledgement section of this article.
53	Prof. John P. Nielsen lived at 1 Washington Square Vlg # 1, New York, NY, 10012-1632 until his death.
54	http://www.nytimes.com/1989/08/15/obituaries/john-p-nielsen-77-professor-ofmetallurgy.html?pagewanted=1.
55	Phillips Laboratory is the old name for the Air Force Research Lab (AFRL), which is currently located at Wright–Patterson AFB.
56	While this may seem like an awful lot of assumptions, it is the best we can do at the present time with the limited information the NYU can provide, and a reasonable one at that.
57	Probably using the latest technology developed by the Battelle Memorial Institute.
58	See Appendix L.
59	Adenstedt & Freeman April 1953, p.1 (PDF p.8).
60	The March 1949 U.S. Navy publication *Titanium—ONR Symposium Report, December 16, 1948* confirms for the existence of a Dr. John P. Nielsen at NYU on page 47 and says that he contributed Paper 17 titled "First Progress Report on Titanium-Carbon and Titanium-Nitrogen Phase Diagrams", pp.153-157, to the symposium.
61	Was Dr. Nielsen trying to find ways to increase the hardness of NiTi by adding carbon (which the other impurities of oxygen and nitrogen could do) just like iron hardens considerably when carbon is added to produce steel? It is possible the USAF (and certainly the U.S. Army did express an interest in this area) was looking into this aspect to help explain one feature of the Roswell foil that other witnesses have noted and mentioned to current researchers—namely the ability to withstand cutting and piercing with a sharp instrument. A hardened alloy is the only way to account for this observation.
62	Ogden & Jaffee 1955, p.1.
63	Ogden & Jaffee 1955, Abstract section.
64	Margolin & Farra 1969, p.329. Also, any titanium alloy with both alpha and beta phases present can potentially move between the alpha and beta phases at the right transformation temperature, meaning it could display a shape-memory effect. Anything that would affect one of these phases, such as the additional of carbon, oxygen, and nitrogen as contaminants, could prevent the shape-memory effect from showing up. Or it could maintain or enhance the effect, but simply adjust the transformation temperature (perhaps to a lower or higher temperature range). As Margolin and Farrar stated:

> "The effects of carbon, oxygen, and nitrogen on the stability of alpha-beta alloys are directly

related to their effects on beta-decomposition reaction kinetics. It is known that interstitial elements dissolved in the beta phase increase the rate of formation of alpha (and omega) from beta, indicating that their presence would be detrimental to stability. This is true for single-phase beta alloys or beta-quenched alpha-beta alloys. However, when alpha-beta alloys are fully annealed (or stabilized), the interstitials partition predominantly to the alpha phase, leaving the beta phase relatively free of interstitials."

65   Ogden & Jaffee 1955, p.20.

66   In fact, titanium nitride (TiN) and titanium carbide (TiC) are noted for being hard materials of which TiC is exceptional in this regard. Due to its similar hardness to tungsten carbide (WC), TiC is often used as a coating for cutting instruments to improve durability and toughness. TiC has the classic NaCl or CsCl-type of cubic crystalline structure, just like the equiatomic NiTi alloy system, which naturally raises the question as to whether Dr. Nielsen was trying to figure out how to harden NiTi with small additions of carbon as part of his work to assist the USAF. Did the USAF discover or predict a particularly hard material would be possible with NiTi-C and wanted Nielsen to investigate this type of alloy? Further details about the physical properties of TiN and TiC can be found at http://www.eifeler.com/portfolio-2/cvd-layer-systems/tintic-titanium-nitride-titanium-carbide/.

67   The Battelle Memorial Institute also looked at adding carbon to the surface of titanium through a process called *carburizing and induction heat treatment*. The final Technical Report covering this work for the period September 1, 1952 to August 31, 1953 can be found under report number NP-5074 and part of contract number DA-33-019-ORD-942. Authors of this report were P. E. Moorhead, A. J. Griest, P. D. Frost, and J. H. Jackson.

68   Report produced under Power Plant Laboratory Contract No. AF 33(038)-8879 (E.O. No. R-506-67).

69   Yurko et al. 1959, p.909.

70   Otsuka & Ren 2005, p.518.

71   Confirmed by Otsuka & Ren 2005, p.518.

72   This would provide further evidence to question whether the USAF could have produced a highly pure titanium-based SMA in 1947. If the phase diagram for NiTi was still not accurate by 1950, how would the USAF know how SMAs work at the fundamental crystalline level? Without an accurate phase diagram for NiTi, it would be next to impossible to predict the existence of a titanium-based SMA before 1950.

73   Otsuka & Ren 2005, p.518.

74   Otsuka & Ren 2005, p.519.

75   Otsuka & Ren 2005, p.519.

76   It is possible this request from the U.S. Army had some ties with other important work being carried out by the USAF at Wright-Patterson AFB concerning the Roswell foil. One of the notable observations made by witnesses of the Roswell foil was its incredible hardness. Some witnesses claimed heavy blows from a sledgehammer would not leave a mark on the surface or leave a permanent dent and cutting with a pair of scissors or piercing it with the tip of the scissors was virtually impossible,. This would require a high level of hardness. But since the U.S. Army was asking about how to make a hard titanium-based alloy to act as a form of armour, it is possible the USAF was also trying to figure out the same problem. Perhaps the USAF knew the situation and wanted the U.S. Army to ask if Nielsen could come up with solutions quickly. The USAF may well have been looking to Dr. Nielsen in the early stages of his study into NiTi for an answer to this "hardness" problem.

77   Perhaps it was realized that data on the phase diagram of NiTi were appearing in the scientific literature, so why not allow Dr. Margolin and his assistant to contribute in this area by publishing their own results without any direct involvement by the USAF? It would be seen as just a natural advancement of scientific knowledge and nothing else. And would avoid any questions being asked about why the USAF was interested in NiTi.

78   Dr. Harold Margolin reviewed a book published in 1982 while working at NYU. The book review was for J. C. Williams and A. F. Belov's, *Proceedings of the Third International Conference on Titanium*.

79   The name of this scientist first cropped up in an article published online by Tony Bragalia in August 2009. He first noted the possibility that his knowledge of chemical bonds in intermetallic materials (i.e., alloys) may have assisted some Battelle staff with their research, some of which were classified at the time, including the work into NiTi. However, Bragalia added sensationalism to his findings when he stated that Pauling had allegedly established a belief in the existence of aliens after secretly studying UFOs in his own time not long after his involvement with Battelle. For example, in his article titled "UFOs and VITAMIN C – Linus Pauling's Flying Saucer Secret" (https://ufocon.blogspot.com/2009/08/ufos-vitamin-c-linus-paulings-flying.html) he wrote:

"...his [Pauling's] belief in ET clearly established..."

On checking these claims, the Special Collections and Archives Research Center at Oregon State University does provide a detailed day-by-day look at the activities of Pauling throughout most of his life based on the numerous memos, letters and books he and his secretary Beatrice Wulf retained throughout his illustrious career as a chemist. Full details of his activities can be found at http://scarc.library.oregonstate.edu/coll/pauling/calendar/1951/01/11.html.

On examining his activities, we do find that the then-director of the Battelle Institute, Clyde Williams, had invited Pauling to give a lecture on February 7, 2016 on a topic of his own choosing. Pauling decided he would talk about the latest work on metals and alloys in terms of his new theory on chemical bonds. The importance of this work can be seen three years later when Pauling was awarded the Nobel Prize in Chemistry for his "research into the nature of the chemical bond and its application in the elucidation of complex substances". Many of the Battelle staff working on various projects at the time were present to listen to Pauling's lecture, and all, including Mr. Williams, expressed their deepest gratitude for his insights.

Beyond that, Dr. Pauling contributed other useful information relating to titanium from a chemical perspective throughout much of the 1950s and early 1960s that could have helped Battelle staff with their own alloy research.

As for claims that Dr. Pauling had secretly studied UFOs, this is true. However, his view appears to be relegated to the area of conspiracy theories or simply natural or man-made phenomena. For example, Dr. Pauling did receive a letter from Gordon L. Jamieson of the Queensland Flying Saucer Research Bureau dated January 2, 1961. Dr. Pauling replied in a letter dated January 11, 1961.

Also, as part of his personal effort to understand the controversial topic that was sweeping his nation in the 1960s, Pauling read *The Scientific Study of Unidentified Flying Objects* (i.e., the Condon Report) commissioned by the USAF and described at the time as "the complete report" on the subject. This, together with another text titled *NASA's Space Science and Applications Program* (in an attempt to better understand spacecraft technology and what could be possible in the future), and two popular UFO books — Brinsley Trench's *The Flying Saucer Story*, and John G. Fuller's *Incident at Exeter* — had effectively formed his complete picture on UFOs at the time. And given the conclusion of Project Blue Book in 1969 and his various copious handwritten question marks and other notes written into Brinsley's and Fuller's books, one can quickly see that he became a skeptic.

Pauling returned to the UFO scene in the 1980s. It was reported that he heard about the Roswell case from Charles Berlitz and William Moore's book and later from other authors. Pauling wrote a letter to the President of the New Mexico Institute of Mining and Technology in which he stated, "There are such difficult cases as the rancher near Roswell, New Mexico...". The letter was found folded in one of the pages of Pauling's personal copy he later acquired of the 1966 book titled *Flying Saucers: Serious Business*.

Otherwise, Pauling's only claim to fame in the Roswell case — if one could call it that — was to impart his expert knowledge of chemical bonds in metals and alloys to the people at Battelle in early 1951, and later his other work relating to titanium, and his reading of books on the Roswell case in the 1980s. Other than this, there are no indications in the archives to suggest that Pauling was directly involved in any secret work relating to the original Roswell materials at the present time.

80   Originally published in French as "La Valence des métaux et la structure des composés intermétalliques" (translated as "The valence of metals and the structure of intermetallic compounds") in Volume 46 of the *Journal de Chimie Physique*, 1949, pp.276-287. An English version appeared under the title, "Principles Determining the Structure of Intermetallic Compounds", published in the Meeting of the American Chemical Society, Chicago, Illinois, September 6, 1950.

81   Titanium is more difficult to machine than aluminium or steel. Special carbide tools are needed to cut titanium. The size of the cutting equipment also restricted the size of the sheets of titanium, which was considered by 1962 to be small compared to the Mercury Space Capsulte, thereby necessitating triple sheet welding techniques to be incorporated into the capsule's construction. Compared this to the dark-grey 9-meter diameter disk found near the Socorro region with no signs of welding or nuts and bolts used on the surface, this indicates a very high level of technological achievement by the USAF for the 1947 period. At any rate, the Battelle Memorial Institute was contracted by the USAF to look at ways of machining titanium alloys, such as cutting and rolling out thin sheets. The earliest document discussing this is titled, "Machining of Titanium Alloys" from the Defense Metals Information Center at Battelle, dated February 2, 1965.

As for machining thin sheets into a reasonable size (although no larger than those produced by 1962), U.S. Patent No.2,651,099 dated November 17, 1950 and issued to the inventors at Mallory-Sharon Titanium Corporation in Niles, Ohio, shows the best that can be achieved with titanium sheets. The question one should ask is why was this invention not made in the first half of 1947 if the Roswell foil is a titanium-based SMA and could be machined to a very thin sheet by the USAF covering a size of at least 9 meters in diameter without signs of welding or other means of connecting titanium sheets together?

82   Given the way the object was struck down by lightning and the possibility it moved across the sky with a glowing surface reminiscent of an electric "light bulb" turned on would suggest the metal was mainly used for the skin (perhaps to hold an oscillating electric charge for some specific purpose). Therefore, structural supports were likely to have been made with a plastic or some tough polymer material.

83   In 1974, NOL merged with Dahlgren's Naval Weapons Laboratory (NWL) to form the Naval Surface Weapons Center. In July 1997. the military R&D activities performed at this site were formerly relocated to other Naval Surface Warfare Center sites, such as Dahlgren in Virginia, and Carderock's Indian Head division in Maryland. The old decommissioned NOL site will be redeveloped to consolidate all research and development offices of the U.S. Food and Drug Administration.

84   *Omega Science Digest* 1984 (Australian edition), p.11.

85   Kauffman, G.B. & Mayo, I. (in press). Nitinol, "the memory metal" has multifarious applications.

86   The term "shape-memory" was not officially coined by Buehler. His preferred term was "mechanical memory". Later, as more scientists learned of NiTi's interesting and highly pronounced elastic property, someone in the scientific community decided "shape-memory" sounded like a better term to use and could be rolled off the tongue with greater ease, so the term has stuck with everyone ever since. From there, SME (Shape-Memory Effect) and SMA (Shape-Memory Alloy) became commonplace acronyms for classifying the elastic effect and the class of alloys having this effect in metallurgical departments. And it was only in the 1970s did scientists see a need for a further classification to handle those SMAs with exceptional superelastic properties. Hence the term superelastic (as opposed to pseudoelastic used prior to 1958 to represent the weaker form of the elastic effect), although some scientists may choose to see the two terms as essentially the same in order to keep things simple. Thus, in recent times, shape-memory is seen as more to do with a heating of the alloy to cause an elastic response, whereas pseudoelastic and superelastic may be used to describe the stress-induced elastic response.

87   http://robots.rutgers.edu/papers/NDE.pdf (Information about Brussels World's Fair and Au-Cd appearing online in 2005).

88   W. J. Buehler, J. V. Gilfrich and R. C. Wiley, "Effect of Low-Temperature Phase Changes on the Mechanical Properties of Alloys near Composition TiNi": *Journal of Applied Physics*. Volume 34, Issue 5 (1963), p.1475 (pp.1475-1477).

89   Otsuka & Wayman 1998, p.2.

90   http://www.americanheritage.com/articles/magazine/it/1993/2/1993_2_18.shtml (Request for information about nitinol by Congress members and high-ranking officials).

91   Personal statement obtained from http://www.wolaa.org/files/Nitinol_Oral_History.pdf.

92   K. N. Melton and O. Mercier, "Fatigue of NiTi thermoelastic martensites": *Acta Metallurgica*. Volume 27, Issue 1, pp.137-144. Published in January 1979.

93   R. H. Bricknell, R. N. Melton, and O. Mercier. "The Structure of NiTiCu Shape Memory Alloys": *Metallurgical Transactions A*. Volume 10, Issue 6, pp.693-697. Published in June 1979.

94   Rachinger 1958, p.250.

95   The strain limit will go down with each thermal cycling between the martensite to austenite phase and back again. After about a few hundred thermal cycles, NiTi quickly settles down to a strain limit of 6% to 8%. After about 500,000 thermal cycles, the strain limit will drop further. After a million thermal cycles, NiTi is extremely hard and tough to bend. However, if the strain limit is exceeded (which requires considerable force), the alloy will snap or break, mainly across a crystalline face where the unit cells have developed imperfections and/or contain impurities.

96   Teramoto 2010, p.2.

97   Otsuka et al. 1974, p.913.

98   Teramoto 2010, p.4.

99   Saunders 1984, p.12.

Chapter 8

1   *An Encyclopedia of Metallurgy and Materials* 1984, p.291.
2   Craighead et al. 1950, p.485.
3   Grayson 1983, p.98.
4   http://metals-history.blogspot.com/2010/04/titanium-new-metal-for-aerospace-age.html as of June 22, 2010.
5   Augenstein, Bruno W. & Murray, Bruce. 2004, *Mert Davies: A RAND Pioneer in Earth Reconnaissance and Planetary Mapping from Spacecraft*. California, USA: RAND Corporation.
6   Craighead et al. 1950, p.485.
7   Craighead et al. 1950, p.486.
8   Craighead et al. 1950, p.486.
9   Craighead et al. 1950, p.486.
10 *Encyclopedia Britannica* Volume 21, p.485.
11 Official quote from the U.S. Army Material Command Historical Office obtained from http://www.amc.army.mil/amc/ho/studies/titanium.html. Dr. John P. Nielsen at NYU would study TiN and TiC as well as TiNi on behalf of the U.S. Army.
12 Official quote from the U.S. Army Material Command Historical Office obtained from http://www.amc.army.mil/amc/ho/studies/titanium.html.
13 Adenstedt & Freeman April 1953, p.1 (PDF p.8).
14 N. N. Thadhani's current field of expertise at the George Institute of Technology is quoted from http://www.mse.gatech.edu/faculty/thadhani as of May 2015.
15 Craighead et al. 1950, p.485.
16 Battelle was essentially looking at iodine as a viable replacement for the chlorine atoms in the titanium tetrachloride structure in the hope of creating a more robust and chemically inert structure before reducing it with magnesium through the Kroll method.
17 The U.S. investigators have a web site at http://www.roswellinvestigators.com/.
18 Mr. Carey, a former USAF crypto-security expert, turned his attention to UFOs and later specialized in the Roswell case in the mid-1980s where he teamed up with Donald Schmitt, a lawyer by trade. Mr. Schmitt had already worked with a number of other investigators in the Roswell field and had co-authored several books on the subject before working with Mr. Carey.
19 SUNRISE was not informed by Mr. Bragalia of his link to his friend Mr. Carey. His early interest at the time was determining the name of the Australian researcher (similar to the way the USAF wanted the name of the late Steven Schiff through the GAO) who gathered the link between the USAF at Wright–Patterson AFB, Battelle and NiTi through a scientific footnote. SUNRISE could not obtain any information online about Mr. Bragalia at the time to show his connection to his friend other than his contact details in the Boston area. Since 2010, he has published online several articles on NiTi and the links to the Roswell case. None of the articles acknowledges the work conducted by SUNRISE unless a name is given.
20 See Appendix F for first page details.
21 SUNRISE learned at this point that Mr. Bragalia assists Mr. Carey in gathering information on the Roswell incident, including the names of international researchers and investigators involved in the study of the Roswell case.
22 A copy of this article can be downloaded from http://www.aimehq.org/search/docs/Volume%20236/236-2 09.pdf.
23 In the "Fourth Quarterly Progress Report" for the period December 1, 1961 to February 1, 1962 prepared by senior engineering metallurgist P. Bergstedt under contract number AF 44(616)-7984 Task no. 73812, it is reported "that the zirconium alloying [to titanium] may be responsible for the low temperature embrittlement" (PDF p.34). This would occur at 5 percent zirconium. For high zirconium content in titanium, the investigators were definite about the low temperature embrittlement caused by the element. The only alloy recommended for low temperature use is one without zirconium. In this contract, the USAF was interested in finding suitable tough titanium-based alloys at cryogenic temperatures.
24 See Appendix F.
25 Rapid solidification processing is reminiscent of how amorphous alloys are created.
26 Wang 1972, p.1.
27 Eutectic and eutectoid temperatures are essentially the toughs and crests in a phase diagram showing the minimum or maximum temperatures for crystalline structures can remain unchanged. Scientists know a different crystalline structure can be formed by the combining of two or more elements and allowed to solidify, but the structure can change across different percentages of the composition and at different temperatures. A phase diagram is the best way to visualize these different crystalline structures by showing the regions where these structures remain stable.

28   Adenstedt & Freeman April 1953, Foreword section (PDF p.4).
29   Wang 1972, p.ii.
30   Teeple 1950, pp.1990-2001.
31   Adenstedt and Freeman published the latest phase relationship diagram for TiAg in their WADC Technical Report 53-109 dated April 1953. It shows the alpha and beta dual crystalline structures within a phase region in a narrow temperature band of around 1,600°F and between 5 and 18 weight percent silver (see page 9, PDF p.16).
32   A nickel-rich (gamma) phase means the nickel atoms are clustering or sticking together within the titanium matrix structure. Higher temperatures help to eliminate the gamma phase.
33   Columbium and niobium are the same element.
34   U.S. Patent No. 3,351,463.
35   *Chemical Abstract* 1949, Volume 43, Number 2, 3766c-d.
36   This report is mentioned in the recently unclassified U.S. military document titled "Some Mechanical and Ballistic Properties of Titanium and Titanium Alloys" by R.KK. Pitler and A. Hurlich. The report has been released and was obtained in March 2010.
37   See Appendix E.
38   *Encyclopedia Britannica* Volume 21, p.485.
39   Crain 1951, Column 9441d.
40   John.Powell@f4.n1010.z9.FIDONET.ORG (Quote from *The Research Airplane Program* obtained from http://www.ufo.net/ufodocs/text.documents/n/nasa-xplanes).
41   Mesthene January 1962, p.v.
42   Mesthene January 1962, p.3.
43   Mesthene January 1962, p.3.
44   Mesthene January 1962, p.3.
45   Only the gold-cadmium (Au-Cd) and copper-zinc (Cu-Zn) alloys showed a pseudoelastic (or shape-memory) effect before 1947, but not the superelastic property. If the USAF had realized any of these alloys were superelastic, Au-Cd would not be dark-grey in color to match the Roswell foil. Furthermore, the lowest martensite temperature transformation starts at 30°C and would require higher than body temperature to reveal any possible superelastic property. In fact, at 47.5 at.% composition for Au that was studied prior to 1947, the martensite temperature would have been well above 40°C, making it harder to reveal a superelastic property. For Cu-Zn, the austenite transformation temperature is well below -10°C, which is too low to reveal a superelastic property if it had one. It would appear that the USAF had discovered superelasticity in another alloy, and one that contained titanium as its major constituent.

Chapter 9

1   http://www.dupont.com/content/dam/dupont/products-and-services/membranes-and-films/polyimde-films/documents/DEC-Kapton-summary-of-properties.pdf
2   The physical properties were obtained from http://www.bopetfilms.com.
3   http://usa.dupontteijinfilms.com/informationcenter/downloads/Physical_And_Thermal_Properties.pdf as of August 2016.
4   It has been estimated that this form of toughened rubber is 10 times stronger than natural rubber.
5   Earliest mention of anodizing titanium (and its alloys) is the First Progress Report by Charles Levy for the U.S. Army Watertown Arsenal, titled, "Electroplating on Titanium", dated July 8, 1952.
6   Li 1996, p.iii.
7   Y. H. Wen, H. S. Peng, D. Raabe, I. Gutierrez-Urrutia, J. Chen and Y. Y. Du. "Large Recovery Strain in Fe-Mn-Si-based Shape Memory Steels Obtained by Engineering Annealing Twin Boundaries": *Nature Communications*. September 17, 2014, Article number 4964.

Chapter 10

1   SUNRISE has published *Can UFOs Advance Science? A New Look at the Evidence*, which includes statistics and examples of man-made and natural phenomena commonly mistaken for UFOs. Also included, for the first time in any book on the subject, is a technological explanation to support the observations of genuine UFOs involving disc-shaped metallic and glowing objects seen by witnesses.

2   Question sent to http://www.madsci.org/ask.html on April 23, 2012. Response received by email on April 24, 2012.

3   Swords 1998, p.124.

4   Data obtained from various online sources. As the figures do vary with different online sources, the lowest empty weights are shown where they were observed at time the data was gathered.

5   *USAF Executive Summary: Report of Air Force Research Regarding the 'Roswell Incident',* September 8, 1994, pp.12. *The Roswell Report* 1995, pp.19.

6   The alternative would have to be a large airship and with the right weather conditions as well as a large cabin space for multiple passengers and plenty of food to make the long arduous journey to the US.

7   Paul et al. 2002, p.21.

8   A modern-day lightweight aircraft such as the Adam A500 weighs in at 3,450kg through the use of a combination of lightweight materials such as titanium, aluminium and other materials.

9   http://www.titaniumeden.com/about_ti.php (3 tons of titanium in 1948).

10   *USAF Executive Summary Report of Air Force Research regarding the 'Roswell Incident',* September 8, 1994, p.12. *USAF Roswell Report – Fact or Fiction?* 1995, p.20.

11   *USAF Executive Summary: Report of Air Force Research Regarding the 'Roswell Incident',* September 8, 1994, p.12. *The Roswell Report* 1995, p.20.

12   *USAF Executive Summary: Report of Air Force Research Regarding the 'Roswell Incident',* September 8, 1994, p.12. *The Roswell Report* 1995, p.20.

13   Just to add to the probable explanation from the witnesses, it has been noted the type of materials that were used in the construction of the Roswell object. In particular, the "electrically-insulating" plastic-like structural beams and some dark-colored plastic sheets, and the outer shape-memory metal skin. Man-made aircraft in the late 1940s (and even to this day) usually have their structural beams made of metal. Not so for this Roswell object. The only other known man-made object with the insulating beams and electrically conducting outer skin combination is a weather balloon. However, the toughness of the materials tell us that we are dealing with a high speed flying object, not a weather balloon or airship. When combined with the unusual way the lightning strikes were repeated until the "odd explosion" was heard and everything died down again, it suggests the Roswell object could be one where an electric charge had been used on the metal skin for some specific purpose. As soon as the electric charge was reduced significantly, the lightning strikes diminished dramatically. This is something that the scientific community will need to keep in mind when determining what it was the USAF had recovered in early July 1947.

Chapter 11

1   Email received by SUNRISE.

2   http://www.abovetopsecret.com/forum/thread35945/pg1.

3   http://www.virtuallystrange.net/ufo/updates/1998/sep/m17-025.shtml.

PICTURE CREDITS

Chapter 1
Roswell Daily Record   front page spread for July 9, 1947: Public domain image.

Chapter 2
Colonel William H. Blanchard: USAF http://www.af.mil/information/bios/bio.asp?bioID=4708; Copy of original news release from Second Lieutenant Walter Haut: USAF (digitally enhanced by SUNRISE); Various newspaper articles for the July 1947 period: Public domain images; Major Clement McMullen: USAF; Photo of Rawin sonde weather balloon: New Mexico Institute of Mining and Technology in Socorro; Photo of Rawin target weather balloon: U.S. Meteorological Service; Photos of substituted wreckage in General Ramey's office: James Bond Johnson and Major Charles A. Cashon: USAF and University of Texas Arlington Library Archive; General Nathan F. Twining: USAF http://www.jcs.mil/cjs/history_files/bios/twining_bio.pdf; Photos of secret memo close-up: Public domain; Photo of Fort Worth AAF airmen with weather balloon: *Fort-Worth Star Telegram* Photograph Collection at the University of Texas Arlington Library; World War II aerial photograph of Wright-Patterson AAF: USAF Archive. U.S. President Harry S. Truman: U.S. Department of Defense (ca. 1974-1984), public domain image.

Chapter 3
Brigadier General Arthur E. Exon: NICAP.

Chapter 4
Book cover images of Roswell Report: USAF; Prof. Charles B. Moore holding a Mogul balloon: NYU and shared under the Creative Commons (CC) Attribution Share Alike 4.0 International License; Colonel John Haynes: CNN.

Chapter 5
Photographs of Miracle Spoon and packaging: SUNRISE; Cover of now defunct *Omega Science Digest* magazine containing article on nitinol: Scanned by SUNRISE.

Chapter 6, 7, 8
Scanned images of various scientific and military abstracts, graphs etc.: Obtained from various metal journals (mostly published from the 1940s to the 1960s) and declassified USAF/Battelle Reports on titanium alloy studies after 1947; Other images created by SUNRISE; Photo of Dr. John Nielsen: New York University (NYU) Archives; U.S. Naval Ordnance Laboratory: Washington Section Archives, Washington, D.C: William J. Buehler: Unknown; Battelle Memorial Institute: Analogue Kid, July 28, 2007, under the CC 3.0 License, available from Wikipedia.

Chapter 9
Photographs of polymers: Various sources from unknown photographers.

Appendices
Several U.S. Government documents digitally re-created from originals and related to the Roswell case: Various sources (originals available for download from the SUNRISE web site).

Although every effort has been made to trace the copyright holder, we apologize in advance for any unintended omissions and would be pleased to insert the appropriate acknowledgements in any subsequent edition of this book.

ABOUT THE AUTHOR

SUNRISE Information Services (SUNRISE) is a private research center aimed at producing original, stable, interesting and easy-to-read educational and research information for the global community, while uncovering new and original knowledge.

⌐

www.ingramcontent.com/pod-product-compliance
Lightning Source LLC
Chambersburg PA
CBHW060421220326
41598CB00021BA/2252